Soil Health on the Farm, Ranch, and in the Garden

Kenneth E. Spaeth Jr.

Soil Health on the Farm, Ranch, and in the Garden

 Springer

Kenneth E. Spaeth Jr.
Hayley Drive
Weatherford, TX, USA

ISBN 978-3-030-40397-3   ISBN 978-3-030-40398-0   (eBook)
https://doi.org/10.1007/978-3-030-40398-0

This Springer imprint is published by the registered company Springer Nature Switzerland AG
The registered company address is: Gewerbestrasse 11, 6330 Cham, Switzerland

Preface

Soil health is undoubtedly a "hot topic" in agriculture; although not a new idea, the resurgence of soil health principles is producing positive results throughout the agricultural community with emphasis on cultural acceptance, improving hydrology and reducing erosion, planting cover crops, and restoring and maintaining soil organic matter, microbial populations, fertility, and soil physical properties. Soil health is a concept that is important across all land uses, including cropland, rangeland, pastureland, forestland, and garden environments. Why write a book on soil health? As a scientist and conservation agent, this is the book I want as a reference to quickly obtain information on a myriad of soil health subjects. This book is not intended as a comprehensive textbook; the content of the material would need to be greatly expanded in size, scope, and content; however, *Soil Health on the Farm, Ranch, and in the Garden* can augment the study of soil health in the college curriculum and/or provide reference material for use in practical situations (teaching, use by landowners, managers, and professional agents). The objectives of this book are 1) to convey a conceptual framework i.e., ecological site concept, which can serve as a reference for soil health attributes, 2) clarify the aspects of how environmental variables interact with the biotic and abiotic factors in the ecosystem; 3) differentiate soil health dynamics between different land uses (cropland vs. grazinglands), 4) highlight the importance of hydrology and hydrologic function with soil health, 5) present examples of how cover crops and natural vegetation protect the soil surface and, after decomposition become organic matter, 6) display and discuss the organic matter cycle perceptions from several authors; and provide some realistic examples of organic matter development in the soil, and 7) present some concepts that are useful in assessing soil health on site for cropland, pastureland, rangeland, and gardens.

There are many viewpoints as to the definition of soil health and what constitutes soil health in the field. Chapter 1 provides a historical context of taming the land and current definitions of soil quality and health. Before soil health can be ascertained on the local field level, an understanding of the ecological site is imperative so that baselines can be established as to site and soil potential (Chap. 2). Ecological sites are dynamic and unique across the landscape, and awareness and use of ecological site descriptions on the lands we live on or manage, i.e., in a farm, ranch, or garden setting, is the first step in maintaining soil health and continued prosperity of any agricultural endeavor. State and transition models and narratives associated with ecological sites are the culmination of a collection of knowledge about ecosystem dynamics and changes in response to natural events and management applications.

There are many inherent environmental variables associated with soil health. The interactions between climate, soil physical and chemical properties, and historic plant community and existing plant composition, soil flora and fauna, hydrology, and past and current management are complex and dynamic. When one considers the most limiting factor in terrestrial environments, i.e., "water," it should be no surprise that hydrology is related to almost every aspect of terrestrial ecosystems. The hydrologic cycle and its function are integral to the health of the soil. Soil health is dependent on proper hydrologic function; organic matter cycling cannot proceed without water and associated hydrologic dynamics. Hydrology and interactions with agricultural land use are covered in detail in Chaps. 3, 4, and 5.

Organic matter is the core of soil health as it is involved in many environmental processes related to atmospheric carbon, climate, soil, plants, and microorganisms. The soil organic matter and carbon cycle are important aspects of understanding the interactions of abiotic and biotic variables in the environment. Why Chap. 6 heading "Organic Matter: The Whole Truth and Nothing but the Truth"? There are many misconceptions and exaggerations regarding the organic matter and soil health. This chapter is intended to provide the reader with a relevant collection of reference material.

The final chapter, "Soil-Hydrology-Plant Assessment Technologies for Cropland, Rangeland, Pastureland, and Gardens," provides a summary of some of the "state-of-the-art" assessment technologies. One novel approach in Chap. 7 encourages the customization of visual land assessments in cropland, grazingland, and garden settings to address local concerns.

Weatherford, TX, USA   Kenneth E. Spaeth Jr.

Contents

Chapter 1
Taming the Land: A Historical Perspective

Abstract Soil health and quality are not new ideas but have gained a resurgence among land users. Many popular publications and scientific studies published in peer-reviewed journals address new ideas and research on cover crops, applying manure and other soil amendments, practicing crop rotations, and using minimum tillage or no-till cropping systems. On cropland, soil health dynamics are quite different from grazinglands as the soils are cultivated and fertilized with various amendments (synthetic and/or organic). On grazinglands, ranchers manage soil health through grazing practices and managing vegetation. Therein lies the foundation of this book: current soil health emphasis is based on conservation, sustaining the soil resource and maintaining production, and restoration, where economically possible and feasible.

Questions

- Differentiate the concepts of soil quality and soil health.
- Throughout history, what have been the key challenges in cropland agriculture?
- How has the pristine land use concept affected the use and management of lands in the United States?
- Discuss how conservation practices are related to soil health.

I have surveyed vast landscapes from precipices, gorges, and mountains and marveled at the respective landscapes across the United States. The beauty and the complexity of geological influences, topography, and vegetation are incomprehensible when one considers all the environmental factors and evolutionary events that were responsible for what we observe today. One can only wonder what first impressions indigenous peoples had as they explored and settled in unknown lands throughout the world—awe, and amazement perhaps? Our ancient ancestors who lived in greater intimacy with nature realized the importance and state of the resources around them because of their direct contact with their environment for their livelihood. Various estimates of when and where the earliest systems of cultivation and animal husbandry vary, but 10,000 years ago is a reasonable estimate.

Agricultural practitioners, those involved in farming, ranching, or gardening, often ask questions and wonder about what the land was like before human

© Springer Nature Switzerland AG 2020
K. E. Spaeth Jr., *Soil Health on the Farm, Ranch, and in the Garden*,
https://doi.org/10.1007/978-3-030-40398-0_1

Fig. 1.1 Soil health principles are ubiquitous in crop, pasture, range, and garden settings. There are, however, different dynamics related to soil organisms and the amount of organic matter turnover among the land uses. These dynamics are discussed in the chapters in this book. Photos: Palouse wheat field (photo K1441-5 courtesy USDA-ARS), Fort Keogh livestock and range research station in southeastern Montana (photo K3908-1, Jack Dykinga, courtesy USDA-ARS), vegetable garden (photo by author)

settlement—romanticization about an unspoiled landscape.[1] If land managers wish to understand local dynamics of soil health, it is important to be aware of the unique connections between climate, soils, and native vegetation. The driving forces that formed local soils and vegetation are unique and are different from place to place. This connection is the foundation of ecology, a branch of biology that studies the interrelationships of organisms and their environments. In order to understand soil health, we must also consider other aspects of our environment, such as climate, plants, microorganisms, hydrology, nutrient cycles, and soil dynamics. All these factors are intertwined and are the drivers of the particular ecosystems where we live.

---

[1] Plant communities are unique in that one set of factors or variables may be driving the ecosystem and a different set in a different ecosystem. There are commonalities to be sure, the function of organic matter, nutrient uptake, energy cycles and photosynthesis, plant adaptation to pH, hydrology, but the finer points or a universal model that identifies key plant community attributes is elusive (e.g., plant succession, plant competition, diversity models, ecosystem resistance, and resilience to invasive spp.).

Fig. 1.2 Lewis and Clark, American explorers who traveled across prairies, through the Rocky Mountains, Intermountain West, and finally to the Pacific Coast during 1804–1806. Painting by Edgar Paxson, Lewis and Clark at Three Forks, mural in lobby of Montana House of Representatives, courtesy of Wikipedia

In the United States, we, fortunately, have a "glimpse into history" concerning the vast expanses throughout the Great Plains and Intermountain West. Lewis and Clark and the Corps of Discovery[2] were the first American explorers who traversed the uncharted American interior during 1804–1806 (Fig. 1.2). The 12,874 km (8000 miles) expedition started on the Ohio River, on to the Missouri River, through the Northern Great Plains, across the Continental Divide, into the Pacific Northwest, and finally reaching the Pacific Ocean. Members of the Corps of Discovery recorded their impressions in journals[3]:

> The country on both sides of the missouri from the tops of the river hills, is one continued level fertile plain as far as the eye can reach, in which there is not even a solitary tree or shrub to be seen except such as from their moist situations or the steep declivities of hills are sheltered from the ravages of the fire –Captain. M. Lewis, April 10, 1805 (mile 1669; Little Basin, McLean Co, North Dakota).

---

[2] The Corps of Discovery was commissioned by President Thomas Jefferson in 1803; he appointed his personal secretary US Army Captain, Meriwether Lewis, who selected Second Lieutenant William Clark (had simulated rank of Captain equal to Lewis) as his partner. Together with 33 people (mostly volunteer soldiers) and 1 dog (Seaman), and a Shoshoni teenager Sacagawea and baby acting as interpreter embarked on the journey from Fort Mandan, North Dakota, up the Missouri River on May 14, 1804, to explore and record their observations and find an all-water route to the Pacific Ocean.

[3] Approximate location designation in miles from starting point, Camp Wood (River Dubois), about 15 miles north of St. Lewis, at the junction of the Mississippi and Missouri rivers in Illinois Territory (Plamondon 2000).

I ascended to the top of the cut bluff this morning, from whence I had a most delightful view of the country, the whole of which except the valley formed by the Missouri is void of timber or underbrush, exposing to the first glance of the spectator immence herds of Buffaloe, Elk, deer, & Antelopes feeding in one common and boundless pasture –Captain. M. Lewis, April 22, 1805 (mile 1832; "Cut Bluff," Williams Co, North Dakota).

After dinner Capt. Clark walked out a fiew miles to view the Country, which he found verry rich Soil produceing but little vigitation of any kind except the prickly pairs but little grass & that verry low. a great deal of Scatering pine on the Lard. Side & Some fiew on the Stard. Side –Sergeant John Ordway, Wednesday, May 22, 1805 (mile 2294, "Teapot Creek," Phillips Co, Montana).

In the after part of the day I also walked out and ascended the river hills which I found sufficiently fortiegueing. On arriving to the summit [of] one of the highest points in the neighborhood I thought myself well repaid for my labor: as from this point I beheld the Rocky Mountains for the first time –Captain. M. Lewis, Sunday, May 26, 1805 (mile 2375, "Turtle Creek," Blaine Co, Montana).

the land is tolerably fertile, consisting of a black or dark yellow loam, and covered with grass from 9 Inches to 2 feet high. the plain ascends gradually on either side of the river to the bases of two ranges of mountains which lay parrallel to the river and which erminate ⟨it's⟩ the width of the vally. the tops of these mountains were yet partially covered with snow while we in the valley –Captain. M. Lewis, August 2, 1805 (mile 2876, "Frazier Creek," Jefferson Co, Montana).

the soil as you leave the hights of the mountains becomes gradually more fertile. the land through which we passed this evening is of an excellent quality tho very broken, it is a vary grey soil. a grey free stone appearing in large masses above the earth in many places – Captain. M. Lewis, Wednesday, September 20, 1805 (near mile 3345, Ant Creek, Idaho Co, Idaho).

As Lewis and Clark and the Corps of Discovery explored the Great Plains and Pacific Northwest, they saw landscapes "in a light" that no descended European Americans had seen before. As Stephen Ambrose stated in *Undaunted Courage*: "Lewis was stepping into the unknown … he was entering a completely unknown territory, nearly a half-continent wide, as any explorer ever was … he was entering a heart of darkness. Deserts, mountains, great cataracts, warlike Indian tribes—he could not imagine them, because no American had ever seen them … he turned his face west. He wound not turn it around until he reached the Pacific Ocean. He stepped forward into paradise." Was Lewis stepping into a land of paradise? Realistically, no, but as Ambrose said, "one would hardly know it from Lewis's descriptions, which were accurate enough but always given a positive slant."

Other observers have documented their impressions of the American landscape before widespread settlement. Judge James Hall in Plumbe's Sketches (1839) gave this description of Iowa prairies: "the scenery of the prairie is striking, and never fails to cause an exclamation of surprise. The extent of the prospect is exhilarating. The outline of the landscape is sloping and graceful. The verdure and the flowers are beautiful; the absence of shade, and consequent appearance of profusion of light, produce gaiety which animates the beholder … it is impossible to conceive a more infinite diversity, or a richer profusion of hues or to detect the predominating tint, except the green, which forms the beautiful ground, and relieves the exquisite brilliancy of all the others." In a book on Illinois, Gerhard (1857) provided this observation: "Immense prairies of grass, interlaced with groves and stretching, principally along the water courses, cover two-thirds of the entire state in the north ... Some extend in the immense level plains. Others are rolling, others again broken by hills ... on the lower, humid prairies, where the clayey stratum lies close to the surface, the middle or principal stalk of grass, bearing the seed, grows very thick, having long and course leaves, and attaining a height of 9 feet, so that the traveler on horseback will frequently find it higher than his head ... on the undulating prairies the grass is finer, and exhibits more leaves, its roots are interlaced so as to form a compact mass, and its leaves spread in a dense sod, which rarely exceeds the height of 18 inches, until late in the season, when the seed-stalk shoots up." Interestingly, not all observers had an appreciation for the virgin landscape, in an essay by Atwater (1818), he said this about western Ohio: "On the prairies and barrens of the West ... the traveler who for several days traverses these prairies and barrens, their appearance is quite uninviting, and even disagreeable. He may travel from morning until night, and make good speed, but on looking around him, he fancies himself at the very spot whence he started. No pleasant variety of hill and dale, no rapidly running brook delights the eye, and no sound of woodland music strikes the ear; but, in their stead, a dull uniformity of prospect spread out immense." Was the American continent the "Land of Milk and Honey"? In one respect it was, the abundance of untouched natural resources and wildlife, and opportunities for America seemed limitless. However, to portray and romanticize the land as celestial, pristine, and undisturbed by humans is a false notion. This idea was perpetuated by primitivist early American writers (Thoreau, Longfellow) and landscape artists (e.g., Thomas Cole, 1801–1848; Albert Bierstadt, 1830–1902).

Some of the lands that Lewis and Clark traveled across were not "pristine or untouched" by humans. Lewis and Clark saw firsthand how Native Americans used fire to manipulate or manage lands to their advantage. Captain Clark observed and recorded several entries about prairie fires in his journal: March 6, 1805 "a Cloudy morning & Smokey all Day from the burning of the plains, which was Set on fire by the Minetarries [Hidatsa tribe] for an early crop of Grass as an endusement for the Buffalow to feed on"; July 20, 1804 "Plains interspesed with Small Groves of Timber on the branches and Some Scattering trees about the heads of the runs, I Killed a Verry large yellow wolf, The Soil of Those Praries appears rich but much Parched with the frequent fires."

For at least 10,000 years, Native Americans of the Great Plains used fire as a tool to manage nearby buffalo herds as the shoots of grasses are tender and very palatable after a burn. Scientific research suggests that there is no substantial evidence that indigenous people wantonly burned large areas of the prairie; however, they certainly increased fire frequency compared to natural fires caused by lightning strikes. In the mixed-grass prairies of North Dakota, natural fire frequencies vary but on average occurred every 7–12 years (Frost 1998). Lewis (1973) mentions 70 reasons why Native Americans throughout the continent used fire; however, this list can be condensed to 11 primary uses: "hunting, managing crops, improving growth and yields in wild plants, fireproofing areas, collecting insects, managing pests, fighting wars and sending signals, extorting goods from other people, clearing areas for travel, felling trees, and clearing riparian areas" (Brauneis 2004). However, during the timeframe of the Corps of Discovery, there were undoubtedly expansive areas in the west that were uninhabited or rarely-to-occasionally visited by Native Americans. These expansive land areas between tribal boundaries were most likely influenced more by natural disturbance events.

Who were and when did the first Native Americans settle in the present-day United States? There was a time in North America's past when no humans occupied the continent, but at some point, in time, adventurous humans made the journey to the New World. There is considerable debate about who, when, and where this colonization took place. The scientific consensus is that the first ancestors of modern Native Americans traveled from Siberia from the Bering Land Bridge (Beringia—a land bridge from Siberia to Alaska) (Raghavan et al. 2015). However, the timing of the migration remains controversial according to newly published anatomical and genetic data and studies. Recent studies find that ancestors of present-day Native Americans migrated as a single wave from Siberia and entered the American continent no earlier than 23 thousand years ago (Raghavan et al. 2015). After an 8000-year sequestration period in Beringia, the first Native Americans traveled to the North American continent about 13 thousand years ago and eventually diversified into two branches, one group across North and South America, the second group colonized North America. Undoubtedly, there is a long history of people inhabiting the American continent, and they had an impact on the land over thousands of years.

The lands that agricultural producers work today, whether it is growing crops, managing livestock on range and pasture, managing forests, or working in your home garden, most likely have a long history of agrarian use and have been subjected to a variety of disturbances (natural and anthropogenic). Some lands settled by colonial settlers in the eastern United States have been in different kinds of agricultural production for three centuries. If we are concerned about soil health, a historic reference of the soil and vegetation is required, and then we can proceed to understand inherent capabilities and what is needed to maintain sustainability and productivity and, if needed, a practical guide to remediation. This concept is a major theme throughout this book.

Taming the Land and Its Consequences

In the journal entry by Captain Meriwether Lewis on April 10, 1805 (see above), he mentions a treeless plain. Captain Lewis assumed that the treeless plain was due to wildfires that eliminated tree species; his record mentions the "solitary tree or shrub to be seen except such as from their moist situations or the steep declivities of hills are sheltered from the ravages of the fire." This journal entry was recorded in McLean Co, North Dakota, where average annual precipitation is 43 cm (17 in) per year. Captain Lewis did not have time to collect detailed climate data and observe seasonal patterns as he traveled west; his climatological perception was his birthplace, Albemarle County, Colony of Virginia (present-day Ivy Virginia) where the average annual precipitation is 114 cm (44.8 in) per year. Lewis grew up in the Blue Ridge Mountains with forest vegetation dominated by oaks, pines, and the American chestnut.[4] I am sure that the members of the Corps of Discovery had their biases about land, productivity, and agricultural potentials,[5] after all, they were born and lived in a humid eastern climate with abundant rainfall and vegetation.[6] Captain Lewis's observations were partially correct in thinking that trees and brush in the wetter draws enabled these plants to avoid the effects of frequent fires, become established, and grow. However, there was another ecological factor that Lewis perhaps did not "recon with—drought." The northern plains of North Dakota experiences periodic droughts, severe enough to limit deep recharge soil moisture during the winter. Trees with deep roots depend upon deep soil moisture during the summer, whereas the dominant prairie grasses such as western wheatgrass (*Pascopyrum smithii*), needle-and-thread grass (*Hesperostipa comata*), blue grama (*Bouteloua gracilis*), green needlegrass (*Nassella viridula*), prairie June grass (*Koeleria macrantha*), plains muhly (*Muhlenbergia cuspidata*), little bluestem (*Schizachyrium scoparium*), and sideoats grama (*Bouteloua curtipendula*) can tolerate the dry climate. What plant ecologists realize now but was not realized by Lewis at the time was that there are many interrelated environmental factors that maintain prairie or grassland vegetation (discussed in later chapters). The Northern Great Plains climate has inherent cyclic droughts (occasionally severe, see discussion below) that maintain grassland vegetation and limit the extent of trees. Grasslands are dominated by grasses and forbs (non-woody herbaceous plants) for a reason. For example, the historical origin of grasslands in North America can be traced to the Miocene-Pliocene epoch, about 7–5 M years before the present. During that time, a drying trend was occurring from the Miocene uplift of the Rocky Mountains, which created a rain shadow effect from the moist Pacific air masses. Grasses, which are more adapted to drought than trees, flourished and became the dominant vegetation. Grazing and browsing animals then flourished with the transition and expansion of grassland vegetation (Stebbins 1981; Axelrod 1985).

---

[4] These native tree species in Albemarle County forests are the same as in Meriwether Lewis day, with the exception of American chestnut, which have succumbed to the chestnut blight (*Cryphonectria parasitica*).

[5] Case in point, many pioneers who took advantage of the Homesteading Act of 1862 (160 acres of public land) settled on unproductive lands that could not sustain crops, and life was extremely harsh with many trials and tribulations. It was not long until many abandoned their claims.

[6] As I reflect upon my own experiences as a new graduate student at the University of Wyoming, I was quite familiar with the eastern deciduous forests and Flint Hills tall grass prairie of Kansas. Upon arriving at the University of Wyoming, I was immediately sent out to the western part of the state in the Red Desert to do field work. I was shocked when I saw the barren landscape and depauperate vegetation. If fact, I seriously contemplated switching from rangeland management to forestry—my initial impressions of the desert vegetation were dubious. However, by the end of my 2-year master's program, I developed a love for Wyoming and the ecological intricacies of the cold desert vegetation, sagebrush steppe, and juniper woodlands.

The Advent of Agriculture

The agricultural transition from nomadic societies to the advent of agriculture, the Neolithic Revolution in the Near East 10,000 years ago, was due to the domestication of various plant and animal species. The African bush dwellers and Australian aborigines were probably some of the last of dominant hunter-gatherer societies. Food supplies began to be more plentiful and secure, which facilitated permanent settlements, communities, and, eventually, cities. Slash-and-burn agriculture was used toward the end of the Neolithic Era, where forests were cut and burned for planting; however, many of these efforts were temporary, as the American pioneer settlers also discovered. Slash-and-burn methods are still used today, mainly in tropical forests in South American, Asia, and Africa. During the agricultural transformation, almost every aspect of domestic life changed. Food production and the specialization of activities changed the social role of men, women, and children (Tyldesley 1995). With the advent of increasing populations and cities, governments were established; religions emerged with different aspects of mythologies, traditions, rituals, morals, and behavior standards; architecture, art, the building up of armies, and world conquests were all tied to the success or failure of agriculture. The development of agriculture also ushered in the demise of the land, primarily caused by farming, grazing, and harvesting of trees in forests. Erosion and salinization of farmlands occurred rapidly. The once fertile soils of the Fertile Crescent associated with the Nile, Euphrates, and Tigris rivers soon degraded due to human agricultural activities. On more sloping lands, the erosive activities of water and wind erosion on cropland and the denuding of vegetation cover on grazinglands eventually took its toll where the soil was transported down the valleys and eventually to the sea. It was no coincidence that deterioration of the soil resource coincided with the movement and recolonizing activities of the Greeks, Romans, Carthaginians, and Phoenicians. Some societies were more successful than others with respect to land management, e.g., the terrace builders in Asia, Mesoamerica, and South America.

Throughout ancient history, civilizations and population expansion occurred independently in at least six different regions of the world (Fig. 1.3) (Mazoyer and Roudart 2006). Agriculture, and on even a more fundamental level, the soil, was the common denominator to the survival and perpetuation of humans. Early humans were hunter-gatherers, and eventually, the nomadic lifestyle gradually gave way to more permanent settlements that depended on agricultural pursuits. These early settlements began in broad alluvial plains near larger rivers where soils were fertile and supported crops and herding animals. People worked with what they had, so naturally fertile lands were a necessity. Ancient farmer relied on annual or occasional flooding to replenish eroded sediments and plant nutrients. The first crops were limited to cereals planted in continuous succession until productivity suffered, which necessitated the abandonment of the field and moving to new land. These eroded lands were replenished by alluvial sediments or were permanently

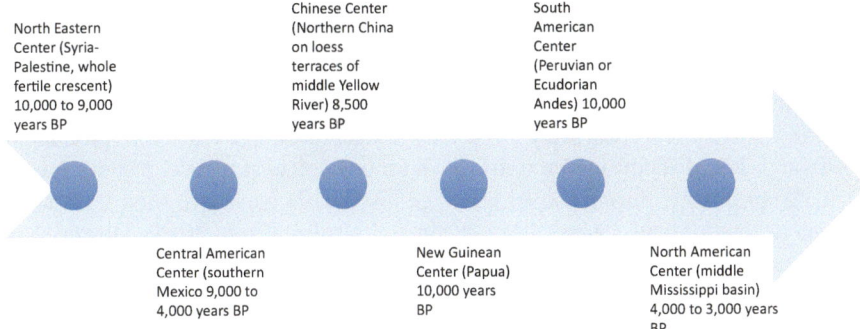

Fig. 1.3 Timeline of important centers where agriculture emerged

Fig. 1.4 Ancient farmer
with plow and oxen, from
burial chamber in the tomb
of Sennedjem, Egypt.
(Courtesy of Wikipedia)

abandoned if significant erosion had occurred. These early agricultural settlements evolved into sophisticated societies, e.g., Ancient Egypt thrived along the Nile, India along the Ganges River, and China on loess terraces of the middle Yellow River (Fig. 1.4). Eventually, farming became more refined and successful, allowing others in the population to concentrate on developing arts, crafts, education, science, astronomy, medicine, religion, and developing governments. Any prosperous civilization can trace its roots to healthy, productive soils and agriculture. As civilizations expanded, settlements migrated away from soils along flood plains to more inland upland environments (e.g., loess soils in China, and soils formed from glacial deposits in North America). Note that the North American Center originating 4000 and 3000 years before present was the most recent agricultural settlement center on the timeline (Fig. 1.4).

Knowledge of rudimentary soil conservation practices and basic soil health principles have been known for millennia. New studies report that 8000 years ago, manure was used to fertilize crops (Bogaard et al. 2013) across European climate zones (sites in the United Kingdom, Denmark, Germany, Hungary, Bulgaria, and Greece). Observant farmers noticed that plants grew better where animals congregated and dung accumulated. Ancient farmers found that by applying manure, cultivating plants and providing water by irrigation improved crop production. Although it is difficult to determine from archeological evidence, Middle Eastern and Asian farmers during the Bronze Age (3000 BC) may have used leguminous crops in rotation with cereals to maintain soil fertility (Selin 2013). However, the value of lime, manures, and cover crops (green manure) was not fully realized until 2000 years ago by the Romans (books written by Pliny 23 AD–79 AD; Varro 116 BC–27 BC; Columella 4–c. 70 AD).

Three thousand years ago, Chinese farmers developed a ridged tillage system and the development of fallowing crops (1-year fallow, 2 years cultivated crops). By 500 BC, the fallowing system also incorporated crop rotations, and by 200 BC intercropping and multiple cropping emerged (Xu et al. 2000). By 200 to 580 AD in the Yellow River Valley, plows and harrows were used to roughen the soil surface to conserve moisture during fallow periods. In a book published at this time, *Important Means of Subsistence for Common People*, the modern ideas of crop rotations with legumes and cereal crops (legumes should be planted before cereal crops) and green manuring were advocated (Miao 1981). During the rule of Charlemagne (814), crop rotations were used in Europe and consisted of a two-field rotation alternating between cropping and fallowing (Butt 2002). During the Carolingian Empire (800–888 in Western and Central Europe), a three-field rotation system was developed, which usually consisted of cropping with a winter cereal crop, a second field in annual legumes, and the third field fallowed. The subsequent year, the pattern was reversed (Bruns 2012). Later, around 1730, Charles Townshend 2nd Viscount Townshend of Raynham (Turnip Townshend) introduced a four-field rotation to the Waasland region of England (Ashton 1948). This system consisted of a rotation of a root crop [turnips (*Brassica rapa* var. rapa), wheat (*Triticum aestivum*), barley (*Hordeum vulgare*), and clover (*Trifolium* spp.) followed by fallow (Bruns 2012)].

The early agriculturists did have an appreciation and understanding of the underlying factors associated with productive soils. The writings of the Greek philosopher Plato (437–327 BC) indicate that he was aware of the values of soil health and productivity; Xenophon, the Greek historian and essayist (434–355 BC), observed ruination about a farm he knew: "the estate has gone to ruin" because "someone didn't know it well to manure the land." The story of land degradation and erosion continues throughout history and in our modern era (see Warkentin 2006; Mazoyer and Roudart 2006). As then, and now, humans have not always taken proper care of the soil. Mismanagement can be attributed to several common factors, lack of knowledge and ignorance; there was a time when land was plentiful, and natural resources seemed limitless—when a field or farm was worn out, there was always another; and economics and/or the desire for immediate gain resulted in management decisions that caused degradation of the soil and other resources such as forests and water.

Text Box 1.1 (extract from Hillel 2005)
Two ancient civilizations, Mesopotamia (today occupying Iraq, Kuwait, Eastern Syria, Eastern Iran, and southeastern Turkey) and Egypt, experienced different levels of success and environmental problems. The soils of the Mesopotamian region were deep and fertile and with moderate growing climate and available irrigation water from the Tigris and Euphrates. The upper watersheds of these lands due to overgrazing and deforestation eroded transported sediment onto the plains. Spring flooding was more frequent due to accelerated runoff, which caused the river channels to change course, and summer evaporation of the floodwaters aggravated the salinity of the soil. The incoming silt from the uplands inundated the irrigation channels. Over time, seepage from the rivers, irrigation canals, and flood irrigation practices also caused salts to accumulate in the soil and groundwater. Various remedial measures were practiced, such as fallowing and planting barley (a more salt-tolerant grain); however, over time the population centers outstripped resources, and the population centers moved from the lower-lying valleys to the uplands of the Tigris-Euphrates valley. In contrast, the Egyptians had more long-term success because of different soils and hydrologic regimes. Runoff carrying silt from the Blue Nile combined with organic matter from the White Nile added nutrients to the farmlands. Late summer-to-fall flooding and transported nutrients coincided with fall planting of grain crops. As the Nile River flooding receded, the water table dropped, and the land drained along with leaching accumulated salts back to the Nile.

Between 1620 and 1860, an era of expansion and the first semi-modern agricultural revolution coincided with the industrial revolution. The rotation of annual grain crops with perennial fodder crops began, which increased crop production and surpluses of food. Pioneers and settlers who immigrated to the New World also brought their Old World farming practices. Many of the European farms were situated on relatively flatlands, and climates and rainstorms were generally more moderate in intensity. When European immigrants established farms in North America, they found a complex landscape. Many farms were established on steeper lands, and high-intensity convective rainstorms are a common occurrence. The inevitable result was that soil was transported via runoff into nearby water bodies and if connected to major rivers. Many farmers did not realize the severity of the erosion until it was too late—thus the migration to other fertile lands west and south. Interestingly, there were individuals in colonial times that we're aware of the effects of soil and resource degradation. Jared Eliot (1685–1763), farmer, minister, and physician in Guilford, Connecticut, wrote six essays that were later combined and published as *Essays Upon Field Husbandry* in 1748. His writings addressed various subjects from reclaiming swampland, draining waterlogged soils, alternative crops, use of fertilization to improve hay production, and the invention of the drill plow. Another early agriculturalist, Samuel Deane (1733–1814), a Congregational clergyman and farmer, advocated soil conservation and the use of crop rotations, use of manure, and contour plowing on steeper soils to lessen the impact of sheet-rill and gully erosion. I believe Deane was a visionary man as he proposed that "white crops" such as oats, corn, flax, rye, and barley should be followed by "green crops" such as grasses and legumes.

In America, early colonists and later pioneers encountered vast frontiers as they migrated east to west and north to south. The virgin lands they cleared and tilled were either covered by forests or grasslands.[7] Colonial settlement started with the Plymouth Colony in the present-day modern town of Plymouth, Massachusetts, during November of 1620. Their subsequent farming activities were at first limited to barley and peas, but their initial farming efforts were largely a failure due to the seeds they brought from England. The Patuxet, a Native North American tribe, showed the settlers how to grow corn and fertilize the soil with shads or herrings. At that time, agricultural pursuits occupied 90% of the population, and most efforts consisted of subsistence production for family use. Subsequently, more ships arrived, and by 1678, about 60,000 people lived in New England. During the 1700s, plantation agriculture migrated south to Virginia, Maryland, and South Carolina. After 1800, cotton became a major crop in North Carolina down through Texas. By 1850, there were 1.4 M farms, 4 M by 1880, and 6.4 M by 1910. After pioneer farmers settled in the Midwest and Northern Michigan, Wisconsin, and Minnesota, the Great Plains attracted the sodbusters, who plowed the lush prairies. As pioneers moved west through the Dakotas and Montana, they discovered that erratic and low rainfall was not conducive to dependable and profitable farming (given the technology at the time). However, the rangelands of the western Great Plains were suitable for livestock.

Patrick Henry (1736–1799) reportedly commented after the American Revolution that "since the achievement of our independence, he is the greatest patriot who stops the most gullies" (Helms 1991). Henry's quote illustrates early in America's history the concern about soil erosion and its economic ramifications. Almost a century later, the Homestead Act of 1862 opened millions of acres of public lands, and by 1934, the federal government had granted 1.6 M homesteads and distributed 109 M ha (270 M acres; 420,000 mi^2) of federal land for private ownership (10% of the land area of the United States). Railroads advertised the lands open to homesteading as the land of "milk and honey"; however, many found that soil quality was poor and the landscape harsh and barren. In addition, natural events such as winter blizzards, tornados, insect infestations, and prairie fires could destroy years of work. Many homesteads failed because of a lack of knowledge and technology to control erosion and restore fertility. Cultivation of the land and the lack of conservation measures led to subsequent erosion and loss of topsoil—a tragedy of monumental proportions in the United States. Even today, an experienced eye of agricultural specialists and soil scientists can spot abandoned homesteaded lands. As productivity declined,

---

[7]There are three basic land types in the continental United States, forestland (includes conifer, deciduous forests), rangeland (includes grasslands, cold and warm deserts), and alpine-above timberline. Tundra occurs in high latitudes of Canada and Alaska.

many marginal croplands were converted to pasture or simply abandoned (termed "go-back lands" usually consisting of weedy species).

Today, for the farmer, rancher, and homeowner gardener, there is a high probability that a portion or all of your land has been altered and disturbed and/or has in the past undergone significantly accelerated erosion. I use the term accelerated erosion to denote that which is caused by human activities such as removal or disturbance of natural vegetation, farming, forestry, and construction. Soil erosion is a natural process; mountains, valleys, canyons, gorges, bottomlands, and stream channels were formed by geologic water-induced soil erosion. Wind erosion typically occurs in drier, more arid areas, while water erosion is more common in humid areas. The physical properties of soil erosion depend on a combination of factors, slope of land, inherent soil physical properties, vegetation, and timing, amount, and intensity of rainfall.

The worst modern erosion event was the Dust Bowl of the Southern Great Plains of North America (western Kansas, southeastern Colorado, Oklahoma Panhandle, the northern two-thirds of the Texas Panhandle, and Northeastern New Mexico) (Fig. 1.5). The 1934 *Yearbook of Agriculture* stated: "Approximately 35 M acres of formerly cultivated land have essentially been destroyed for crop production ... 100 M acres now in crops have lost all or most of the topsoil; 125 M acres of land now in crops are rapidly losing topsoil ...". What was the cause of the 1930's Dust Bowl? Many factors were responsible; the drought, although not unusual or extreme from a climatological perspective in the Great Plains, initiated a cascading set of disastrous events (Lee and Gill 2015). A combination of drought, primitive farming methods, lack of vegetative cover, intensive cultivation, fallowing, and overgrazing were responsible. Farming practices were primitive, and dry farming was the standard farming practice of the day. Dry farming of nonirrigated lands (irrigation was not widespread and was limited to floodplains where diverted water from streams was used to flow into fields and exited at a point downstream) consisted of moldboard plowing or use of a lister, which buried most of the crop litter. On fallow fields, tillage created a loose layer of soil (soil mulch or dust mulch) that was intended to reduce evaporation of water at the soil surface. If the soil mulch was too fine, it was subject to accelerated wind erosion. If rain compacted the soil mulch, the fields were tilled again to rebuild it (Hargreaves 1957).

In response to the Dust Bowl days, Hugh Hammond Bennett (1881–1960, the Father of Soil Conservation) helped establish the Soil Erosion Service in the Department of the Interior and was the first director in September 1933. Bennett was an advocate for soil conservation around the country and gave speeches at farm meetings and to Congress. During the height of the Dust Bowl in 1935, dust storms reached as far as Washington DC. Bennett anticipated a storm and testified before a Congressional committee[8] on the bill that officially created the United States Department of Agriculture—Soil Conservation Service (USDA-SCS; Soil

---

[8] I have heard stories from old SCS conservationists that Bennett used to throw handfuls of soil in Congress to make his point.

Fig. 1.5 (**a**). Farmer Arthur Coble and sons walking in the face of a dust storm, Cimarron County, Oklahoma. Arthur Rothstein, photographer, April 1936 (Library of Congress). (**b**). Dust storm approaching Stratford, Texas. Dustbowl surveying in Texas Image ID: theb1365, Historic C&GS Collection Location: Stratford, Texas Photo Date: April 18, 1935 Credit: NOAA George E. Marsh Album. (**c**). Dust storm from the Midwest blew into Washington in 1935, darkening the skies over the Lincoln Memorial. (USDA-NRCS photo archive)

Conservation Act of April 27, 1935).[9] In response to the Dust Bowl, in the 1930s, the Roosevelt Administration also purchased lands to help preserve prairie ecosystems and soils. The USDA—National Forest Service's Cimarron, Comanche, Kiowa, and Rita Blanca National Grasslands in the Great Plains were some of the most severely wind-eroded areas.

Drought is inherent in the Great Plains. Winter count data was traditionally collected by Native American winter count keepers (tribal historian—a lifetime job and handed down to relatives) who consulted with tribal elders and leaders (Fig. 1.6). Winter count data from 1777 to 1869 shows that ten droughts were recorded during this period, five being extreme, one severe, one moderate, and three mild (Gallo and

---

[9] In 1994, SCS's name was changed to the Natural Resources Conservation Service, which represents a more broadened scope of the agency's activities (soil, water, animals, plants, air).

Fig. 1.6 Native American paints a winter count on a buffalo skin. Each individual drawing or pictograph represents a significant event that occurred in the past year. National Anthropological Archives. Item naa inv. 03494000. Smithsonian Institution, Washington DC

Wood 2015). Will a soil erosion event occur again that was as devastating as the Dust Bowl? Severe droughts have occurred since and will continue to occur periodically; however, modern conservation practices can greatly diminish the environmental consequences and soil erosion levels. Since the Dust Bowl days, droughts have occurred in each decade throughout the United States. Another severe drought (1952–1957) in the Great Plains had serious consequences for agriculture, and by the end of that drought, a national emergency was declared. Another major drought occurred in the Great Plains during 2012. That summer, dust storms closed sections of US Interstate 70.[10] In the Great Plains, summer rainfall generally occurs mostly from May to August. However, in 2012, the rains failed, and the usual abundance of slow-soaking precipitation-frontal systems and thunderstorms that characterize the Great Plains' climate did not provide needed moisture. The 2012 drought developed rapidly from May, reaching its peak intensity in August and continuing through the fall. Records show that the 4-month cumulative rainfall deficit (averaged over a six-state area of the Central Great Plains) was the greatest since record-keeping began in 1895 (Hoerling et al. 2014). This drought is ranked as the most severe summertime drought for 117 years, surpassing the droughts of 1988, 1934, and 1936—the cause of the 2012 drought was meteorological.

Significant advances in agriculture began in the 1950s and late 1960s. The Green Revolution introduced new agricultural technologies that increased agricultural

---

[10] We must be ever vigilant as wind erosion is still a lingering threat to agricultural pursuits.

production worldwide. The Green Revolution resulted in the use of high-yielding varieties, especially dwarf wheat and rice varieties, improved hybrids, and more widespread use of chemical fertilizers and agrochemicals, improved irrigation, more conservation-minded cultivation, and mechanization. Norman Borlaug was the "Father of the Green Revolution," and he received the Nobel Peace Prize in 1970 and was credited with saving over a billion people worldwide from starvation. In the 1980s, technologies derived from the Green Revolution reached a plateau. During the 1990s, the focus was on biotechnology, altering the genetics of plants to thwart pests and weeds (Parayil 2003).

Land Statistics: Area and Ownership in the United States and Erosion Trends

As a point of reference, there is approximately 0.92 B ha (2.27 B ac) in the United States (50 states), 259 M ha (640 M ac) are federal and 0.6 B ha (1.63 B ac) are private. Federal lands are largely administered by the Bureau of Land Management (100.5 M ha; 248.3 M ac), US Forest Service (78 M ha; 192.9 M ac), Fish and Wildlife Service (36 M ha; 89.1 M ac; 76.6 M ac in Alaska), National Park Service (31.9 M ha; 78.8 M ac; 52.4 M ac in Alaska), and Department of Defense (4.6 M ha; 11.4 M ac) (Vincent et al. 2017). The amount of federally owned land varies across the United States, but most federal land is in the Western United States.

On private land in the conterminous 48 states, range and forestland occupy the largest land areas. Cropland and pastureland follow in order of acreage, followed by other land uses such as developed land, other rural, water, and Conservation Reserve Land (CRP) (Fig.1.7).

What is the status of soil erosion today? Averages on natural geologic erosion vary according to landscape position. Geologic erosion rates increase from gentle sloping landscapes (0.0001–0.01 mm year^{-1}) to moderate hillslopes (0.001–1 mm year^{-1}, to steep mountainous alpine topography (0.1–10 mm year^{-1}). In contrast, erosion under conventional agricultural practices generally exceeds 0.1 mm year^{-1} (Montgomery 2007). Pimentel (2006) estimates that soil loss is 10 to 40 times greater than average rates of soil formation.[11]

Some global statistics concerning land degradation (Secretariat of the United Nations Convention to Combat Desertification 2014):

- 70.9% of the Earth's surface is water (about 3% fresh), and 29.1% is land area (about 25% of is either highly degraded or experiencing high rates of degradation).
- As of 2008, there was 1386 M ha (3425 M ac) of arable land in the world, and 52% used by agriculture is moderately to severely impacted by land degradation.

---

[11] Termed soil loss tolerance level "T," NRCS. Every soil component has an assigned soil loss tolerance level.

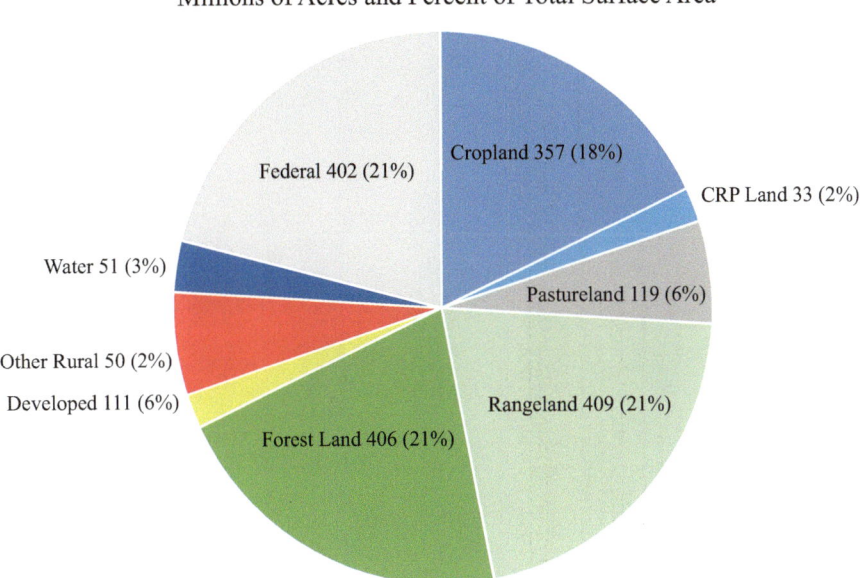

Fig. 1.7 Surface area of land cover/use in conterminous 48 states (USDA-NRCS 2007)

- The Food and Agriculture Organization of the United Nations (FAO 2011, 2015) estimates that over the last 40 years, about 33% of the world's arable land has had significant erosion and continues at a rate of 10 m ha^{-1} year^{-1} (24.7 M ac).
- About 60% of the agricultural land in the Mediterranean Basin is degraded, 30% in the European Mediterranean, and 83% of the African Mediterranean (Safriel 2009).
- On a global level, it is estimated that human land use activities have degraded about 43% of the Earth's vegetated lands (Weil and Brady 2017).

Soil loss from cropland in the United States is decreasing each decade due to conservation measures. Current estimates of soil loss are 1.5 B ha year^{-1} [1.7 B tons year^{-1}; 870 Mg (959.9 M tons)] from water erosion; and 694 Mg (765.1 M tons) from wind erosion (USDA-NRCS 2007). Soil loss estimates on cropland are derived from computer models such as the Revised Universal Soil Loss Equation.[12] It is interesting to note that the Roman poet Virgil (70 BC–19 BC) advised farmers: "study the ways of the winds" before farming in a new land (see Virgil 2005). About 69 Mg (75 B tons) of soil is eroded worldwide each year; about two-thirds comes from agricultural land. If we assume a cost of $3 per ton of soil for nutrients, $2 per ton for water loss, and $3 per ton for off-site impacts, this massive soil loss costs the

---

[12] The Revised Universal Soil Loss Model (RUSLE2) is the erosion prediction technology that uses the familiar USLE empirical structure.

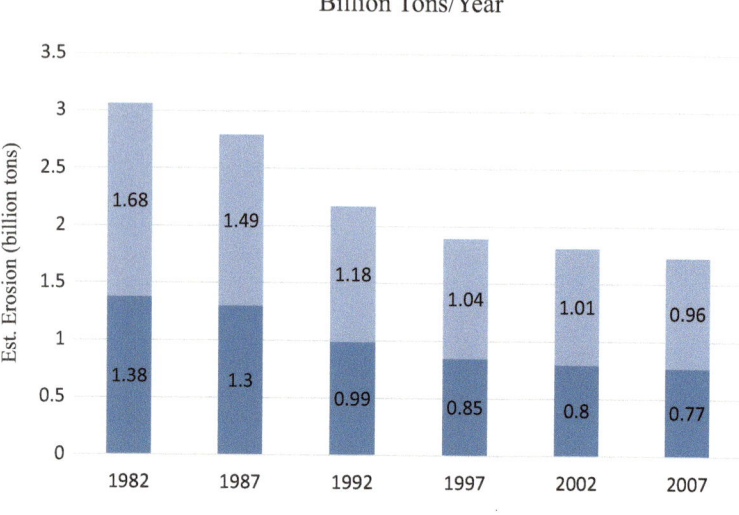

Fig. 1.8 USDA—Natural Resource Conservation Service estimates of soil loss from wind and water erosion on cropland (1982–2007) (USDA-NRCS 2007; English units)

world about $400 B per year, or about $67 per person per year" (Pimentel et al. 1995; Pimentel 2006) (Fig. 1.8).

Degradation of the land has been a serious problem in the past, but it is still an important global issue today because:

> of its adverse impact on agronomic productivity, the environment, and its effect on food security and the quality of life. Productivity impacts of land degradation are due to a decline in land quality on site where degradation occurs (e.g., erosion) and off-site where sediments are deposited. However, the on-site impacts of land degradation on productivity are easily masked due to the use of additional inputs and the adoption of improved technology and have led some to question the negative effects of desertification. The relative magnitude of economic losses due to productivity decline versus environmental deterioration also has created a debate. Some economists argue that the on-site impact of soil erosion and other degradative processes are not severe enough to warrant implementing any action plan at a national or an international level. Land managers (farmers), they argue, should take care of the restorative inputs needed to enhance productivity. Agronomists and soil scientists, on the other hand, argue that land is a non-renewable resource at a human time-scale, and some adverse effects of degradative processes on land quality are irreversible, e.g., reduction in effective rooting depth. The masking effect of improved technology provides a false sense of security. (Eswaran et al. 2001).

In Jared Diamond's book, *Collapse: How Societies Choose to Fail or Succeed*, the author poses the question: "What is the single most important environmental population problem facing the world today? A flip answer would be, the single most important problem is our misguided focus on identifying the single most important problem"! I think it is interesting to contemplate what agriculture would be today if soil conservation knowledge and practices were employed by early and current agriculturalists on a consistent basis. A rudimentary knowledge of conservation practices was known by our ancestors as discussed above, but no formal organization of technical assistance to landowners and operators began until 1905 (USFS), 1935 (SCS), and 1934 (BLM). As Eswaran et al. (2001) point out in the quote above, debates throughout the world continue regarding the importance and implementation of conservation soil erosion action plans. Where debates and doubt exist throughout the world, and conservation programs are ignored or delayed, it is indeed a sad day for those societies.

Gifford Pinchot (1865–1946) was an American visionary, forester, and politician who was the first Chief of the newly created US Forest Service by President Theodore Roosevelt in 1905.[13] Pinchot's definition of conservation was "the wise use of the earth and its resources for the lasting good of men." The Nature Conservancy also has a similar vision of conservation: "Protecting land is where our story began, and it will always be a part of our DNA. Today, we know we must use, conserve, and restore land at an unprecedented scale—for the benefit of people, wildlife, and our climate." The USDA–Natural Resources Conservation Service (USDA-NRCS), instructed by Congress, is mandated with providing conservation management programs on non-federal lands. The USDA-NRCS does not administer and manage public lands; the agency offers conservation assistance to landowners on a voluntary basis. The USDA-NRCS mission is about "helping people help the land … to provide resources to farmers and landowners to aid them with conservation. Ensuring productive lands in harmony with a healthy environment is our priority."[14]

The US Forest Service is also dedicated to healthy productive soils, and its mission is to "sustain the health, diversity, and productivity of the nations' forests and grasslands to meet the needs of future and present generations. Many of the forests and grasslands we manage today were created as part of a national effort to protect soil and water resource degradation and restore forests and ecosystems. The original forest reserves were identified to protect and secure favorable flows of water and timber (Organic Act) to reduce or minimize soil erosion" (USFS 2015). The US

---

[13] Theodore Roosevelt's conservation group, Boone and Crockett Club, created the concept of National Forests in 1875, as a result of environmental concerns regarding Yellowstone National Park (the first National park established on March 1, 1872, during the administration of Ulysses S. Grant). In 1876, Congress created the office of Special Agent in the Department of Agriculture with Franklin B. Hough as head, for the purpose of assessing forest conditions in the United States. In 1881, the office was merged into the Division of Forestry (renamed the Bureau of Forestry in 1901). The Transfer Act of 1905 moved the management of forest reserves from the General Land Office of the Interior Department to the Bureau of Forestry and created the United States Forest Service (Gifford Pinchot was named United States Chief Forester during the Presidency of Theodore Roosevelt).

[14] The NRCS is the USDA's primary agency to assist, on a voluntary basis, private landowners, conservation districts, tribes, other organizations, and interested citizens. Crossover to federal lands occurs in cooperative efforts, e.g., US Forest Service, USDI-Bureau of Land Management, US Fish and Wildlife Service, and vice-versa.

Department of Interior, Bureau of Land Management (USDI-BLM) administers more land in the United States than any other government agency (247 M surface acres of public land).[15] The USDI-BLM is also committed to managing and conserving resources for multiple use and sustained yield. The BLM mission is: "sustain the health, diversity, and productivity of the public lands for the use and enjoyment of present and future generations" (USDI-BLM 2017).

What Is Soil Health: Is It a New Idea?

Fortunately, the idea of soil health is now making it to national news; although the concept of soil health is not a new idea, the subject has resurged and is now highlighted in newspapers, agricultural magazines, scientific journals, gardening clubs, cooperative extension bulletins, and government agency policies. The resurgence of interest in soil health most likely is in response to a variety of reasons:

- Exponential growth and interest in organic food products and farming ($43 B according to the Organic Trade Association's industry; global sales of organic foods amounted to about 81.6 B US dollars in 2015) (Statista 2018).[16]
- The US Department of Agriculture (USDA) reported 19,474 organic farms, ranches, and processing facilities in 2015, up more than 5% from the previous year and 250% from 2002, when record-keeping began. Throughout the world, there are more than 27,800 organic producers.[17]
- Impetus from the FAO International Year of Soils (2015) by the United Nations Food and Agriculture Organization. Interest worldwide on landscape sustainability.
- Earths' population growth.
- Increasing recognition of degraded resources throughout the world.
- Interest in sustainable crop production systems that reduce tillage operations and wind and water erosion, use of no-till and reduced tillage to reduce soil disturbance.

---

[15] The history of the Bureau of Land Management extends as far back as 1785 with the Land Ordinance and NW Ordinance of 1787, created to survey and usher settlement of lands in the original 13 US colonies after the American Revolution. In 1812, Congress established the General Land Office (in the Dept. of Treasury) to administer the disposition of lands sold by the federal government (i.e., federal lands). Subsequent disposal of federal lands occurred over many years (beginning with the Homestead Act of 1862 and others, homesteading was discontinued in 1976 in the conterminous United States, and Alaska in 1986). During 1934, there was an important event in the history of US rangelands occurred: the Taylor Grazing Act, which established the US Grazing Service and initiated formal management of public rangelands through organized grazing advisory boards and grazing fees. The Grazing Service merged with the General Land Office in 1946, to form the Bureau of Land Management within the US Dept. of Interior, which then began to focus more on land management and less on historic land disposal.

[16] https://www.statista.com/topics/1047/organic-food-industry/.

[17] https://www.nass.usda.gov/Surveys/Guide_to_NASS_Surveys/Organic_Production/FAQ/index.php.

- The correlation between soil health and human health.
- Greater affluence throughout the world and desire for higher-protein diets.
- Desire to improve water quality.
- Maintaining productive farmland for future generations.
- The economics of reducing fuel, soil amendments, and wear and tear on equipment.
- Improve and sustain crop yields and productivity.
- Maximize the local water budget, improve hydrologic function, and manage soil moisture.
- Enhancing wildlife habitat.
- New science and better information about the quantitative effects of soil health.

The USDA-NRCS created a Soil Health Division in 2016 because of "the increasing interest in soil health by farmers, ranchers, and landowners across the country. USDA-NRCS responded by creating a Division of Soil Health. The new Division, staffed by soil health experts located strategically across the country, will provide direct training and technical assistance. Goals of the new Division focus on soil health training, assessment, planning, and implementation … direct assistance [will be provided] to NRCS field staff, landowners, and others seeking to improve the health of their soils."

Other organizations are active in promoting soil health. In 2013, the Samuel Roberts Noble Foundation (Ardmore, OK) and the Farm Foundation along with leaders from government and non-government agencies, industry, and private land-owners to evaluate the current statistics on soil health throughout the world. The Soil Health Institute[18] was created to foster science-based information about soil health. Their mission and vision is to: "safeguard and enhance the vitality and pro-ductivity of soil through scientific research and advancement … we believe that healthy soil is the foundation of life and society, and should be treated as an irre-placeable resource. We believe that knowledge advancement and research will improve soil health and sustain the earth's most valuable asset … now is the time for farmers, ranchers, researchers, academia, legislators, government agencies, indus-try, and environmental groups to invest in our soil like never before. Together, we can unify, restore and protect this land, so it may enrich our lives" (Soil Health Institute 2018).

In 2013, the Food and Agriculture Organization of the United Nations (FAO) emphasized a greater focus on soil health (FAO 2013). FAO stated:

> More attention to the health and management of the planet's soils will be needed to meet the challenge of feeding a growing world population the importance of soil for food secu-rity should be obvious. From the origins of civilization in early farming communities up through today, we can see how societies have prospered thanks to healthy soils and declined when their lands became degraded or infertile …

> Healthy soil is not only the foundation of food production, but serves other functions … for example, soil is critical to the health of ground and surface waters and ecosystem health,

---

[18] https://soilhealthinstitute.org/.

and sequesters twice as much carbon as is found in the atmosphere … until recently, soils were the most overlooked and widely degraded natural resource. Today that state of affairs that has at last begun to change, with World Soil Day poised to be recognized by the United Nations and a new, international Global Soil Partnership.

Often the terms soil health and soil quality intertwine. Often the concepts of soil health and quality are used synonymously; however, soil quality denotes a soil's natural composition and quantifiable natural properties that are inherent for a particular soil type, e.g., soil physical/chemical characteristics and historical soil-forming factors, which are fixed by nature. Soil quality factors "generally cannot be influenced by human management" (Moebius-Clune et al. 2016). Karlen et al. 1997 defined soil quality as: "The capacity of a specific kind of soil to function, within natural or managed ecosystem boundaries, to sustain plant and animal productivity, maintain or enhance water quality and air quality, and support human health and habitation." Soil functions in various capacities:

1. Soil sustains biological components, their activity, diversity, and ultimately productivity.
2. Soils regulate and partition water as a function of the hydrologic cycle.
3. Soils serve as an environmental filter; they are involved in buffering and the degradation of organic and inorganic matter.
4. Soil stores and cycles nutrients and other elements in the environment.
5. Soil provides the foundation and support for the infrastructure of civilization.

There are numerous definitions of soil health: "the continued capacity of the soil to function as a vital living ecosystem that sustains plants, animals and humans" (USDA-NRCS 2012); "the capacity of a soil to function, within ecosystem and land-use boundaries, to sustain productivity, maintain environmental quality, and promote plant and animal health" (Doran and Parkin 1994); and FAO defines soil health as "the capacity of soil to function as a living system, with ecosystem and land-use boundaries, to sustain plant and animal productivity, maintain or enhance water and air quality, and promote plant and animal health. Healthy soils maintain a diverse community of soil organisms that help to control plant disease, insect and weed pests, form beneficial symbiotic associations with plant roots; recycle essential plant nutrients; improve soil structure with positive repercussions for soil water and nutrient holding capacity, and ultimately improve crop production" (FAO 2008). The USDA—Natural Resources Conservation Service defines soil health as "the capacity of the soil to function as a vital living ecosystem to sustain plants, animals, and humans. Six key soil physical and biological processes were identified that must function well in a healthy soil, and therefore would especially benefit from measurement methods standardization: (1) organic matter dynamics and carbon sequestration, (2) soil structural stability, (3) general microbial activity, (4) C food source, (5) bioavailable N, and (6) microbial community diversity (USDA-NRCS 2019)." In summary, soil health "refers to soil properties that change as a result of soil use and management over time" (Moebius-Clune et al. 2016).

In order to accomplish goals of improving soil health, an integrated approach is needed where several practices are used in tandem. For example, cover crops, residue, and tillage management, reduced and no-till, and conservation crop rotation are some of the principal conservation management practices used in sustainable cropping systems. Their collective use in a "soil health management system" can provide for more long-term soil health benefits. The overall goal of the conservation management practices listed above are to (1) minimize soil disturbance, (2) maximize soil cover, (3) maximize biodiversity, and (4) maximize and enhance living roots—microorganisms are most active in and around plant roots and in healthy cropping environments; the increase in diversity of beneficial microbia can deter pathogenic organisms and enhance nutrient cycle in soils. In addition, aggregate stability and porosity in the soil also enhance hydrologic function.

All soils (i.e., not degraded soils) have a natural level of soil quality and health, which are both unique to that particular soil in its ecosystem setting. Soils develop and evolve over long periods of time in conjunction with climate, geology, parent materials, vegetation, and local microbial and animal floras. "Soils—in the pedological [science that deals with formation, biotic and abiotic components and classification] sense—are born, mature, and die. They are born when sediments are deposited. They mature after sedimentation stops and soil horizons develop. And they die when erosion strips away the soil horizons and exposes underlying parent material" (USDA-NRCS 2016b). Early Russian soil scientists (pedologists) developed a metaphor called "soil memory"; soils have a past history and so-called memory dictated by climate, vegetation, and land use—these properties are discernable by the properties and stratigraphy of the soil (scientific study of rock strata or layers, especially the distribution, deposition, correlation, and age of sedimentary rocks).

Where levels of soil organic matter may be naturally low (e.g., semiarid and arid environments), this property is representative of that soil type, which was influenced by the historic soil-forming factors and the soil's evolution over time. Another example may be soil pH; soils developing with forest vegetation are typically acid; however, soils with high carbonates and limestone-derived parent material are typically alkaline. Both these conditions are representative and cannot be altered easily without high agricultural inputs. Inherent natural levels of organic matter, pH, carbonates, sodium, sand, silt, clay, and soil structure constitute "natural soil quality." Some of these soil physical and chemical attributes may not be conducive to agricultural or construction pursuits (or to your liking), but they still represent the particular qualities of that soil. Because a soil may not have abundant organic matter, or a pH level that is desirable for a particular crop, or maybe relatively low in plant essential nutrients, that does not mean that the soil is "bad" or "unhealthy." It just means that that soil will not meet your needs or agricultural objective, but regardless, it is healthy from a natural point of view. The critical point regarding soil quality or health, a reference condition is required for each soil type so that comparisons of existing conditions can be made. Without a reference for each particular soil, soil quality or health determinations are not possible. This is an extremely important point, which is discussed in the next chapter "The Ecological Connection," where an ecological site defines the reference standard as well as deviations caused by perturbations to the system.

Objectives of This Book

Soil health is now a prominent topic on farms, ranches, and homeowner's gardens. Nature takes thousands of years to create fertile soil out of sterile bedrock. Soils are born when sediments are deposited; mature after sedimentation stops, and soils develop in conjunction with climate and vegetation and die when erosion removes topsoil and exposes the underlying geologic parent material. Soil health and sustainable agriculture are fundamental to our very existence and survival. In the United States, there are currently 3.2 M agricultural producers operating 2.1 M farms and ranches covering 370 M ha (915 M ac) that produce food, fuel, and fiber for Americans and people around the world. In addition, 35% of all households (42 M, up 17% the last 5 years) in America are growing food at home or in a community garden. Soil properties and related health dynamics are unique from place to place. New research and concepts concerning soil health are focusing on several issues: (a) the connection of hydrology and soil health; (b) the role of soil microbes; (c) the importance of organic matter, (d) cover crops, green manuring, and permanent vegetation relationships to carbon cycling; (e) the effects of organic and inorganic soil amendments; and (f) improving soil health and evaluating the effectiveness of conservation practices. The objectives of this book are to:

- Introduce hydrologic and soil science principles and attributes that are important in understanding and maintaining soil health in farming, ranching, and gardening. Each of these land uses is unique, and the book will address soil science and attributes pertinent to cropland, range, and gardens.
- Provide multiple views of carbon cycling and soil organic carbon reserves and renewal in the soil.
- Identify challenges and pragmatic avenues of management to restore and maintain soil health.
- Encourage a science-based approach to soil health and include science notes from the scientific literature that summarizes new technology and/or refutes certain exaggerated popularized notions of soil health.

References

Ashton, T.S. 1948. Some statistics of the industrial revolution in Britain. *The Manchester School* 16: 214–234.

Atwater, C. 1818. On the prairies and barrens of the west. *American Journal Science and Arts* 1: 116–125.

Axelrod, D.I. 1985. Rise of the grassland biome, central North America. *The Botanical Review* 51: 163–201.

Bogaard, A., R. Fraser, T.H. Heaton, M. Wallace, P. Vaiglova, M. Charles, and R.M. Arbogast. 2013. Crop manuring and intensive land management by Europe's first farmers. *Proceedings of the National Academy of Sciences* 110: 12589–12594.

Brauneis, K. 2004. Fire use during the Great Sioux War. *Fire Management Today* 64: 4–9.

Bruns, H.A. 2012. Concepts in Crop Rotations, Agricultural Science. Agricultural Science. Rijeka, Croatia: Intech Publishers 25–48.

Butt, J.J. 2002. *Daily life in the age of Charlemagne*, 82–83. Westport: Greenwood Publishing Group.

Doran, J.W., and T.B. Parkin. 1994. Defining and assessing soil quality. In *Defining soil quality for a sustainable environment*, ed. J.W. Doran et al. , 3–21. Madison: Soil Science Society America Inc. and American Society Agronomy Inc.SSSA Spec. Publ. No. 35

Eswaran, H., R. Lal, and P.F. Reich. 2001. Land degradation: An overview. Responses to land degradation. In *Proc. 2nd. International conference on land degradation and desertification*, ed. E.M. Bridges, I.D. Hannam, L.R. Oldeman, F.W.T. Pening de Vries, S.J. Scherr, and S. Sompatpanit. Khon Kaen/New Delhi: Oxford Press.

Food and Agriculture Organization (FAO). 2008. Soil health definitions. http://www.fao.org/agriculture/crops/thematic-sitemap/theme/spi/soil-biodiversity/the-nature-of-soil/what-is-a-healthy-soil/en/

———. 2011. Scarcity and abundance of land resources: Competing uses and shrinking land resource base, SOLAW TR02.

———. 2013. http://www.fao.org/news/story/en/item/209369/icode/

———. 2015. International year of soils. Food and Agriculture Organization of the United Nations, Rome Italy. soils-2015@fao.org.

Frost, C.C. 1998. Presettlement fire frequency regimes of the United States: A first approximation. Fire in ecosystem management: Shifting the paradigm from suppression to prescription. In *Tall timbers fire ecology conference proceedings*, vol. 20, 70–81. Tallahassee: Tall Timbers Research Station.

Gallo, K., and E. Wood. 2015. Historical drought events of the Great Plains recorded by native Americans. *Great Plains Research* 25: 151–158.

Gerhard, F. 1857. *Illinois as it is*. Chicago: Keen and Lee.

Hargreaves, M.W.M. 1957. *Dry farming in the northern Great Plains, 1900–1925*. Cambridge, MA: Harvard University Press.

Helms, D. 1991. Readings in the history of the soil conservation service – NRCS – USDA. Two centuries of soil conservation. OAH Magazine of History (Winter 1991), 24–28.

Hillel, D. 2005. Civilization, role of soils. In: Encyclopedia of soils in the environment (Vol. 1). New York: Elsevier.

Hoerling, M., J. Eischeid, A. Kumar, R. Leung, A. Mariotti, K. Mo, S. Schubert, and R. Seager. 2014. Causes and predictability of the 2012 Great Plains drought. *American Meteorological Society* 95: 269–282.

Karlen, D.L., M.J. Mausbach, J.W. Doran, R.G. Cline, R.F. Harris, and G.E. Schuman. 1997. Soil quality: A concept, definition, and framework for evaluation (a guest editorial). *Soil Science Society of America Journal* 61: 4–10.

Lee, J.A., and T.E. Gill. 2015. Multiple causes of wind erosion in the Dust Bowl. *Aeolian Research* 19: 15–36.

Lewis, H.T. 1973. *Patterns of Indian burning in California: Ecology and ethnohistory (No. 1)*. Ramona: Ballena Press.

Mazoyer, M., and L. Roudart. 2006. *A history of world agriculture: From the neolithic age to the current crisis*. New York: NYU Press.

Miao, Q. 2018. The selected readings from "Agricultural Treatise of Chen Fu". Beijing: Agricultural Press: 1–10.

Moebius-Clune, B.N., D.J. Moebius-Clune, B.K. Gugino, O.J. Idowu, R.R. Schindelbeck, A.J. Ristow, H.M. van Es, J.E. Thies, H.A. Shayler, M.B. McBride, K.S.M Kurtz, D.W. Wolfe, and G.S. Abawi. 2016. Comprehensive Assessment of Soil Health – The Cornell Framework, Edition 3.2, Geneva, NY: Cornell University.

Montgomery, D.R. 2007. Soil erosion and agricultural sustainability. *Proceedings of the National Academy of Sciences* 104: 13268–13272.

Parayil, G. 2003. Mapping technological trajectories of the Green Revolution and the Gene Revolution from modernization to globalization. *Research Policy* 32: 971–990.

Pimentel, D., C. Harvey, P. Resosudarmo, K. Sinclair, D. Kurz, M. McNair, and R. Blair. 1995. Environmental and economic costs of soil erosion and conservation benefits. *Science-AAAS-Weekly Paper Edition* 267 (5201): 1117–1122.

Plamondon, M. 2000. Lewis and Clark Trail Maps: Missouri River Between Camp River Dubois (Illinois) and Fort Mandan (North Dakota)-Outbound 1804; Return 1806. Pullman, Washington: Washington State University Press.

Plumbe, J. 1839. Sketches of Iowa and Wisconsin. *Annals of IA* 14 (483–531): 595–619.

Raghavan, M., M. Steinrücken, K. Harris, S. Schiffels, S. Rasmussen, M. DeGiorgio, and A. Eriksson. 2015. Genomic evidence for the Pleistocene and recent population history of native Americans. *Science* 349 (6250): aab3884.

Safriel, U.N. 2009. Status of desertification in the Mediterranean region. In *Water scarcity, land degradation and desertification in the Mediterranean region*, 33–73. Dordrecht: Springer.

Secretariat of the United Nations Convention to Combat Desertification. 2014. UN Campus, Bonn, Germany. *Platz der Vereinten Nationen* 1: 53113.

Selin, H. (ed.). 2013. Encyclopedia of the history of science, technology, and medicine in non-western cultures. Berlin, Germany: Springer Science and Business Media.

Soil Health Institute. 2018. https://soilhealthinstitute.org/

Statista. 2018. Organic food and non-food sales in the United States from 2008 to 2018. https://www.statista.com/statistics/244394/organic-sales-in-the-united-states/

Stebbins, G.L. 1981. Coevolution of grasses and herbivores. *Annals of the Missouri Botanical Garden* 68: 75–86.

Tyldesley, J. 1995. *Daughters of Isis: Women of ancient Egypt*. London: Penguin Books.

USDA-National Forest Service. 2015. https://www.usda.gov/media/blog/2015/08/6/healthy-soils-provide-foundation-healthy-life-national-forests-and-grasslands

USDA-NRCS. 2007. Soil erosion on cropland. https://www.nrcs.usda.gov/wps/portal/nrcs/detail/national/technical/nra/nri/?cid=stelprdb1041887

———. 2012. Natural resources conservation services: Soil health. http://www.nrcs.usda.gov/wps/portal/nrcs/main/soils/health/

———. 2016a. USDA-NRCS. Soil Health Division. https://www.nrcs.usda.gov/wps/portal/nrcs/main/national/about/history/

———. 2016b. PA#2205, 2017. Soil Planner. Washington, DC.

———. 2019. Recommended soil health indicators and associated laboratory procedures. June 5, 2019, Tech. Note 450-TCH-3. Washington DC.

USDI-BLM. 2017. https://www.doi.gov/sites/doi.gov/files/uploads/FY2017_BIB_BH007.pdf

Vincent, C.H., L.A. Hanson, and C.N. Argueta. 2017. Federal land ownership: overview and data. Congressional Research Service 42346. https://fas.org/sgp/crs/misc/R42346.pdf

Virgil. 2005. *The georgics of Virgil: A translation by David ferry*. New York: Farrar, Straus and Giroux.

Warkentin, B.P. 2006. *Footprints in the soil: People and ideas in soil history*. Amsterdam: Elsevier.

Weil, R.R., and N.C. Brady. 2017. *Nature and properties of soils*. 15th ed. New York: Pearson.

Xu, H.L., H. Umemura, and J.F. Parr Jr. 2000. Nature farming and microbial applications. (Vol. 1). Boca Raton FL: CRC Press.

Chapter 2
The Ecological Connection in Farming, Ranching, and Gardening

Abstract Understanding the ecological connection about the lands we farm, graze, and plant gardens are important because the historic or natural plant communities that existed before cultivation are inextricably linked to what our lands are capable of today (potentials, productivity, sustainability, and resistance and resilence to disturbances). Many US government land management agencies are now using the concept of ecological site or similar models of classification in land management. State and transition models associated with ecological sites are the culmination of a collection of knowledge about ecosystem dynamics and changes in response to natural events and management applications. They are diagrammatic portrayals with narratives and identification of specific environmental drivers—states can change because of natural or anthropogenic disturbance events or lack of natural events. Altered states such as cropland and pasturelands emerged at some point in history from clearing timber and/or plowing grasslands and are now identified in state and transition models. Nature often imposes its will where some cropland and pastureland sites, when abandoned, have a tendency to return to brush and trees (often referred to as "go-back land." Note the connotation to some cropland and pasturelands—the use of ecological sites and inclusion of altered states identifies if this dynamic is operative. Ecological sites are dynamic and unique across the landscape of the United States, and awareness and use of ecological site descriptions on the lands we live on, either on a farm, ranch, or garden, is the first step in maintaining soil health, sustained productivity, and continued prosperity of any agricultural endeavor. The objectives of this chapter are (1) introduce mainstream ecological classification concepts, (2) discuss the connection and importance of ecological sites in addressing soil and plant health, and (3) provide an example of ecological site description with state and transition model with natural vegetation states and altered states (pastureland, cropland, and agroforestry).

Questions

- How can the ecological site concept assist in understanding the ecological dynamics of soil health for cropland, rangeland, and pastureland and in the garden?
- What is the value of ecological sites to agricultural producers (farming, ranching, forestry, and gardening)?
- What is the purpose of an ecological site description?

© Springer Nature Switzerland AG 2020   27
K. E. Spaeth Jr., *Soil Health on the Farm, Ranch, and in the Garden*,
https://doi.org/10.1007/978-3-030-40398-0_2

- What are the pertinent points of interest to farmers, ranchers, and gardeners? What sections in ESD's may be of particular interest to academics?
- In cropland systems, how can the state and transition model depicting reference and successive states assist in maintaining and improving soil health?
- What are the limitations of the state and transition concept in explaining plant successional dynamics in terrestrial plant communities?
- Is successional theory clear and consistent across the gradient of forest and range-land types? How are the dynamics of plant succession related to maintaining soil health? Insert two more questions 1) Discuss the advantages of identifying states and transitions in state and transition models for developing managment plans. 2) How can ecological state and transition models in ecological sites assist communication with land users (including the status of the ecological condition, the path leading to current conditions, options for restoration, and avenues for management)?

Introduction

Pioneers in the United States found that lands that were historically grasslands with deep topsoil and void of boulders were more productive and better suited for crop-land agriculture than forestlands. There are always exceptions, but knowing the eco-logical potential of a particular soil and it's associated historic vegetation are key topics in understanding and managing soil health. Various land management agen-cies have developed various land classification systems to facilitate management and use of different land types. None of these hierarchical systems are perfect; however, these classification schemes have a major function in providing information about soils, management potentials (cropping, grazing, and wildlife), nutrient manage-ment, construction capabilities and, ultimately, maintaining or improving soil health. Each of us has a different familiarity, understanding, and/or perspective of the eco-systems we live in – the specific ecological patterns and variables that affect our agricultural pursuits. One of the main objectives of this book is to explore, in logical order, the complexities of the environments we live in and develop an enhanced understanding of how they function. What are the possibilities and limits with respect to agricultural activities? Ecologically sound information about land and site potentials can be provided by different levels of land classification methodologies.

In *Soil Health on the Farm, Ranch, and in the Garden*, I discuss various ecologi-cal and biological attributes that are important to various land uses. Here I submit land use definitions from different sources to clarify the differences, especially on grazinglands. Even within Federal agencies, more than one definition may exist, and in professional circles, some land use definitions, e.g., rangeland, pastureland, grassland, and prairie can be confusing[1, 2] Definitions of various land uses can be tautological by nature, and ecologists seem to use some terms interchangeably. For example, in *Grasses and Grassland Ecology*, Gibson (2009) mentions that the defi-nition of grassland can be elusive and that many experts do not bother at definition because "we know a grassland when we see one."

---

[1] Food and Agriculture Organization of the United Nations has compiled a list of definitions for rangeland-grazinglands, cropland, etc. http://www.fao.org/docrep/005/Y4171E/Y4171E37.htm
[2] Gibson (2009) provides a thorough discourse on land type definitions.

Text Box 2.1

Various sources of land use are cited (more than one example is given for some land uses as they provide additional insight. See Lund (2006) for an exhaustive treatise on definitions of various land uses).

Land Use: "The degree to which the land reflects human activities (e.g., residential and industrial development, roads, mining, timber harvesting, agriculture, grazing, etc.). Land use describes how a piece of land is managed or used by humans. Land use is generally locally regulated in the U.S. based on zoning and other regulations. Land use mapping differs from land cover mapping in that it is not always obvious what the land use is from visual inspection" (USA-FED-EPA 2006).

Land Use: "The purpose to which land is put by humans (e.g., protected areas, forestry for timber products, plantations, row-crop agriculture, pastures, or human settlements) (Turner and Meyer 1994). Change in land use may or may not cause a significant change in land cover. For example, change from selectively harvested forest to protected forest will not cause much discernible cover change in the short term, but change to cultivated land will cause a large change in cover" (Lund 2006).

Land Use: "The description of the human cultural activities on the land and water" (SRM 1998).

Cropland: "includes areas used for the production of adapted crops for harvest. Two subcategories of cropland are recognized: cultivated and noncultivated. Cultivated cropland comprises land in row crops or close-grown crops and also other cultivated cropland, for example, hayland or pastureland that is in a rotation with row or close-grown crops. Non-cultivated cropland includes permanent hayland and horticultural cropland" (USDA-NRCS NRI Handbook 2018).

Arable and Permanent Cropland: "Arable and permanent cropland is the total of 'arable land' and 'land under permanent crops'. Arable land is the land under temporary crops, temporary meadows for mowing or pasture, land under market and kitchen gardens and land temporarily fallow (for less than 5 years); and land under permanent crops is the land cultivated with crops that occupy the land for long periods and need not be replanted after each harvest" (FAO 2005).

Grazingland: "Grazing lands exist in every state, but the amounts and kinds of land and the uses, products, and values from grazing lands vary from state to state. Examples of grazing land include--annual grasslands of California; hot deserts in the southwestern states and cold deserts in the Great Basin; shrub-grasslands throughout the western states; prairie grasslands of the Great Plains and Corn Belt; humid grasslands of the eastern United States and Hawaii; tundra rangelands of Alaska; improved pasture and hay lands throughout the Intermountain West, Northern Great Plains, Great Lakes, Northeast,

(continued)

and South; wetlands and riparian areas in every state; and grazed forests in all states adjacent to and east of the Mississippi River and in the mountain states of the West" (USDA-NRCS 1995).

Rangeland: Rangelands are those areas of the world, which by reason of physical limitations—low and erratic precipitation, rough topography, poor drainage, or cold temperatures—are unsuited to cultivation and which are a source of forage for free-ranging native and domestic animals, as well as a source of wood products, water, and wildlife (Stoddard et al. 1975).

Rangeland: "Land on which the indigenous vegetation (climax or natural potential) is predominantly grasses, grass-like plants, forbs, or shrubs and is managed as a natural ecosystem. If plants are introduced, they are managed similarly. Rangeland includes natural grasslands, savannas, shrublands, many deserts, tundra, alpine communities, marshes and meadows" (SRM 1998).

Rangeland: "A land cover/use category on which the climax[3] or potential plant cover is composed principally of native grasses, grasslike plants, forbs or shrubs suitable for grazing and browsing, and introduced forage species that are managed like rangeland. This would include areas where introduced hardy and persistent grasses, such as crested wheatgrass, are planted and such practices as deferred grazing, burning, chaining, and rotational grazing are used, with little or no chemicals or fertilizer being applied. Grasslands, savannas, many wetlands, some deserts, and tundra are considered rangeland. Certain communities of low forbs and shrubs, such as mesquite, chaparral, mountain shrub, and pinyon-juniper, are also included as rangeland" (USDA-NRCS NRI Handbook 2018).

Rangeland: "A kind of land on which the historic climax vegetation was predominantly grasses, grasslike plants, forbs, or shrubs. Rangeland includes land revegetated naturally or artificially to provide a plant cover that is managed like native vegetation. Rangelands include natural grasslands, savannas, most deserts, tundra, alpine plant communities, coastal and freshwater marshes, and wet meadows" (FAO 2018).

Rangeland: "American term for land on which the indigenous vegetation is predominantly grasses, grass-like plants, forbs, or shrubs and is managed as a natural ecosystem" (Forage and Grazing Terminology Committee 1992).

---

[3] The term climax, referring to an end point in plant succession (Whittaker 1975). In the United States, nonnative invasive plants have disrupted plant successional pathways that have existed in the evolution and development of an indigenous vegetation. For example, exotic annual grasses are very competitive on western rangelands and have altered the energy and water dynamics in many rangeland plant community types. Perennial grasses and forb seedlings have been unable to compete with annual species that grow and mature quickly, usurp moisture, and prevent perennial seedlings survival (Harris 1977).

(continued)

Rangeland: "Uncultivated land that will provide the necessities of life for grazing and browsing animals" (Holechek et al. 2004).

Rangeland Management: "is the manipulation of rangeland components to obtain the optimum combination of goods and services for society on a sustained basis … range management has two basic components: (1) protecting and enhancing the soil and vegetation complex, and (2) maintain or improving the output of consumable range products, such as red meat, fiber, wood, water, and wildlife" (Holechek et al. 2004).

Pastureland: "Grazing lands, planted primarily to introduced or domesticated native forage species, which receive periodic renovation and/or cultural treatments such as tillage, fertilization, mowing, weed control and irrigation" (SRM 1998).

Pastureland: "A land cover/use category of land managed primarily for the production of introduced forage plants for livestock grazing. Pastureland cover may consist of a single species in a pure stand, a grass mixture, or a grass-legume mixture. Management usually consists of cultural treatments: fertilization, weed control, reseeding or renovation, and control of grazing . . . land that has a vegetative cover of grasses, legumes, and/or forbs, regardless of whether or not it is being grazed by livestock" (USDA-NRCS NRI Handbook 2019).

Pastureland: "Pasturelands are distinguished from rangelands by the fact that periodic cultivation is used to maintain introduced (non-native) forage species, and agronomic inputs such as irrigation and fertilization are applied annually" (Holechek et al. 2004).

Pasture: "A type of grazing management unit enclosed and separated from other areas by fencing or other barriers and devoted to the production of forage for harvest primarily by grazing" (Forage and Grazing Terminology Committee 1992).

Permanent Pasture: "Pastureland composed of perennial or self-seeding annual plants that are grazed annually, generally for 10 or more successive years" (Barnes and Nelson 2003).

Hayland: "A subcategory of cropland managed for the production of forage crops that are machine harvested. These crops may be grasses, legumes, or a combination. Hayland also includes land in set-aside or other short-term agricultural programs" (USDA-NRCS NRI Handbook 2019).

Meadow: "A tract of grassland where productivity of indigenous or introduced forage is modified due to characteristics of the landscape position or hydrology, e.g., hay meadow, wet meadow" (Forage and Grazing Terminology Committee 1992).

Grassland: "Land on which the vegetation is dominated by grasses, grass-like plants, and/or forbs (cf. dominant). Lands not presently grassland that were originally or could become grassland through natural succession may be classified as potential natural grassland" (SRM 1998).

(continued)

Grassland: "Types of vegetation that are subject to periodic drought, that have a canopy dominated by grass and grass-like species, and that grow where there are fewer than 10–15 trees per hectare" (Risser 1988).

Prairie: "French term for grassland, used now to describe the grasslands of North American Great Plains. Defined in the USA as nearly level or rolling grassland, originally treeless, and usually characterized by fertile soil" (Forage and Grazing Terminology Committee 1992).

Forest: "Forests (according to the U.S. National Vegetation Classification system) consist of trees with overlapping crowns forming 60–100% cover. Woodlands are more open, with 25–60% cover" (ISM 2018).

Forestland: "Land that has at least 10% crown cover by live tally trees of any size or has had at least 10% canopy cover of live tally species in the past, based on the presence of stumps, snags, or other evidence. To qualify, the area must be at least 0.4 ha (1.0 acre) in size and 36.5 m (120 ft) wide. Forest land includes transition zones, such as areas between forest and non-forest lands that meet the minimal tree stocking/cover and forest areas adjacent to urban and built—up lands. Roadside, streamside, and shelterbelt strips of trees must have a width of at least 36.5 m (120 ft) and continuous length of at least 110.6 m (363 ft) to qualify as forest land. Unimproved roads and trails, streams, and clearings in forest areas are classified as forest if they are less than 36.5 m (120 ft) wide or less than an acre in size. Tree-covered areas in agricultural production settings, such as fruit orchards, or tree—covered areas in urban settings, such as city parks, are not considered forest land" (USDA-USFS 2016).

Forestland: "Land on which the vegetation is dominated by trees or, if trees are lacking, the land shows historic evidence of former forest and has not been converted to other uses" (SRM 1998).

Forestland: "A land cover/use category that is at least 10% stocked by single-stemmed woody species of any size that will be at least 4 meters (13 ft) tall at maturity. Also included is land bearing evidence of natural regeneration of tree cover (cutover forest or abandoned farmland) and not currently developed for non-forest use. Ten percent stocked, when viewed from a vertical direction is a canopy cover of leaves and branches of 25% or greater. The minimum area for classification of forest land is 1 acre, and the area must be at least 30.5 m (100 ft) wide" (USDA-NRCS NRI Handbook 2019).

Forestland, grazed: "A category under forest land cover/use that includes forest land that is being grazed by livestock and managed using range or pasture management principles and practices adapted to the forest ecosystem" (USDA-NRCS NRI Handbook 2019).

Forestland, not grazed: "A land cover/use category that includes areas meeting the definition of forest land [at least 10% stocked by single-stemmed woody species of any size that will be at least 4 meters (13 ft) tall at maturity] that is not visibly grazed or otherwise disturbed by domestic livestock" (USDA-NRCS NRI Handbook 2019).

Classifying Terrestrial Ecosystems

Ecosystems are tremendously complex, and only a rudimentary understanding of ecological dynamics is possible given all the interactions between all the living and nonliving variables. An ecosystem includes living organisms (animals, plants, microorganisms) living and interacting with abiotic components (nonliving; air, water, soil) that are affected by disturbances over time and are linked to the environment. The environment is the complex of abiotic and other biotic factors that have influenced living organisms at time scales from millennia to present. Figure 2.1 gives a basic representation of the driving factors in an ecosystem. Ecosystems can be viewed on global (the biosphere), continental, regional, local, and individual scale (Fig. 2.1). On the farm, ranch, or in the garden, you are probably more concerned with ecosystem dynamics of an agricultural field, pasture, or garden. Concerns about climate change often are examined at regional to global scales.

No aspect of the environment, including plant communities and soils, is independent of each other. The environment includes the realm of all outside forces in nature that influence and affect the life cycle of all living organisms. In order to understand soil dynamics and health, it is important to identify the role of environmental factors and how plant communities, from a historic and current perspective, have affected the "ecological cycle of events." In any agricultural endeavor, greater knowledge of your immediate environment can only be beneficial to success. Consider a coniferous forest, a prairie, or deciduous forest and the diversity of living organisms living and linked together, how they affect each other, and how they respond to different environmental factors that are unique to the area. In each of these regions, all living organisms are linked in a myriad of ways and interact with climatic and other environmental cycles. If you are crop farming a field or tending a garden, this land was originally forest or rangeland (grasslands, savannahs, warm or cold deserts, shrublands, etc.). In farming, I think it is important to know the history of the land you are working with. For example, soil development under forest or rangeland environments have very different histories with respect to pH, organic matter, soil morphology, and development of the upper layer (topsoil) and subsoil characteristics. This is directly related to understanding soil health. As discussed in Chap. 1, every soil has inherent quality characteristics (some good and bad from a farming perspective). The link to soil health is sometimes confusing, but in my mind, soil health is related to maintenance or sustaining those qualities. A soil may have organic matter content of 3%, which is a quality factor, whereas, soil health is all about maintaining or slightly improving that attribute. The loss of organic matter during cultivation of farmland will be discussed in more detail, but almost all cultivated land initially lost organic matter and most likely has reached a lower, but more stable level of organic matter. In the past, if significant erosion has occurred with organic matter loss, most likely the field has or needs to be re-established with tree and/or grass plantings—crop production will not be lucrative from an economic point of view.

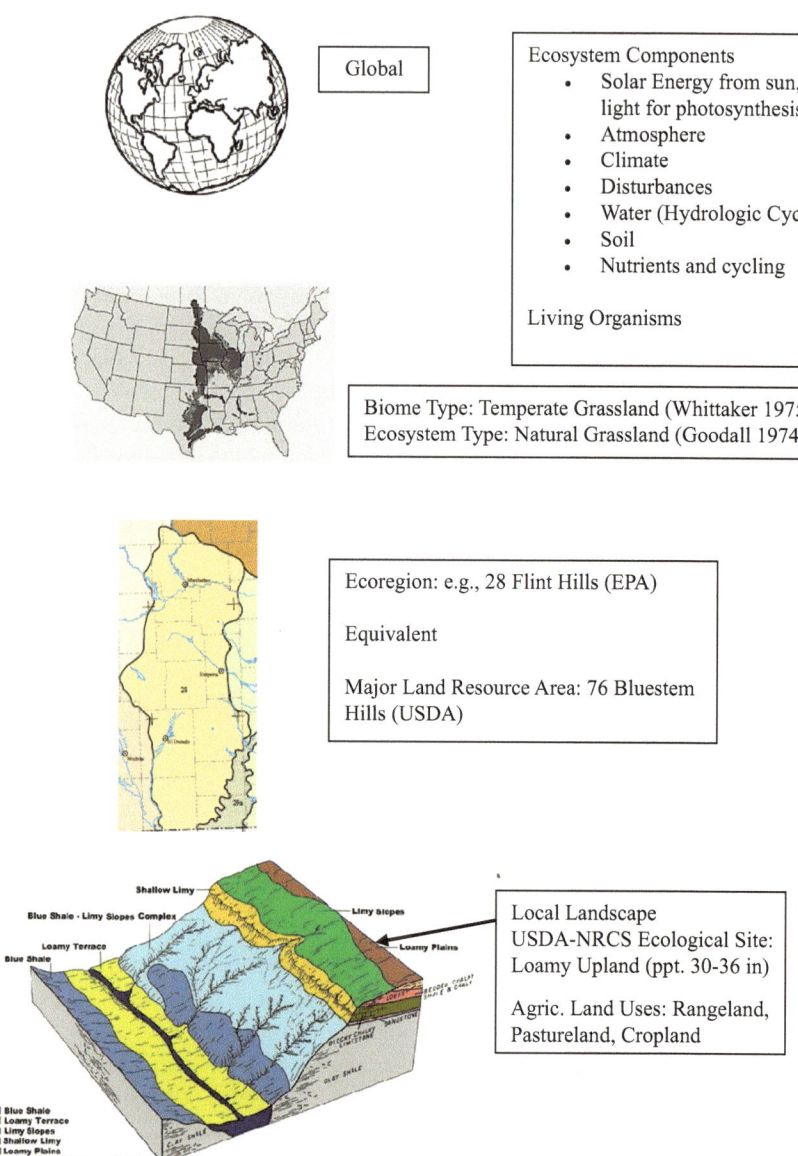

	Ecosystem Components
Global	• Solar Energy from sun, light for photosynthesis • Atmosphere • Climate • Disturbances • Water (Hydrologic Cycle) • Soil • Nutrients and cycling Living Organisms

Biome Type: Temperate Grassland (Whittaker 1975)
Ecosystem Type: Natural Grassland (Goodall 1974)

Ecoregion: e.g., 28 Flint Hills (EPA)

Equivalent

Major Land Resource Area: 76 Bluestem Hills (USDA)

Local Landscape
USDA-NRCS Ecological Site: Loamy Upland (ppt. 30-36 in)

Agric. Land Uses: Rangeland, Pastureland, Cropland

Fig. 2.1 Hierarchy of ecosystem components from a global to a local scale

How do soils affect communities of living organisms and vice versa; how do organisms influence the soil (the physical and chemical structure, hydrology, nutrient cycles, energy cycles, and microclimate)? Energy and nutrients are used in the environment to sustain community vital functions, which are used by organisms and then released back to the environment. In a community, all interacting organisms, biotic factors and pools, and ecosystem processes such as the circulation of energy, matter, and nutrients form an ecosystem. Aggregations of living organisms that are temporally adapted (a community of plants and animals) utilize energy and specific resources within specific environments.

It is interesting to note that basic ecological principles were recognized by Theophrastus of Eresus (372–288 BCE) where he noticed that animal and plant species adapt to specific environments. Ecology as a formal science was established by the German zoologist Ernst Haeckel (1834–1919). He coined the neologism "ecology" [from Greek word oikos (house or dwelling place); logos meaning study, origin; thus "ecology is the scientific study of the earthly dwelling place, or home"] (Keller and Golley 2000). From a contemporary perspective, ecology is a complex science that encompasses biology, botany, evolutionary concepts, chemistry, physics, geology, genetics, and molecular biology. True to Haeckel's view, the modern definition of ecology is the scientific study and conceptual[4] understanding of interactions between organisms and their unique environments (the ecosystem).

Terrestrial ecosystems can be represented on broad and finite landscape scales. The use of terms such as ecosystem, biome, plant association, community types, plant community, and ecological site can be confusing. Even professional ecologists are prone to synonymy and loose interpretations of key ecological terminology (Morin 2011).[5] Ecologists use a variety of terms that represent different scales of classification.[6] Why are terrestrial plant communities classified into various grouping levels and how are they used by land managers? Classification of plant communities is an abstract concept, and each community has a particular characterization based on life/growth forms, species dominance, strata or layers of vegetation (overstory, understory plants), and unique assemblages and composition of those species. For example, as you fly across the United States at 9144 m (30,000 ft), you will see forests. You may be able to differentiate a deciduous from an evergreen coniferous forest, but not identify individual species. At the ground level, the tree species are identifiable and occur in small-to-large tracts of land that reappear on the landscape at various intervals due to aspect of the land, elevation, and local geological and soil differences. For example, in Idaho—Northwestern United States, various dominant tree species may be present as you travel along a highway in the mountains. Ponderosa pine (*Pinus ponderosa*) may be present at lower drier elevations, followed by Douglas fir (*Pseudotsuga menziesii*) and interspersed western larch (*Larix occidentalis*), then grand fir (*Abies grandis*), at still higher elevations, and finally subalpine fir (*Abies lasiocarpa*), which occurs at the highest elevation (Fig. 2.2). In areas with very high elevation gradients above timberline, you may observe an alpine ecosystem with an assortment of different alpine community types or ecological sites. This example transitions three forest types and culminated in a

---

[4] The science of ecology does incorporate factual information about environment and species interactions; however, many dynamics in ecology such as species diversity, competition, resilience of native species to invasive species, and succession rely on model and theories.

[5] In my synecology class, my professor handed out a collection of definitions of community. To this day, I do not know if that was to assist or confuse us more.

[6] The terminology is often not strict in usage and application; often there is overlap in the concepts. There is no single or stringently correct way to classify communities.

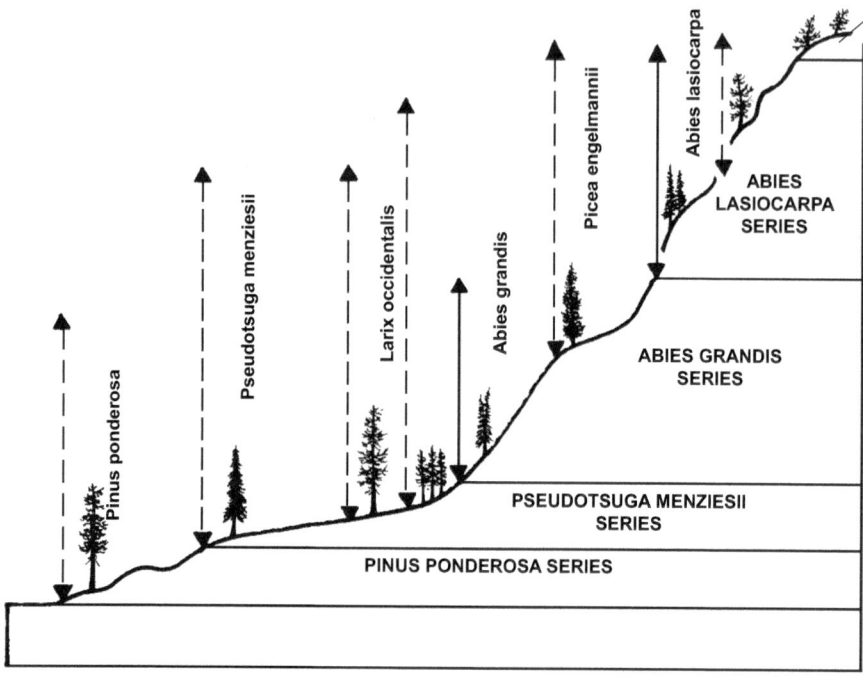

Fig. 2.2 Elevation, moisture, and climate gradient with generalized dominant tree species distribution (northern Idaho). (Courtesy USFS, Cooper et al. 1991)

rangeland type (alpine). This gradient of forest and rangeland types is associated
with elevation, moisture, temperature, and geological and soil changes. Each of
these coniferous tree species grow in specific areas (broad plant associations to
discrete localized sites; habitat types, ecological sites) with characteristic climate,
elevation gradients, and other indigenous plants, animals, and microorganisms.
Geology and soil types found in each of these coniferous ecosystems can also be
variable. The ecology of these four community types is unique and quite different;
therefore, management of these forest types as a single unit would not be efficient
or pragmatic. All these characteristics, which may be unique or common across
various coniferous forest types and for all plant communities (forest or rangeland),
are the basis of terrestrial ecology.

Large-scale terrestrial classification is based on physiognomy (vegetation structure), which can be reduced to six broad physiognomic types called biome types.
Biome types can be continental or semi-continental in scope: forest, grassland,
woodland, shrubland, semidesert shrub, and desert. Biome types can intergrade or
mix due to a wide range of environments and major differences in climate. Biomes
are recognized within biome types and have similar physiognomy, e.g., broadleaf
deciduous forest, western coniferous forest, and temperate grasslands (prairies). For
example, the biomes tallgrass, mixed-grass, and shortgrass prairies are part of the

grassland biome type. Community types (CT) within biomes are more definitive in the approach to land characterization, and there are several approaches and concepts. Dominance types are CTs that specify dominant plant species such as oak-hickory, maple-basswood, Douglas fir, ponderosa pine, or Engelmann spruce (*Picea engelmannii*) forests. This classification approach is overstory species-centric, whereas, a site type also recognizes understory or undergrowth plants. Forest habitat types or ecological sites identify the potential climax tree species, e.g., subalpine fir, which is usually the most adapted shade-tolerant tree species followed by the dominant or understory indicator species such as common beargrass (*Xerophyllum tenax*). A third species name could also be included to denote a phase such as grouse whortleberry (*Vaccinium scoparium*). A phase designation can represent a third understory layer, a transition between two adjacent HTs, or a minor variation within the HT. The US Forest Service has used HTs for several decades to designate the potential climax plant community[7] (Daubenmire and Daubenmire 1968); however, since the publication of the *Interagency Ecological Site Handbook for Rangelands* (USDA 2013), the US Department of Interior Bureau of Land Management (USDI-BLM), US Department of Agriculture Forest Service (USFS), and US Department of Agriculture Natural Resources Conservation Service (USDA-NRCS) have cooperatively agreed to use the concept of rangeland ecological sites to identify and describe rangelands in the United States. This manual set the stage for anticipated future versions of the handbook to address forest, riparian, and culturally managed lands (altered lands such as cropland, agroforestry, and pastureland).

Examining the Landscape: Concepts of Land Classification

The science of vegetation ecology is relatively young, and in the early 1900s, it seems that from the very beginning, the conceptualization of plant communities began with controversy. Two underlying themes (or hypotheses) emerged that categorized vegetation pattern across landscapes: the community unit theory and the individualistic-continuum concept. The debate regarding the nature of community organization has been discussed for almost a century now (Whittaker 1962; Shipley and Keddy 1987a, b; Austin and Smith 1990; McIntosh 1995; Callaway 1997; Reinhart 2012). During the beginning of the twentieth century, the debate started with a basic question: were plant communities an organized system of co-occurring species or an assemblage of a random collection of individualistic species arriving on a site that varied continuously with environmental change along the landscape?

Frederick E. Clements (1874–1945) was an American plant ecologist who presented the view of organismic concept of communities also called the community-unit concept and proposed that plant communities were holistic and interdependent (Clements 1916). Plant communities were likened to a facsimile of an individual

---

[7] Potential natural plant community is the biotic community that would develop if all sucessional sequences in the ecosystem progressed without human-caused distrubances under present environmental conditions.

organism (growth, maturation, and death), visualized as natural units of coevolved species populations forming homogeneous, discrete, and recognizable vegetation units. In contrast, Henry Allan Gleason (1882–1975) an American botanist advocated the individualistic continuum concept or individualistic concept of community organization where communities were a collection of species that had commonality with respect to adaptations to the abiotic[8] environment (Gleason 1926, 1939). Gleason (1917, 1926) deviated from Clement's hypothesis and argued that Clements' monoclimax theory did not account for adjustments of individual plant population dynamics.

Gleason (1917) stated: "Vegetation, in its broader aspects, is composed of a number of plant individuals. The development and maintenance of vegetation is therefore merely the result of the development and maintenance of the component individuals, and is favored, modified, retarded, or inhibited by all causes which influence the component plants. According to this view, the phenomena of vegetation depend completely upon the phenomena of the individual." Here the suggestion of plants being modified by "causes which influence the component plants may be the recognition of environmental effects along the continuum". The transition to the individualistic viewpoint gained momentum when Whittaker (1967) used more sophisticated gradient analyses, which showed patterns of species replacements along a gradient representing the continuum. Ecologists now recognize that species dynamics (existence, composition, fitness, distribution), are not wholly dependent on abiotic conditions and competition, but are highly affected by complex interactions within the plant community, mutualists, and consumers (Callaway 1997).

In reality, vegetation and species populations in plant communities continuously intergrade along environmental gradients—the continuum; however, plant communities with similar species assemblages are also repetitive and recognizable on the landscape. As Whittaker (1975) later stated: "classifications of communities are often needed. There is no real conflict between the principle that communities are generally (but not universally) continuous with one another, and the practice of classifying these communities is a means of communication about them" (Whittaker 1975). I also add you cannot effectively manage a continuum. Ecologists recognize that plant species are distributed in space and respond according to unique genetic, physiological, life-cycle characteristics and physical and environmental factors. Alternatively, community units are characterized as homogenous, discrete community units organized in a hierarchical structure (e.g., plant communities, cover types, habitat types, and ecological sites). Although vegetation occurs along a continuum, ecological understanding and land management are facilitated by forming homogeneous recognizable groups such as the ecological site.

Landscapes can be classified and grouped into recognizable land units for scientific study, evaluation, and land management and planning activities. As shown in Fig. 2.1, common regional conceptualizations among US land management agencies include Omernik ecoregions and Major Land Resource Areas (MLRA) (Omernik 1987, 1995; USDA-NRCS 2006). Many other systems of classifying geographic areas have been used in the United States (Powell 1895, identification of the

---

[8] Abiotic-nonliving basic elements of the environment e.g., rainfall, temperature, wind, soil componentns, minerals etc.

major physiographic provinces within the United States; Fenneman 1928, 1931, regions and subregions of physiographic provinces based on geologic and landform attributes; Dice 1943, biotic provinces characterized by flora-fauna, climate, physiography, soil, and vegetation types; Kuchler 1964, classified ecological provinces and vegetation types; Bailey 1976, 1980, 1983, 1995, ecoregions).

In the United States, Omernik ecoregions (Levels I, II, III, and IV) and MLRAs can be further grouped into more discrete plant community types associated with soil map units called ecological sites. The ecological site concept has now been accepted and used by the major US land management agencies. The *Interagency Ecological Site Handbook for Rangelands* (USDA 2013) was developed to outline and implement policy for the Bureau of Land Management (BLM), US Forest Service (USFS), and Natural Resources Conservation Service (NRCS) for identifying and describing rangeland ecological sites for use in inventories, monitoring, and managing rangelands. On rangelands and forestlands, the USDA-NRCS classifies these land types into ecological sites (USDA-NRCS 2014). Riparian ecological sites are also recognized but need to be correlated with conterminous range or forest upland sites to address the overall hydrologic aspects of the system. Ecological site descriptions document site characteristics and scientific information pertaining to ecological attributes associated with the site. This information is fundamental to USDA-NRCS conservation planning and management and communication with other agencies and land users.

An ecological site is a conceptual classification of the landscape. It is a distinctive land type based on a recurring landform with distinct soils (chemical, physical, and biologic attributes), kinds and amounts of vegetation, hydrology, geology, climatic characteristics, ecological resistance and resiliency, successional dynamics and pathways, natural disturbance regimes, geologic and evolutionary history including herbivore and other animal impacts, and response to particular management actions. These discrete characteristics separate one ecological site from another and are inherent with respect to geological and evolutionary development (Fig. 2.3).

Two questions are relevant: (1) What is the importance of land classification and recognition of ecological sites to soil health? (2) How are ecological sites used? The physiognomy and biological components of terrestrial lands vary tremendously. Crop producers recognize that different soils have varying capacities to grow crops due to physical and chemical soil attributes. Soil amendments and fertilizer inputs vary according to soil type, as does productivity of crops and effects on microorganism dynamics. Familiarization and knowledge about what types of soils and their intrinsic characteristics are necessary for successful farming and maintaining soil health.

On rangeland, many different life/growth forms (trees, shrubs, grasses, forbs, biological soil crusts) can occur. In addition, there are annual and perennial plants occupying different strata (canopy layers). Grass species or graminoids (includes grasses, rushes, and sedges) also have different reproductive strategies and growth habits. Perennial grasses can exhibit rhizomes (underground branched stems) or stolons (aboveground runners that tend to be prostrate on and in the soil surface), which can form a sod appearance, whereas grasses with erect culms that branch extensively from the plant base produce a bunch, tufted, or caespitose appearance.[9]

---

[9] This phenomenon has many ramifications for management and hydrologic responses. Sod vs. bunch grasses have also developed with different disturbance regimes and herbivory.

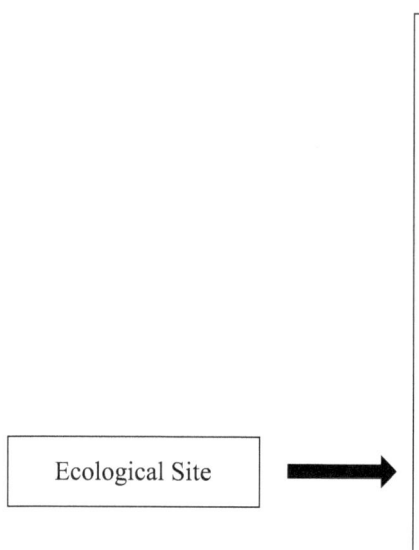

Fig. 2.3 Environmental and ecological factors associated with ecological sites

Fertilizers and soil amendments are not traditionally or typically used on range-lands; therefore, management and sustainability of soil health are largely dependent on natural management options alone. Proper or prescribed grazing management, brush and invasive plant control, and prescribed burning (where applicable to sites and site conditions) are some of the main tools to maintain soil health on rangelands. In the garden, management of soils is similar to large-scale cropping systems; however, the use of pesticides is usually much reduced, and there are many more options using soil amendments to maintain and increase soil organic matter and health.

There are many common uses of ecological sites and descriptions that are related to land management:

- Provide ecological site information to NRCS customers at a finite scale of land classification—the ecological site
- Communication tool with customers and other users

- Document ecological dynamics of rangeland, forestland, and riparian sites
- Conservation planning and application
- Watershed-scale modeling
- Database for ecological site information
- Use in GIS level modeling tools
- Serves as the ecological basis for conducting rangeland health assessments and monitoring
- Provide baseline ecological information for modeling plant growth, hydrology, and erosion (e.g., Rangeland Hydrology and Erosion Model (RHEM)
- Plant community baseline data for scientific research and experimental studies

Examples of Two Hierarchical Classifications of Terrestrial Plant Communities

Why explore the hierarchical structure of terrestrial ecosystems? The reason is to set the stage for you the reader to ponder the complexities and the diversity of plants and animals in nature. If you are engaged in raising crops, grazing livestock, or gardening, taking time to consider the historic nature of the land you are working with, and its specific qualities can give you a new perspective and hopefully more success in your endeavors.

The ecosystem can be dissected into finer levels of classification culminating at a local ecological site. It is at the ecological site level where the details and dynamics of soil health can be realized. Our pioneer ancestors could visually determine land capability on a rudimentary level, which was primarily based on landscape position (uplands, lowlands, river bottoms, etc.). Selecting lands for farming was not totally haphazard, but the process did not include much science. Modern land classification offers the agricultural producer science-based information about the dynamics of their lands. Presented here are two regional land resource classifications with supporting information:

I. EPA Ecoregions (ER). An ER is a geographic area representing similar patterns among biotic, abiotic, aquatic, and terrestrial ecosystem components and serves as an organizational classification and tool for providing information relative to ecosystem management.
II. USDA-Land Resource Region (LRR) and Major Land Resource Areas (MLRA).

An MLRA is regional in size, traversing a state or dispersed throughout several states, and represents a level of classification for an area with similar climate, physiography, soils, and vegetation that support certain land uses. The MLRA defines the extent of an area where individual ecological sites are identified for more specific levels of understanding and use in conservation planning. Specific soil map units and components are correlated with discrete ecological sites.

Both ER and MLRAs are used extensively by US land management agencies and others to facilitate communication about land resources for farming, ranching, forestry, engineering, wildlife habitat, recreation, and other uses.

Case Study

Consider the Great Plains region, which includes the greatest expanse of grasslands in the United States. The ten plains states extend east from the Rocky Mountains (originating in Canada) to the Central Lowlands in New Mexico. The vegetation of the Great Plains is highly diverse and is a "land of marked contrasts and limitless variety: canyons carved into solid rock of an arid land by the waters of the Pecos and the Rio Grande; the seemingly endless grain fields of Kansas; the desolation of the Badlands; and the beauty of the Black Hills" (Trimble 1980). The Great Plains region includes three distinct prairies: the tallgrass, mixed-grass, and shortgrass prairies. Although grasslands are the dominant vegetation type in the Northern Great Plains, shrub, forest, and woodland vegetation also exist throughout the region. Within each of these three prairie types, there are a myriad of ecosystems, plant communities, soil types, and ecological sites (Spaeth 2018).

In this example, the Kansas state soil, Harney silt loam, is applicable to many agronomic land uses such as cropping, rangeland, pastureland, and gardening. The Harney silt loam is the most extensive soil in the state covering 1,609,030 ha (3,976,000 ac) in west-central Kansas. A variety of cash crops, irrigated and dryland, are raised on Harney silt loam.

Figure 2.4 shows the map area delineations of ER 27b and MLRA 73. First, an example of some of the pertinent data from the Kansas and Nebraska ecoregion map (Chapman et al. 2001) is shown with information on physiography, geology, soil, climate, and vegetation. Following is information for MLRA 73 (USDA-NRCS

a)   b)

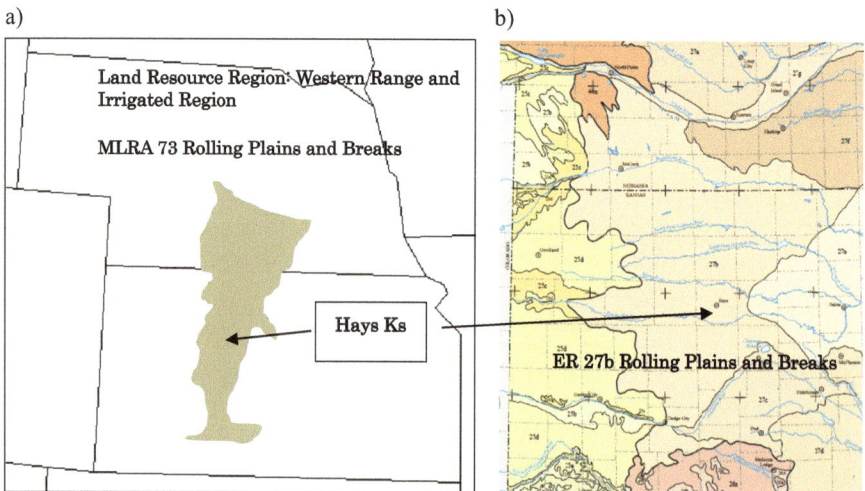

Fig. 2.4 Maps showing two land classification systems: (**a**) Major Land Resource Area and (**b**) ecoregion Level IV. In this example, the names for the designated area "Rolling Plains and Breaks" are the same; however, many times the names are different for an area of interest

2006). Note that not all the information from the EPA and USDA classifications and maps is given; the objective here is to give the reader a basic view of the type of data that is available.[10]

Hierarchical Classification: Ecoregion[11]

Biome Temperate Grassland

Biome Type Tallgrass prairie

Classification EPA—Ecoregions of North America:
Level I, 09 Great Plains

Level II 9.4 South-Central Semiarid Prairies
Level III 27 Central Great Plains
Level IV 27b Rolling Plains and Breaks

Ecoregion 27b Description The Rolling Plains and Breaks ecoregion was historically a mixed-grass prairie. Today, a mosaic of cropland agriculture and rangeland occurs throughout the region. Soils are silty, well-drained, deep, and moderately permeable, formed in loess on uplands. The dissected plains, with broad, undulating to rolling ridgetops, are a contrast to the smoother Western High Plains (25) to the west and the broad, flat regions to the north (27g and 27f). In Kansas, this region contains extensive oil deposits.

Rolling Plains and Breaks

Area: 64,073 square kilometers (24,739 square miles)

Physiography Dissected plains with broad undulating to rolling ridgetops and hilly to steep valley sides.

Elevation/Local Relief 518–975 m (1700–3200 ft)/15–60 m (50–200 ft)

Geology, Surficial Material, and Bedrock Holocene to Illinoian-aged loess on uplands with alluvium in floodplains and stream terraces. Tertiary sandstone (Ogallala Formation) and Cretaceous limestone and shale (Niobrara and Greenhorn formations).

Soil Order (Great Group) Mollisols (Argiustolls, Haplustolls,Calciustolls), Entisols (Ustorthents)

---

[10] Ecoregion data is available at:
    MLRA information is available at: https://www.nrcs.usda.gov/Internet/FSE_DOCUMENTS/nrcs142p2_050898.pdf

[11] Data from Chapman et al. 2001

Common Soil Series Coly, Uly, Harney, Holdredge, Hord

Temperature/Moisture Regimes Mesic/Ustic, Udic[12]

Precipitation 508–609 mm (20–24 in)

Frost Free Mean Annual Days 150–190

Mean Temperature January min/max July min/max −8.8/5.5;14.4/34.4 C (F 16/42;58/94)

Potential Natural Vegetation Tall and mixed-grass prairie: big bluestem (*Andropogon gerardii*), little bluestem (*Schizachyrium scoparium*), blue grama (*Bouteloua gracilis*), needle and thread (*Hesperostipa comata*), sideoats grama (*Bouteloua curtipendula*), and western wheatgrass (*Pascopyrum smithii*). Some areas of floodplain forests are along major riparian corridors.

Land Use and Land Cover Mosaic of predominantly cropland and rangeland. Winter wheat and grain sorghum are the major crops with large areas of corn in the north. Irrigated areas are along the major rivers planted with corn, alfalfa, and small grains. Rangeland on breaks.

Hierarchical Classification: USDA-NRCS-Land Resource Region: Western Range and Irrigated Region and Major Land Resource Area 73 Rolling Plains and Breaks

Data below cited from USDA-NRCS (2006).

This MLRA area is in Kansas (78%) and Nebraska (22%). It makes up about 55,670 square kilometers (21,485 square miles). The towns of Hays, Great Bend, and Dodge City, Kansas, and Alma, Curtis, Holdrege, and McCook, Nebraska, are in this MLRA. The MLRA is bisected by Interstate 70. The Platte River is at the northern edge of the area, and the Arkansas River is at the southern edge.

Physiographic Features
The western half of MLRA 73 and areas along the Arkansas River have remnants of the Tertiary river-laid sediments washed out onto the plains from erosion of the prehistoric Rocky Mountains in Colorado. In the valley of the Arkansas River, the wind reworked these sediments, forming a hummocky dune surface of eolian sand. A loess mantle occurs on the higher ground in the western half of the area. The Tertiary-age Ogallala and White River Formations cover Cretaceous Pierre Shale in the northern part of the area. The Ogallala Formation consists of loose to well-cemented sand and gravel, and the White River Formation consists of ashy

---

[12] Mesic-a soil temperature class with mean annual temperature 8 to 15 deg C. Ustic-soil moisture intermediate between Udic and Aridic regimes, usually relatively dry winters and some plant-available moisture, droughts can be significant. Udic-soil moisture sufficiently high most years for plant growth, common for soils in in humid climatic regions

claystone and sandstone. Pierre Shale and Niobrara Chalk are at the surface in the valleys of the Republican, Smoky Hill, and Saline Rivers. Fort Hays Limestone of the Niobrara Formation and Blue Hill shale of the Carlile Formation are at the surface in the valleys of the Saline and Smoky Hill rivers. Shale is commonly exposed in the eastern half of this MLRA, in Kansas. Quaternary and more recent sand and gravel partially cover the shale in the river valleys.

Climate
The average annual precipitation in this area is 19–30 inches (485–760 millimeters). Most of the rainfall occurs as high-intensity, convective thunderstorms during the growing season. The maximum precipitation occurs from the middle of spring to early autumn. Precipitation in winter occurs as snow. The annual snowfall ranges from about 17 inches (45 centimeters) in the southern part of the area to 24 inches (60 centimeters) in the northern part. The average annual temperature is 48–56 °F (9–14 °C). The freeze-free period averages 180 days and ranges from 145 to 210 days, increasing in length from northwest to southeast.

Water
The total withdrawals average 1400 million gallons per day (5300 million liters per day). About 81% is from groundwater sources, and 19% is from surface water sources. The moderate, erratic precipitation is the source of water for crops and pasture in much of the area. The amount of surface water is limited throughout the area. The Republican and Platte Rivers and their larger tributaries provide surface water for irrigation along their valleys. The surface water is generally of good quality. It is suitable for all in-stream uses and for irrigation. Because of low flows, the rivers in Kansas are limited as sources of irrigation water. Abundant supplies of groundwater for irrigation and other uses are obtained from wells, primarily in the northern part of this MLRA. Alluvium in the valleys along the major rivers and their larger tributaries is one source of groundwater, and Quaternary sand and gravel and the Ogallala Formation within the High Plains aquifer are other sources. The water from the Ogallala Formation is typically low in total dissolved solids but is hard or very hard. The water from the alluvial aquifers is very hard. Shallow groundwater can be contaminated by nitrate and atrazine from agricultural lands. Some deeper wells have been drilled to obtain more saline water from the Dakota Formation. Groundwater becomes scarce in the eastern half of the MLRA, where the High Plains aquifer has been eroded away and Cretaceous Pierre Shale and Niobrara Chalk are at, or close to, the surface.

Biological Resources
This area supports natural prairie vegetation. Little bluestem, big bluestem, switchgrass (*Panicum virgatum*), western wheatgrass, and sideoats grama characterize the vegetation on loamy soils. Blue grama, buffalograss (*Bulbilis dactyloides*), and western wheatgrass characterize the vegetation on clayey soils on uplands. Some of the major wildlife species in this area are mule deer, white-tailed deer, coyote, raccoon, pheasant, bobwhite quail, mourning dove, and meadowlark. The species of fish in the area include bass, bluegill, catfish, and bullhead.

Land Use
Following are the various kinds of land use in this MLRA:

Cropland—private, 55%
Grassland—private, 39%; Federal, 1%
Forest—private, 1%
Urban development—private, 2%
Water—private, 1%
Others—private, 1%

Nearly all of this area is in farms or ranches. More than half of the area is crop-land used for dry-farmed crops. Winter wheat and grain sorghum are the major crops in much of the area. Corn is the main crop in the northern part of the area. Feed grains and hay crops also are widely grown. About 2% of the area, especially on the terraces and narrow bottomland along the Platte and Republican Rivers, is irrigated. Corn, alfalfa, small grains, and grass for hay are grown extensively on the irrigated land. About 40% of the area, consisting of hilly and steep slopes bordering the drainageways, supports native grasses and shrubs used for grazing. The major soil resource concerns are wind erosion, water erosion, maintenance of the content of organic matter in the soils, and surface compaction. Also, soil moisture manage-ment is important in the western part of the area. The resource concerns on range-land are plant productivity, health, and vigor; the spread of noxious and invasive plants; and inadequate wildlife habitat.

Conservation practices on cropland generally include high-residue crops in the cropping system; systems of crop residue management, such as no-till, strip-till, and mulch-till; level terraces in the western part of the area and gradient terraces and grassed waterways in the eastern part; contour farming; conservation crop rotations; irrigation water management; and nutrient and pest management. Conservation practices on rangeland generally include prescribed grazing, brush management, management of upland wildlife habitat, proper distribution of watering facilities, and control of noxious and invasive plant species.

Evaluating the Landscape at the Ecological Site Scale

We step down from the concept of large land uses or regional scales, the Major Land Resource Area (MLRA 73), or ecoregion (27b) to an actual field with individual ecological sites. This step is the basis for conservation planning on farms and ranches, where individual soils are identified and information is given about soil physical, chemical, agricultural potentials, and management. In Fig. 2.5, the field is comprised of several soil components with acres and percentages, e.g., Harney silt loam 0–1% slope; Harney silt loam (1–3% slope); Harney-Uly complex; 3–6% slopes, eroded . . . Ness clay. Even though the soil component, Harney silt loam (1–3% slopes), is commonly cultivated, the historic vegetation associated with this soil was grassland comprised of tallgrasses, a wide diversity of herbaceous plants, with shrubs as a minor component. In Fig. 2.5, Harney silt loam (the Kansas state soil) is highlighted and is correlated with a Loamy Plains (ID = R073XY100KS) ecological site (Table 2.1a, b). USDA-NRCS ecological site descriptions contain the following information:

Fig. 2.5 Aerial photo from USDA-NRCS Web Soil Survey showing field dominated by Harney silt loam. Table 2.1a identifies soil map units with acres in the area of interest (AOI) and percent of the AOI. Table 2.1b defines the map unit components with the correlated ecological site, acres in AOI, and percent of map units in the AOI

Table 2.1 Information associated with Harney silt loam and the corresponding ecological site. Outputs from USDA-NRCS Web Soil Survey

(a)

Map unit symbol	Map unit name	Acres in AOI	Percent of AOI
2612	Harney silt loam, 0–1% slopes	227.5	71.3%
2613	Harney silt loam, 1–3% slopes	64.6	20.2%
2615	Harney silty clay loam, 1–3% slopes, eroded	14.3	4.5%
2630	Harney-Uly complex, 3–6% slopes, eroded	11.8	3.7%
2714	Ness clay	1.0	0.3%
Totals for area of interest		319.1	100.0%

(b)

Map unit symbol	Map unit name	Component name (percent)	Ecological site	Acres in AOI	Percent of AOI
2612	Harney silt loam, 0–1% slopes	Harney (97%)	R073XY100KS — Loamy Plains	227.5	71.3%
		Ness (3%)	R072XY115KS — Closed Upland Depression		
2613	Harney silt loam, 1–3% slopes	Harney (95%)	R073XY100KS — Loamy Plains	64.6	20.2%
		Wakeen (5%)	R073XY101KS — Limy Slopes		
2615	Harney silty clay loam, 1–3% slopes, eroded	Harney, eroded (85%)	R073XY100KS — Loamy Plains	14.3	4.5%
		Coly, eroded (5%)	R073XY101KS — Limy Slopes		
		Uly, eroded (5%)	R073XY100KS — Loamy Plains		
		Wakeen, eroded (5%)	R073XY101KS — Limy Slopes		

(continued)

Table 2.1 (continued)

(b)

Map unit symbol	Map unit name	Component name (percent)	Ecological site	Acres in AOI	Percent of AOI
2630	Harney-Uly complex, 3–6% slopes, eroded	Harney, eroded (65%)	R073XY100KS — Loamy Plains	11.8	3.7%
		Uly, eroded (30%)	R073XY100KS — Loamy Plains		
		Roxbury, frequently flooded (5%)	R073XY108KS — Loamy Floodplain		
2714	Ness clay	Ness (100%)	R072XB011KS — Closed Upland Depression (South) Draft (April 2010) (PE16-20)	1.0	0.3%
Totals for area of interest				319.1	100.0%

(c) Monthly precipitation (mm)

	Jan	Feb	Mar	Apr	May	Jun	Jul	Aug	Sep	Oct	Nov	Dec
High	20.3	26.9	58.9	69.9	110.7	112.8	120.4	99.6	69.3	59.2	32.0	25.7
Med	10.4	13.0	36.1	48.3	74.7	77.5	76.2	66.8	43.7	33.3	16.8	10.7
Low	4.1	5.6	19.1	32.5	49.3	51.1	44.7	41.4	24.6	14.0	7.6	4.1

Monthly Temperature (°C)

	Jan	Feb	Mar	Apr	May	Jun	Jul	Aug	Sep	Oct	Nov	Dec
High	5.7	7.9	13.4	19.1	24.2	29.9	33.4	32.2	27.6	20.6	12.5	5.9
Low	-8.8	-7.0	-2.2	3.3	9.7	15.3	18.3	17.3	11.8	4.4	-2.7	-8.0

Monthly precipitation (Inches)

	Jan	Feb	Mar	Apr	May	Jun	Jul	Aug	Sep	Oct	Nov	Dec
High	0.8	1.06	2.32	2.75	4.36	4.44	4.74	3.92	2.73	2.33	1.26	1.01
Med	0.41	0.51	1.42	1.90	2.94	3.05	3.00	2.63	1.72	1.31	0.66	0.42
Low	0.16	0.22	0.75	1.28	1.94	2.01	1.76	1.63	0.97	0.55	0.30	0.16

Monthly temperature (°F)

	Jan	Feb	Mar	Apr	May	Jun	Jul	Aug	Sep	Oct	Nov	Dec
High	42.2	46.2	56.2	66.3	75.5	85.8	92.1	90.0	81.6	69.0	54.5	42.7
Low	16.2	19.4	28.1	37.9	49.4	59.5	65.0	63.2	53.2	40.0	27.2	17.6

(d) Pertinent soils data

Kind: Loess, calcareous loess
Surface texture: (1) Silt loam, (2) loam, (3) very fine sandy loam
Subsurface texture group: Sandy

	Minimum	Maximum
Surface fragments <=7.6 cm (<=3 in) % cover:	0	0
Surface fragments >7.6 cm (>3 in) (% cover):	0	0
Subsurface fragments <=7.6 cm (<=3 in) (% volume):	0	15
Subsurface fragments >7.6 cm (>3 in) (% volume):	0	0

Drainage class: Well-drained
Permeability class: Moderately slow to moderate

(continued)

Table 2.1 (continued)

(d) Pertinent soils data

	Minimum	Maximum
Depth (cm, inches):	152 (60)	203 (80)
Available water capacity (inches):	11.9 (4.7)	31.4(12.4)
Electrical conductivity (mmhos/cm):	0	1
Sodium adsorption ratio:	0	0
Calcium carbonate equivalent (percent):	0	6
Soil reaction (1:1 water):	6.5	8.3
Soil reaction (0.01M CaCl$_2$):	6.0	7.8

Fig. 2.6 Harney silt loam soil profile (cm). (Courtesy of USDA-NRCS, Kansas)

- Table 2.1a shows the soil map units, acres in the area of interest (AOI), and percent of AOI in Fig. 2.5.
- Table 2.1b shows the soil map units in Fig. 2.5 correlated with the ecological site.
- Table 2.1c shows the average monthly precipitation and temperatures.
- Table 2.1d contains information about soil parent material, surface rock fragments, drainage class, permeability, and basic variables related to soil chemistry. Table 2.1d shows the soil layers or horizons for the Harney soil series. Included in Table 2.1d is information of type location, range of soil characteristics, competing soil series, geographic setting, geographically associated soils, drainage and permeability, agricultural use, and common associated vegetation (Fig. 2.6).
- Table 2.1e Soil layers (horizons) associated with Harney soil series.
- Table 2.1f Ecological dynamics of the site.
- Table 2.1g Reference community (historic plant community) plant species composition by weight. These annual production values represent current year's growth. Shrub production values also represent current year's growth, not the

entire biomass value associated with the woody species. Common name, scientific symbol (first two letters of the genus and species), scientific name, and low and high annual production. Total annual production of plant life forms (forb, graminoids, and shrubs/vines) is summarized.

(e) *Harney Series (USDA-National Cooperative Soil Survey https://soilseries.sc. egov.usda.gov/OSD_Docs/H/HARNEY.html)*

The Harney series consists of deep, well-drained, moderately slowly permeable soils formed in loess. These soils are on uplands on slopes that range from 0 to 8%.

Taxonomic Class: Fine, smectitic, mesic Typic Argiustolls

Typical Pedon: Harney silt loam-in a nearly level cultivated field. (Colors are for dry soil unless otherwise stated.)

Ap—0–23 cm (0–9 inches); dark grayish brown (10YR 4/2) silt loam, very dark grayish brown (10YR 3/2) moist; moderate medium granular structure; slightly hard, very friable; many fine roots; slightly acid; clear smooth boundary 10–36 cm (4–14 inches thick).

AB—23–30.5 cm (9–12 inches); dark grayish brown (10YR 4/2) silt loam, very dark grayish brown (10YR 3/2) moist; moderate fine subangular blocky structure; hard, friable; many fine roots; neutral; clear smooth boundary 0–25 cm (0–10 inches thick).

Bt1—30.5–46 cm (12–18 inches); grayish brown (10YR 5/2) silty clay loam, dark grayish brown (10YR 4/2) moist; moderate medium subangular blocky structure; very hard, very firm; few fine roots; moderately alkaline; clear smooth boundary.

Bt2—46–71 cm (18–28 inches); grayish brown (10YR 5/2) silty clay loam, dark grayish brown (10YR 4/2) moist; strong medium subangular blocky structure; very hard, very firm; few fine roots; moderately alkaline; gradual smooth boundary. (Combined thickness of the Bt horizon is 25–66 cm (10–26 inches.)

BCk—71–89 cm (28–35 inches); brown (10YR 5/3) silty clay loam, brown (10YR 4/3) moist; moderate medium subangular blocky structure; hard, firm; few fine roots; many soft accumulations of carbonates; strong effervescence; moderately alkaline; gradual smooth boundary 0–41 cm (0–16 inches thick).

Ck—89–119 cm (35–47 inches); pale brown (10YR 6/3) silt loam, brown (10YR 5/3) moist; massive; slightly hard, friable; common soft accumulations of carbonates; strong effervescence; moderately alkaline; gradual smooth boundary 0–50 cm (0–20 inches thick).

C—119–152 cm (47–60 inches); pale brown (10YR 6/3) silt loam, brown (10YR 5/3) moist; massive; slightly hard, friable; strong effervescence; moderately alkaline.

Type Location: Pawnee County, Kansas; about 1 ½ miles north of Garfield; 1840 feet east and 100 feet south of the northwest corner of sec. 36, T. 22 S., R. 18 W.

Range in Characteristics: The thickness of the solum ranges from 66 to 127 cm (26–50 inches). Depth to free carbonates ranges from 46 to 76 cm (18–30 inches). The mollic epipedon is 25–50 cm (10–20 inches) thick and includes the upper part of the argillic horizon in some pedons.

Table 2.2 Chemical properties of the Harney soils. Analysis by the Soil Survey Investigations Staff, Natural Resources Conservation Service, Lincoln, Nebraska, Sample # S81KS-165-001. pH 1:1 water method, CEC (cation exchange capacity), and base saturation, NH40AC pH 7.0 method. Ca calcium, Mg magnesium, Na sodium, K potassium

Horizon	Depth(in)	pH 1	Organic C (%)	C.E.C.	Ca	Mg	Na	K	Base Sat.
Ap1	0–18	6.0	1.04	16.8	11.0	3.3	–	1.5	94
Ap2	18–25	6.7	0.80	22.7	16.1	5.0	–	1.0	97
Bt1	25–36	7.0	0.64	23.3	16.8	5.5	TR	1.0	100
Bt2	36–61	7.3	0.44	30.1	21.8	7.8	0.3	1.3	100
BCk	61–79	8.0	0.31	28.3	–	8.0	0.5	1.3	100
Ck	79–122	8.3	0.25	26.1	–	7.7	1.0	1.3	100
C	122–152	8.4	0.25	26.3	–	6.5	2.2	1.3	100

O.C. × 1.728 = % organic matter (Walkley-Black method)

The A horizon has hue of 10YR, value of 4 or 5 and 2 or 3 moist, and chroma of 1–3. It is silt loam or silty clay loam and ranges from medium acid to mildly alkaline.

The Bt horizon below the mollic epipedon has hue of 10YR, value of 5 or 6 and 4 or 5 moist, and chroma of 2 or 3. It is silty clay loam or silty clay; the upper 50 cm (20 inches) averages from 35 to 42% clay. Maximum clay content of any subhorizon within the Bt horizon does not exceed 45%. This horizon ranges from slightly acid to moderately alkaline.

The C horizon has hue of 10YR, value of 5–7 and 4–6 moist, and chroma of 2–4. It is silt loam or silty clay loam and is mildly alkaline or moderately alkaline (Table 2.2).

Competing Series: These are the Beadle, Carlson, Kirley, McClure, Mento, Peno, Raber, Reliance, and Spearville series in the same family and Crete, Detroit, Hastings, Holdrege, Longford, and Richfield series. Beadle, Peno, and Raber soils contain more sand and are formed in glacial till. Carlson soils contain more carbonates in the C horizon and formed in highly calcareous residuum or old alluvium. Kirley, McClure, and Reliance soils have mean annual temperatures lower than 11 °C (52 °F). In addition, Kirley soils have more sand in the solum and formed in alluvium; and McClure soils are slowly permeable in the underlying materials. Mento and Spearville soils have an abrupt or clear boundary between the A and Bt horizon. Mento soils have B and C horizons containing 5–15% exchangeable sodium, and Spearville soils have more than 42% clay in the argillic horizon. Crete and Detroit soils have thicker mollic epipedons. Hastings soils lack carbonates within a depth of 91 cm (36 in). Holdrege soils are fine-silty. Longford soils have redder hues. In addition, Longford soils lack carbonates within a depth of 91 cm (36 in). Richfield soils are drier and depth to carbonates is shallower.

Geographic Setting: Harney soils are on uplands that have slightly concave to convex surfaces. The slope gradient commonly is 0–3%, but the range is 0–8%. The soils formed in loess usually several feet thick. The mean annual temperature varies from 11 to 13.8 °C (52–57 °F), and the mean annual precipitation varies from 48.3 to 68.5 cm (19–27 in).

Geographically Associated Soils: These are the competing Crete, Holdrege, and Spearville soils and the Ness, Penden, and Uly soils. Crete and Spearville soils are on more level ridgetops. Holdrege soils are on steeper slopes. Ness soils are more clayey and are in depressions. Penden and Uly soils lack argillic horizons and are on steeper lower slopes.

Drainage and Permeability: Well-drained. Runoff is slow or medium. Permeability is moderately slow.

Use and Vegetation: Mostly cultivated; wheat and sorghums are the principal crops. Native vegetation is mixed short-, mid-, and tallgrasses.

Distribution and Extent: West-central Kansas. The series is of large extent.

Series Established: Ford County, Kansas, 1962.

(f) *Ecological Dynamics of the Site*

Prairie grasslands in MLRA 73 (Rolling Plains and Breaks) are some of the most productive semiarid grazinglands in the world. The Loamy Plains ecological site receives on the average about 60.9 cm yr^{-1} (24 in yr^{-1}) and occurs on slopes from 1% to 6% slopes. The historic plant community (reference plant community; State 1.1 in Fig. 2.7) was and is dominated by a collection of tall and midgrass warm season grasses that typically occur throughout the tallgrass prairie vegetation type. Overall plant species richness (number of species occurring on site) is high—nearly 60 species. Big bluestem, little bluestem, sideoats grama, Indiangrass (*Sorghastrum nutans*), and switchgrass production make up about 65% of the stand; subdominant grasses such as blue grama, western wheatgrass, buffalograss, prairie June grass (*Koeleria macrantha*), needle-and-thread, and threadleaf sedges (*Carex filifolia*, and other spp.) comprise about 20%. Wildflowers, or non-woody herbaceous plants called forbs, are diverse on this site and comprise up to 13% of the stand production. Some of the common species are Cuman ragweed (*Ambrosia psilostachya*), violet prairie clover (*Dalea purpurea*), prairie fleabane (*Erigeron strigosus*), dotted blazing star (*Liatris punctata*), and common sunflower (*Helianthus annuus*). Shrubs are a minor component of the plant community, contributing about 2% of the production of leadplant (*Amorpha canescens*), broom snakeweed (*Gutierrezia sarothrae*), and plains pricklypear (*Opuntia polyacantha*).

This prairie site is evolved with periodic fire and grazing by large ungulates. Grasses evolving with ungulates commonly display extravaginal tillering and vegetative modes of reproduction in addition to sexual (seed development) such as a rhizomatous habit. Several species in plains grasslands also tiller via stolons [buffalograss, and vine mesquite (*Panicum obtusum*)]. Rhizomes are creeping horizontal underground stems sporting lateral shoots and adventitious roots at nodal intersections, whereas stolon are creeping aboveground horizontal arching stems or runners, which can take root and form new plants (strawberries develop stolons). The bluestems, Indiangrass, switchgrass, and gramas appear as a sea of grass with very little bare ground; however, when viewed in more detail (clipping grasses down to the crown), these species exhibit a clumped or bunch (caespitose) growth

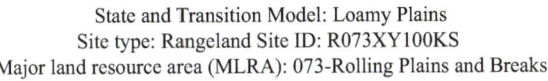

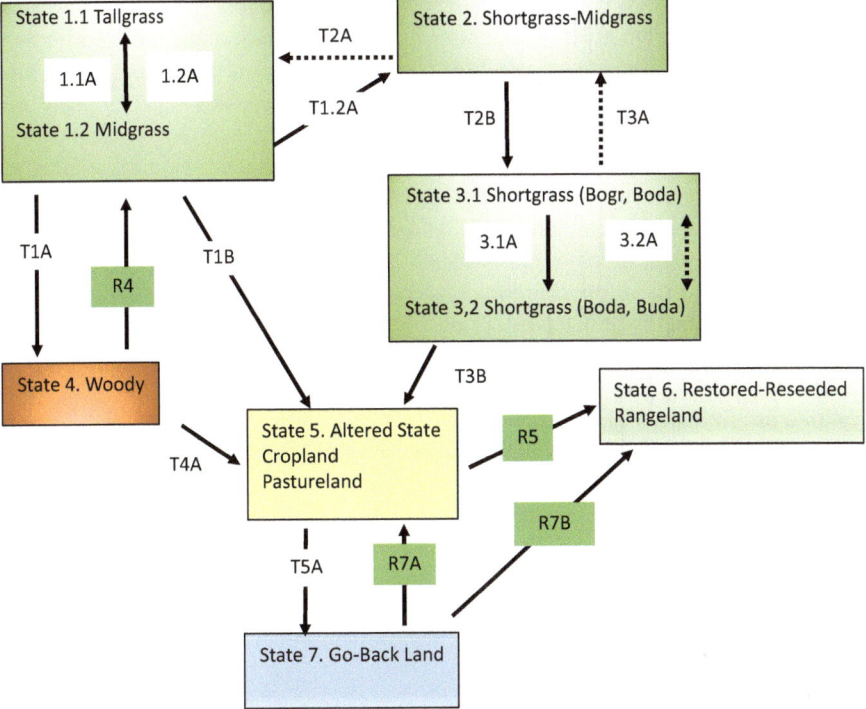

Fig. 2.7 Ecological Site: Loamy Plains, *Andropogon gerardii-Schizachyrium scoparium*. (big bluestem- little bluestem. Site ID: R073XY100KS. Dotted line denotes uncertain or problematic transitions)

form. Mack and Thompson (1982) state: "the morphology of rhizomatous and caespitose grasses reflects the two extremes to which perennial grasses have evolved at least in partial response to continuous high versus low selection pressure by large congregating animals." On the North American steppe east of the Rocky Mountains, rhizomatous warm (C4) and cool season (C3) grasses evolved adaptation mechanisms to grazing by Bison and other grazers (Mack and Thompson 1982).[13] The evolutionary history and expansion of grasses throughout the world occurred in stages beginning in the Paleogene (55–65 Ma), where C3 and C4 grasses appeared in more aridic open habitats (Stromberg 2011). Coevolutionary history between large ungulate grazers and grasses began 45–55 million years ago during the Eocene Epoch (Stebbins 1982, Crepet and Feldman 1991), although Stromberg (2011) states that the evolution of herbivores does not always coincide with evolutionary and ecological events in open-habit grasslands.

---

[13] C4 photosynthesis commonly associated with warm season plants; and C3 photosynthesis associated with cool season plants (see chapter??? for more information).

Fire has been an integral perturbation throughout the evolution of the tallgrass prairie and is necessary to maintain healthy productive stands of native grasses. Natural fire frequency coincided with lightning strikes occurring predominantly during the summer months every 3–10 years on level landscapes and 10–20 years on breaks and stream areas. Indigenous peoples in certain areas initiated burns to facilitate hunting and attract wildlife to areas closer to settlements, which most likely occurred on an annual basis. Fire effects on plant species in the tallgrass prairie depends on soil moisture at the time of the fire, time, and overall frequency of burning. Production of big bluestem, switchgrass, Indiangrass, sideoats grama, and little bluestem generally is not affected by late spring burning (Robocker and Miller 1955; Kucera and Ehrenreich 1962; Anderson et al. 1970; Smith and Owensby 1972). Cool season invasive grasses such as Kentucky bluegrass (*Poa pratensis*), smooth bromegrass (*Bromus inermis*), and annual field bromegrass (*Bromus arvensis*) are detrimentally affected by spring burning. Late spring burning reduces forbs or wildflower species, although stand composition is not significantly affected (Anderson 1965). Eastern red cedar, non-sprouting shrub, is effectively managed with a fire frequency of 20–30 years (Wright and Bailey 1982), although more frequent prescribed burns can eliminate many of the seedlings before they reach 1–1.5 m (4–5 ft) in height. Some shrub species such as snowberry (*Symphoricarpos orbiculatus*) and smoothleaf sumac (*Rhus glabra*) are resprouting species and resilient to fire. Overall, a 3-year burning interval in late spring is recommended to maintain native species dominance and diversity in the tallgrass prairie (Kucera 1970).

(g) *Reference Community Plant Species Composition (by wt.)*

Ecological Site State and Transition Model

Ecological sites contain state and transition models, which are the culmination of a collection of knowledge about ecosystem changes in response to natural events and management applications (Bestelmeyer et al. 2009, 2017). They are diagrammatic portrayals with narratives and specific environmental drivers—states can change as a result of a natural or anthropogenic disturbance event or lack of a natural event. State and transition models are in part the outcome of inadequacies posed in classical successional theories but do not explicitly explain the intricate biological dynamics or theory governing state changes (see discussion Text Box 2.2). Historic plant successional models are simply no longer representative of many plant communities (especially arid ecosystems) where succession and vegetation are discontinuous and exhibit irreversible changes. As a result, "STMs evolved from the recognition that vegetation change was more complex than could be accounted for by classic models of succession (theories and cultural adherence to classical concepts), and could occur along numerous pathways, be discontinuous, and result in multiple stable states in the same environment. Conceptualizing vegetation as discrete states also provides a useful platform for tailoring management actions to the properties and possibilities associated with each state. For rangeland managers, the value of STMs resides both in their flexibility for organizing information and in their ability to foster a general understanding about how rangelands function (Bestelmeyer et al. 2017)."

In contrast to classic plant successional theories, (1) grazing and climatic disturbances do not always produce a linear predictable outcome—variations along a single continuum but may produce multiple stable states as shown in the state and transition model—and (2) state and plant compositional changes may be sudden and not be reversible (where a state threshold is crossed) (Westoby et al. 1989; Bestelmeyer et al. 2017). The premise of state and transition models is to identify natural and anthropogenic disturbances and thresholds, which may or may not be reversible. They are commonly used in conservation planning and for assessment and monitoring vegetation changes and health status of rangeland ecological sites (Carpenter and Brock 2006; Forbis et al. 2006; King and Hobbs, 2006; Bestelmeyer et al. 2004, 2009). For example, what state changes occur in response to common ecological drivers such as drought, fire, or lack of, grazing at different levels, and invasion of exotic annual/perennial plants and woody species? In summary, the STM is a graphical composition that depicts various vegetation outcomes, multiple states, and transitions that occur as a result to disturbance and management (Stringham et al. 2003; Bestelmeyer et al. 2003, 2009; Briske et al. 2005, 2008) (Fig. 2.7). State and transition model diagrams are an effective way to portray and relay information during planning exercises with landowners—the basic causative disturbance factors are provided, but the ecological successional mechanisms and drivers are not exhibited, because either no theory uniformly fits, the variability in the ecosystem is unpredictable, or the complexity of variables in the environment are beyond the scope of the ecological site description. Many rangeland plant communities exhibit very different states and transitions; as a consequence, no universal model relative to ecological successional dynamics exists or is universally applicable.

State 1.1A Reference state representing historic plant community. Figure 2.9 represents a species importance curve where plant species are ranked by production (Whittaker 1965; Tjørve 2003). The shape of the curve is sigmoid, indicating high diversity, where eight species are >100 kg ha^{-1} and 11 species are 8 to 100 kg ha^{-1}. Tall- and midgrass species are dominant (65% composition by weight, big bluestem, sideoats grama, little bluestem, Indiangrass, and switchgrass). Shortgrasses and cool season grass species are subdominant (18%, blue grama, western wheatgrass, buffalograss, and composite dropseed (*Sporobolus compositus*). The remaining vegetation is comprised of other grasses (2%), forbs (13%), and shrubs (2%). The reference state has a diverse population of wildflowers (forbs). This plant community is resistant to short-term stresses (<4 years) such as drought and short-term heavy stocking; however, long-term stresses will result in transitions to State 1.2A and then State 2 and beyond. Ecological resilience is dependent on adequate rest and recovery from grazing (proper grazing use) and prescribed fire.

Fig. 2.8 Loamy Plains ecological site, State 1, reference community. (Photo courtesy of USDA-NRCS)

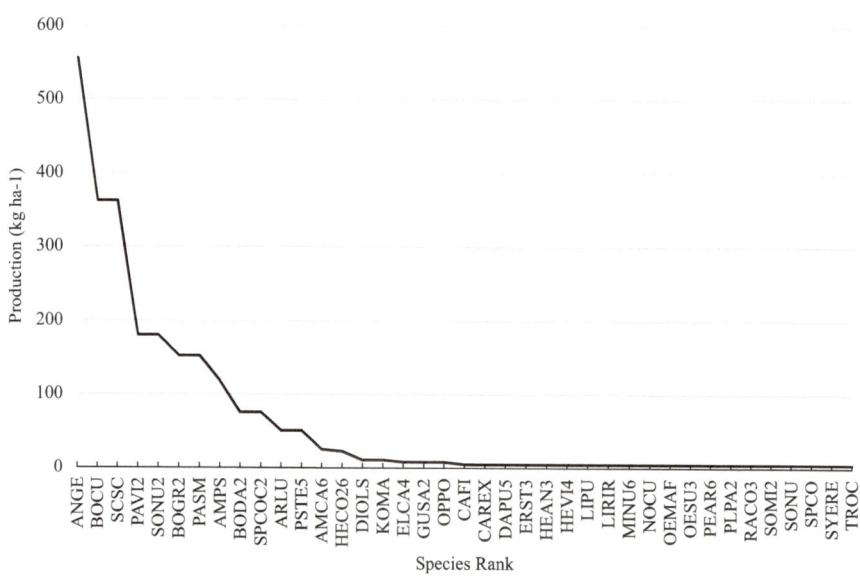

Fig. 2.9 Species importance curve by production rank. Species symbols defined in Table 2.3

Percent organic matter is at maximum in State 1.1A and is approximately 2%. Organic matter content declines significantly upon cultivation. Total annual production ranges from 2750 to 4256 kg ha yr^{-1} (2500–3800 lbs ac yr^{-1}) air-dried vegetation with an average of 3300 kg ha yr^{-1} (3000 lbs ac yr^{-1}). Production varies during favorable, normal, and unfavorable years due to the timing and amount of precipitation and temperature (Fig. 2.8).

Table 2.3 Plant species associated with Loamy Plains ecologial site (ID = R073XY100KS), common name, plant symbol, scientific name, and expected annual production by species (Low-High). Total annual production at bottom of table (Low-Representative value, and High production value)

Common name	Symbol	Scientific name	Annual production (lbs/ac)		Annual production (kg/ha)	
			Low	High	Low	High
Group 1 Tall- and midgrasses (65% by wt.)			970	1950	1087	2186
Big bluestem	ANGE	*Andropogon gerardii*	330	660	370	740
Sideoats grama	BOCU	*Bouteloua curtipendula*	215	430	241	482
Switchgrass	PAVI2	*Panicum virgatum*	105	215	118	241
Little bluestem	SCSC	*Schizachyrium scoparium*	215	430	241	482
Indiangrass	SONU2	*Sorghastrum nutans*	105	215	118	241
Group 2 Subdominant mid-short grasses (18% by wt.)			270	540	303	605
Buffalograss	BODA2	*Bouteloua dactyloides*	45	90	50	101
Blue grama	BOGR2	*Bouteloua gracilis*	90	180	101	202
Western wheatgrass	PASM	*Pascopyrum smithii*	90	180	101	202
Composite dropseed	SPCOC2	*Sporobolus compositus*	45	90	50	101
Group 3 Other grasses (2% by wt.)			0	60	0	67
Threadleaf sedge	CAFI	*Carex filifolia*	0	10	0	11
Sedge	CAREX	*Carex*	0	10	0	11
Scribner's rosette grass	DIOLS	*Dichanthelium oligosanthes*	0	20	0	22
Canada wild rye	ELCA4	*Elymus canadensis*	0	15	0	17
Needle-and-thread	HECO26	*Hesperostipa comata*	0	40	0	45
Prairie June grass	KOMA	*Koeleria macrantha*	0	20	0	22
Group 4 Subdominant forbs (13% by wt.)			175	390	196	437
Cuman ragweed	AMPS	*Ambrosia psilostachya*	75	135	84	151
White sagebrush	ARLU	*Artemisia ludoviciana*	25	65	28	73
Violet prairie clover	DAPU5	*Dalea purpurea*	0	10	0	11
Prairie fleabane	ERST3	*Erigeron strigosus*	0	10	0	11
Common sunflower	HEAN3	*Helianthus annuus*	0	10	0	11
Hairy false goldenaster	HEVI4	*Heterotheca villosa*	0	10	0	11
Dotted blazing star	LIPU	*Liatris punctata*	0	10	0	11
Stiffstem flax	LIRIR	*Linum rigidum*	0	10	0	11
Nuttall's sensitive-briar	MINU6	*Mimosa nuttallii*	0	10	0	11
Prairie false dandelion	NOCU	*Nothocalais cuspidata*	0	10	0	11
Fremont's evening primrose	OEMAF	*Oenothera macrocarpa*	0	10	0	11
Scarlet beeblossom	OESU3	*Oenothera suffrutescens*	0	10	0	11
Silverleaf Indian breadroot	PEAR6	*Pediomelum argophyllum*	0	10	0	11
Woolly plantain	PLPA2	*Plantago patagonica*	0	10	0	11

(continued)

Table 2.3 (continued)

			Annual production (lbs/ac)		Annual production (kg/ha)	
Slimflower scurfpea	PSTE5	*Psoralidium tenuiflorum*	25	65	28	73
Upright prairie coneflower	RACO3	*Ratibida columnifera*	0	10	0	11
Missouri goldenrod	SOMI2	*Solidago missouriensis*	0	10	0	11
Silky sophora	SONU	*Sophora nuttalliana*	0	10	0	11
Scarlet globemallow	SPCO	*Sphaeralcea coccinea*	0	10	0	11
White heath aster	SYERE	*Symphyotrichum ericoides*	0	10	0	11
Prairie spiderwort	TROC	*Tradescantia occidentalis*	0	10	0	11
Group 5 Shrubs (2% by wt.)			15	60	17	67
Leadplant	AMCA6	*Amorpha canescens*	15	30	17	34
Broom snakeweed	GUSA2	*Gutierrezia sarothrae*	0	15	0	17
Plains pricklypear	OPPO	*Opuntia polyacantha*	0	15	0	17

		Annual production (lbs/ac)		
Plant type	*Low*	*Representative value*		*High*
Forb	325	390		494
Graminoids	2125	2550		3230
Shrub/vine	50	60		76
Total:	2500	3000		3800
		Annual production (kg/ha)		
Forb	387.4	437		554
Graminoids	2533	2858		3620
Shrub/vine	59.6	67		85
Total:	2980	3363		4259

State 1.2A Reference state community phase where vegetation composition shifts from tallgrasses species dominance to more midgrasses. Prescribed burning and proper grazing use during normal years rainfall can shift stand composition to more tallgrass species. The time scale of this shift is highly dependent on the degree of change that occurred from 1.1A. Adequate periods of average and above average precipitation will also play an important role in the shift (Fig. 2.10).

T1.2A Transition from midgrass species dominance to more shortgrass species dominance due to long-term heavy continuous grazing and no prescribed fire.

T2A A transition shift from shortgrasses back to midgrasses is possible with prescribed grazing and fire. This transition could take many years-decades-centuries or longer depending on the degree of loss of tallgrasses. Dotted line indicates that this transition could be dubious and depends on how much departure from States 1.1 and 1.2. If significant plant community composition has changed, this transition may

Fig. 2.10 State 1.2 transitioning to State 2. (Photo courtesy of USDA-NRCS)

not be possible. If the hydrologic function has changed significantly in T2A, where higher runoff and erosion has occurred over time, transition to the reference State 1.2 and 1.1 will not be possible.

State 2 Shortgrass-midgrass. This plant community defines a shift in plant composition, function, and structure from a tall- and midgrass community to a predominantly mid- and shortgrass community. Tallgrasses, such as big bluestem, little bluestem, Indiangrass, and switchgrass, will decrease in composition and vigor, with continuous heavy grazing use. This state change is indicative of increased dominance of sideoats grama, blue grama, and western wheatgrass. Western wheatgrass and/or needle-and-thread (*Hesperostipa comata*) will continue to increase in composition with no prescribed fire and/or repetitive late growing season grazing use. Western ragweed (*Ambrosia psilostachya*) and yucca (*Yucca glauca*) are common forbs that increase in the stand. With increasing plant composition of the mid- and shortgrasses, there is a decrease in litter amount and greater potential of soil water evaporation and/or surface runoff resulting in less production. The total average annual production of this site is approximately 2578 kg ha^{-1} (2300 lbs ac^{-1}) (air-dry weight).

States 3.1 and 3.2 Shortgrass. This is a predominantly shortgrass plant community, about 88% graminoids, 2% forbs, and 10% woody plants. Total average annual production is approximately 897 kg/ha (800 lbs ac^{-1}) (air-dry weight) because of the dominance of the less productive shortgrass species.

Blue grama and buffalograss are the dominant grasses and form a dense sod with characteristic dense fibrous root systems in the top 10 cm (4 in) of the soil surface.

Silver bluestem can also be present with interspersed shortgrasses. In addition to the belowground morphological characteristics, these species are also associated with water repellency issues caused by synergistic relationships in the soil rhizosphere that affect hydrologic dynamics. Ritsema et al. (1998) state that water repellency is considered a plant-induced soil property. Sources of water-repellent compounds include accumulated plant-derived organic matter from mulch, decomposing roots and plant material, and root exudates (Doerr et al. 1996; Czarnes et al. 2000). Particulate organic matter contains plant- and microbial-produced compounds such as waxes (Franco et al. 2000; Schlossberg et al. 2005); humic acids (Spaccini et al. 2002); presence of a protective water-repellent lattice of long-chain polymethylene compounds around soil aggregates (Shepherd et al. 2001); aliphatic C present in organic matter (Ellerbrock et al. 2005); mycorrhizal and saprobic soil fungi (Bond and Harris 1964; Paul and Clark 1996; Hallett and Young 1999; White et al. 2000; Rillig 2004); basidiomycete fungi (Bond and Harris 1964; Fidanza 2003); fungal proteins such as hydrophobins (Rillig 2005; Rillig and Mummey 2006); and fatty acids, fulvic acids, extracellular enzymes, and polysaccharides (Bisdom et al. 1993; Kostka 2000; Eynard et al. 2006) (Fig. 2.11).

Blue grama and buffalograss are more drought- and grazing-tolerant than tall-grasses because of short shoot growing points being close to the soil surface (Dahl 1995). Blue grama is typically a bunchgrass but becomes more sod forming with heavy grazing use. A slight increase in plains pricklypear and yucca (*Yucca glauca*) may be observed. Forb diversity declines, with Cuman ragweed (*Ambrosia psilostachya*) and white sagebrush (*Artemisia ludoviciana*) being common. Species diversity declines compared to the reference State 1.1. Declining composition of tall

Fig. 2.11 State 3, shortgrass with silver bluestem increasing. (Photo courtesy of USDA-NRCS)

and mid-warm season native grass, some native cool-season grasses [western wheatgrass (*Pascopyrum smithii*); needle-and-thread, prairie June grass (*Koeleria macrantha*)], and native forbs (including leguminous spp.) is concomitant with changes in energy flow, nutrient cycling, and the hydrologic cycle.

Individual plant species have a profound effect on hydrology and erosion dynamics (USDA-NRCS 2003, Spaeth 1996a, b). Field hydrology studies have documented infiltration capacity with individual species composition (Mazarak and Conard 1959; Dee et al. 1966; Spaeth 1990, 1996a, b; Pierson et al. 2002). Dee et al. (1966) found that water infiltrated three times faster in blue grama and silver bluestem (*Bothriochloa saccharoides*) stands than areas dominated by annual weeds such as summer cypress (*Kochia scoparia*) and windmill grass (*Chloris verticillata*). Blue grama terminal infiltration capacity was about four times higher than buffalograss stands, holding soil type constant.

The primary difference between community phases 3.1 and 3.2 is that the ratio of buffalograss is greater than blue grams in 3.2. This transition results in less infiltration capacity because of the root pan and water-repellent properties in the soil surrounding buffalograss.

State 4 Woody State is representative of an increase and spread of trees resulting from absence of fire. The predominant woody trees are eastern red cedar (*Juniperus virginiana*) and honeylocust (*Gleditsia triacanthos*) with an average canopy cover of >20%. Birds, small mammals, and livestock are the primary distributors of seed, which accelerate the spread of most tree and shrubs common to this site. Many species of wildlife, especially bobwhite quail, turkey, and white-tailed deer, benefit from the cover, and food was provided from the trees and shrubs. Conversely, the presence of trees is considered detrimental to populations of greater prairie chickens and other grassland obligate birds. When management for specific wildlife populations is desirable, these options should be considered in any brush management plan.

The speed of encroachment varies considerably and can occur on both grazed and nongrazed pastures.

Fire is an important factor in maintaining historic open prairie vegetation and relative plant community composition and structure. Bragg and Hulbert (1976) reported that with a lack of fire, which is a normal disturbance component on this site, woody plants can increase up to 34% in contrast to a 1% increase on burned sites. It is worthy to note: not all unburned Loamy Plains sites will have woody invasion to this extent.

Grass production can be reduced by 10–30% due to encroachment of woody species. Hydrologic function and soil/site stability can be diminished with increases of invasion and increased woody canopy cover. Various hypotheses have been posed to explain how juniper can dominate a site: (1) shading by increased canopy cover and interception of rainfall by foliage (Schott and Pieper 1985; Gifford 1970; Wu et al. 2001); (2) allelopathic conditions and deep litter accumulation (Jameson 1966; Horman and Anderson 2003); and (3) changes in soil nutrients (Doescher et al.

1987; Tiedemann 1987) and surface water balance and moisture conditions (Miller et al. 1987; Breshears et al. 1997). As juniper overstory increases, understory vegetation decreases with increases in bare ground in interspaces between trees.

Canopy interception of rainfall can be significant and is affected by precipitation patterns and water holding capacity of plant canopies and litter in the coppice under the trees. Owens et al. (2006) reported that average interception of rainfall by juniper in Ashe juniper studies in Texas was about 35% of the total, 5% was transported down the stem to the soil, and 5% was intercepted by coarse litter under the trees. Some studies have indicated that interception of precipitation in juniper canopies can reach 70–80% (Eddleman 1983). Woody plants in semiarid rangelands also transpire more water than grasses due to deeper rooting structures, which can result in less available water for grass production (Thurow and Hester 1997; Wu et al. 2001). Small rainfall events are usually intercepted and retained in the foliage; thus water does not reach the litter layer at the base of the tree. Only when canopy storage is exceeded does precipitation reach the soil surface. Interception losses associated with the accumulation of the shrub coppice duff layer under juniper (partially decomposed leaves, twigs, and branches at the base of trees usually forming a mound under the tree canopy (coppice)) are considerably higher than losses associated with the canopy. The results affecting soil dynamic properties include infiltrability of water, biological activity, and soil nutrients.

Careful planning is necessary to apply prescribed burning to assure safety and provide for sufficient amounts of fine fuel to carry a fire and consume and control woody species. In some instances, the use of chemicals or mechanical means as a brush management tool may be desirable to initiate and accelerate this transition.

State 5 Altered State Cropland Pastureland

In Ford County, Kansas, where the Harney soil series was established, a famous frontier town, Dodge City (formally called Buffalo City), was established as a frontier town and shipping center. Dodge City was a supply station for soldiers, cowboys, buffalo hunters, and homesteaders from about 1822 to the early 1870s. The Santa Fe railroad extended to Dodge City in 1872, which established the city as a major shipping point for cattle. Cultivation by pioneer farmers most likely began soon after 1822.

Harney soils are extensively farmed; more than half of the land in this MLRA is used for dry-farmed crops. Winter wheat, oats, and grain sorghum are the major crops in much of the area, while corn is grown in the northern part of the area. Feed grains and hay crops also are widely grown. On irrigated land, corn, alfalfa, small grains, and grass for hay are grown. The major resource concerns associated with tillage is wind and water erosion and maintenance of soil organic matter. Soil organic matter decreases with soil depth. Plant litter and organic residues decompose and accumulate mostly at the soil surface. In highly productive grassland soils, SOC extends deeper into the soil profile because of the decomposition of deep extensive grass and herbaceous plant (forbs) root systems compared to less productive arid environments. Soil organic matter decreases after cultivation; the rate is determined by the type of tillage and management. Research shows that cultivation results in soil organic carbon losses between 20% and 40% over a 5- to 20-year period. Many highly productive tallgrass prairie soils contained upward of 7.5% organic matter before cultivation, and after 100 years of cultivation, about 56% of the total was lost. When cultivation starts, the easily decomposed active SOM pool of organic material is rapidly metabolized, which explains the sudden decline in the first few years. Growing and incorporating cover crops, adding manure, and no-till and reduced tillage systems are ways to manage the losses of organic matter in the soil. Cropland conservation management practices include no-till, strip-till, and mulch-till; level terraces in the western part of the area and gradient terraces and grassed waterways in the eastern part of the MLRA, contour farming, conservation crop rotations; irrigation water management; and nutrient and pest management (Fig. 2.12).

State 6 Restored-Reseeded Rangeland

Reseeding the field to native tallgrasses after 100 years of cultivation would slowly and continually begin to build organic matter content over time. Some restoration studies converting marginal cropland to perennial grasses show about a 12% recovery of the original target soil organic carbon over a decade (Garten and Wullschleger 2000). Recovery of soil organic carbon in soil depends upon the soil type, initial carbon inventories, climate, field cultivation practices, and field management. Management practices capable of minimizing erosion and maximizing

Fig. 2.12 State 5, altered state wheat field. (Photo courtesy of USDA-ARS, Stephen Ausmus)

return of residue biomass are fundamental to the maintenance of organic matter in cultivated soils.

This state and established plant community occurs on areas that were formerly farmed. When farming operations are terminated, the area can be seeded and established to a mixture of plants, usually native species common in the historic plant community. Most seeding mixtures consist of a mixture of grasses that include big bluestem, Indiangrass, switchgrass, little bluestem, and western wheatgrass. Depending on seed costs, other grass adapted native grass, and forb species can be added to the mix.

Once native grass stands become fully established (usually 2–5 years), production is comparable to that of the historic plant community. Total annual production ranges from 2980 kg ha^{-1} (2500 lbs ac^{-1}) to 4530 kg ha^{-1} (3800 lbs ac^{-1}) of air-dry vegetation per acre (avg. prod is 3363 kg ha^{-1}; 3000 lbs ac^{-1}). In native reseeded pastures where remnants and areas of original native rangeland exist in the same pasture, they seldom are utilized at the same intensity because domestic livestock usually prefer plants found on the native rangeland areas. For optimum management, reseeded plant communities should be managed as separate pastures. Reseeded pastures can also be managed for native hay production.

State 7 Go-back land. This plant community was created when the soil was tilled (most likely plowed) and farmed (sodbusted) and then abandoned because of exhausted fertility and low crop productivity. Native plants were cultivated and lost, soil organic matter declined, soil structure was altered, and a plow pan or compacted layer can be observed, limiting root penetration and water infiltration. Soil mycorrhizae and other microbial activity are reduced—overall soil health is usually poor. After abandonment, during the early successional stages, this state is not stable as wind and water erosion can be excessive until some form of permanent plant cover is re-established.

Early succession of weedy plants will primarily consist of kochia (*Kochia scoparia*), annual bromes (*Bromus* spp.), pigweed (*Amaranthus albus*), foxtail (*Setaria* spp.), Russian thistle (*Salsola kali*), witchgrass (*Panicum capillare*), and tumblegrass (*Schedonnardus paniculatus*) as well as other annuals. These plants give some protection from erosion and start to rebuild organic matter. The next successionary stage will be grasses such as sand dropseed (*Sporobolus cryptandrus*), threeawn (*Aristida* spp.), silver bluestem, and annuals. Eventually, after decades, blue grama, sideoats grama, and buffalograss may emerge; however, they will not regain in proportions to that of the reference state. Unless this state is reseeded, a stand of low productivity weeds will exist. Erosion from wind and water can stabilize, but productivity and quality of grazing is low.

Text Box 2.2 Basic Ideas Underlying Plant Community Succession and Change

As mentioned in the abstract of this chapter, knowledge of individual ecological site dynamics associated with agricultural fields (cropland, pastureland, rangeland, garden areas) is the primary starting point toward sustainable management and having a plan to maintain and/or improve soil health. Knowledge and understanding of many of the dynamics of natural vegetation systems is often based on theory and supposition, possibly supported by experiments where possible, and/or anecdotal observations. For example, there are many contrasting studies involving plant succession, competition, cyclic events, resilience, and resistance that propose various hypotheses and different conclusions. To make things more complicated, these dynamics often exhibit different characteristics among different plant community types. At best, we can track plant community pattern responses to disturbances via state and transition models, but science has limited knowledge of the discrete drivers concerning these dynamics. State and transition models are an integral part of ecological sites and portray plant successional dynamics associated with an ecological site. Ecologists and land managers can observe responses of vegetation to fire, invasive plants, grazing, and cyclic weather events; however, the ecological intricacies and responses of species (plants and soil microorganisms) with environmental variables can take unexpected turns and be unpredictable. Often management in natural range and forest ecosystems can result in expected results; however, there is always the unexpected. MacArthur (1972) wrote that the main objective of community ecology is to identify patterns and functional rules. State and transition models can identify the patterns and observable changes in plant composition and associated environmental factors (soil erosion, etc.); however, the part that is missing is the repeatable rules or dynamics that govern the ecosystem. Some ecologists have suggested that "community ecology is a mess," with respect to community dynamics, and that general laws consist of relatively few fuzzy generalizations (Lawton 1999). Simberloff (2004) adds that "laws and models in community ecology are highly contingent, and their domain is usually very local": general rules in ecology have been difficult to achieve due to the complex nature of plant communities and their environments. The intent of these statements is not to imply that ecology is a "weak science"—nature is elusive: studies in community ecology are critically important in pursuing knowledge about how and why environmental change occurs, even if we cannot derive sound theories and laws that pertain to main themes in plant ecology.

For example, plant competition is a primary ecological process where plants directly or indirectly impact one another. The main concepts that dominate discussions about plant competition are intraspecific competition, interspecific competition, competitive exclusion, coexistence, and symmetrical or

(continued)

asymmetrical competition (Connell 1990).[14] No general overriding law exists regarding competition of plants in a crop monoculture or in the natural environment.

Plant succession is another major theme for plant ecologists (McIntosh 1981, 1985); it is conceived as an orderly transition and sequence of changes in species occurrences with changing and evolving community processes over time. Pickett et al. (1987) simply define vegetation succession as the process of species replacement over time. Huston and Smith (1987) relate sequential transition as a series of intervals where a "once dominant species or group of species" recedes and other species transition into the community or gain dominance. A former dominant species will remain subdominant or disappear altogether unless a disturbance and/or environmental change occurs. Barbour et al. (1987) stress that time can be highly variable with respect to successional pathways. Primary succession refers to lands that are in very early stages of soil development (e.g., volcanic activity, geologic disturbances, sand dune formation, drying of peat bogs) where no previous vegetation exists can go through series of succession for millennia. Secondary succession refers to lands where vegetation exists, and a series of vegetation transitions occur due to climatic changes and disturbances such as fire, drought, and grazing pressure. Disturbances may cause temporary periods of bare soil; however, if the intensity of the disturbance did not remove plant propagules (seeds, roots, rhizomes), plants will slowly re-establish on the site. Mueller-Dombois and Ellenberg (1974) and Major (1974) estimate that secondary succession can proceed five to ten times faster than primary succession. The sequence of vegetation progression is marked by stages of specific vegetation composition and growth called seres or seral stages.

Many different theories and hypotheses have been introduced in the literature. The traditional ideas about plant succession encompass the works of Whittaker (1953), Egler (1954), Drury and Nisbet (1973), Horn (1974), Connell and Slatyer (1977), McIntosh (1981, 1985), Miles (1987),

---

[14] "Intraspecific competition implies that more closely related species are less likely to coexist; the idea is that individuals of the same species share similar resource requirements. Interspecific competition in natural plant communities focuses on competition between different species. Competitive exclusion considers that two species, evenly distributed and with limited resources, cannot coexist for extended periods. Competitive exclusion may be avoided if the species either develop or have different requirements (e.g., niche differentiation—deeper versus shallower roots, plants requiring high-intensity versus filtered light, larger versus smaller leaves—or differentiation in adapting to varying moisture levels in the soil, or in the ability of a plant to maintain internal water balance). Coexistence is necessary if two or more species are to survive; natural selection in the plant community coincides with maintaining balance for the same contested resources. Niche segregation can be symmetrical or asymmetrical and usually refers to how plant species evolve away from each other and adapt to unique niches or locations in their communities" (Spaeth 2018).

(continued)

Glenn-Lewin and van der Maarel (1992), and Kahmen and Poschlod (2004). Since the early twentieth century, observations and treatises of Clements (1916), Sampson (1917, 1919), Weaver and Clements (1938), and Dyksterhuis (1949) formed the basis of classic plant successional theory used on the rangelands in the United States. Many papers cover the history of successional concepts on rangeland (Westoby et al. 1989; Friedel 1991; Laycock 1991; Joyce 1993; Briske 2017a, b). Frederic Edward Clements (1874–1945) s theory on plant community succession was highly influential and was based on gradual vegetation changes over time, a unidirectional progressional sequence with advancing stability and retrogression of a series of intermediate plant communities (seral stages), eventually culminating with a climax plant community. Clements formulated the monoclimax theory (a self-perpetuating climax community), which included several assumptions: (1) climate of a region is related to the dominant species that adapt and populate a site; (2) succession culminates in climax state of vegetation; and (3) the climax state represents equilibrium with temporary fluxes of compositional change that is cyclic (seasonally, yearly, or longer). Whittaker (1951, 1953, 1967) and Meeker and Merkel (1984) provide a thorough synopses of the climax concept. Clements progressed on the work of Henry Cowles (1869–1939) who described plant succession in a study of sand dune vegetation development on the shores of Lake Michigan (Cowles 1899; Clements 1904). Cowles depicted a chronosequence of vegetation from sand dunes near the shore, progressing more inland from the sandy beach to grassland, to mature forests (Fig. 2.14a, b).

Whittaker (1953) emphasized that natural plant communities are responsive to many different environmental factors such as genetic adaptabilities of plant species, climate, soil, seed and propagule resources, and a variety of disturbances such as herbivory or fire. In reality, several climax community outcomes can develop. Whittaker (1975) stressed that different stable communities can develop on south vs. north slopes, granitic vs. limestone, sandstone, shale thus an area can contain a number of different kinds of climax communities—forming a mosaic (polyclimax). Considering the ecological site concept used today by ecologists and land managers, these polyclimax alternatives would be explained by the environmental specifics of the ecological site: the north and south slope vegetation differences and the soil parent materials would not be considered a polyclimax of one community but would be specific to the individual ecological sites. Egler (1954) discussed the concept of relay floristics, which involved a succession of plants that emerge then recede and in the process prepare the site for another cycle of incoming and outgoing plants.

Relay floristics depicts a succession of incoming and outgoing plants; each group appears along a timeline and then subsequently is replaced by another group of species (Fig. 2.13). Initial floristics composition considers that each

(continued)

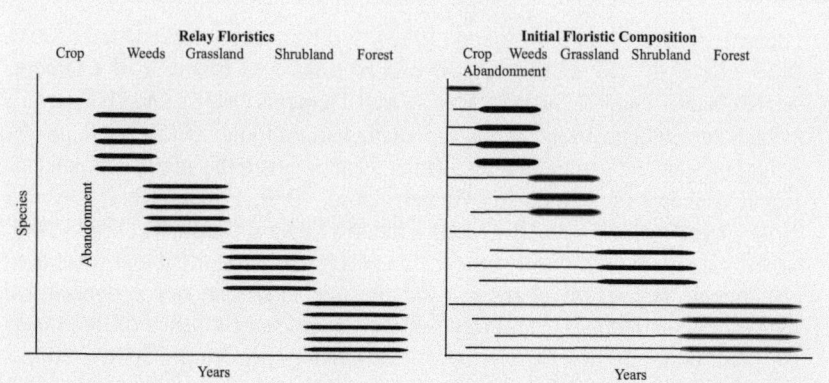

Fig. 2.13 Diagrammatic representation of relay and initial floristic composition. (Adapted from Egler 1954)

Fig. 2.14 (**a**) Low crescentic dune complex, High Island, Michigan. Photo by George Damon Fuller, image apf8-02986, 4-09-1917. Photo courtesy of University of Chicago Department of Botany Records. (**b**) Succession on the dunes, *Populus balsamifera*, *Picea*, *Abies*, *Thuja*, and *Acer*, High Island, Michigan. *Prunus* is also present. Photo by George Damon Fuller, image apf8-02990, 4-09-1917. (Photos courtesy of University of Chicago Department of Botany Records)

species is present as seed or propagules from the onset of cropland abandonment, and then there is a progressive visible successional sequence.

Drury and Nisbet (1973) later challenged studies of succession in forests and concluded that many did not "conform to contemporary generalization." Drury and Nisbet (1973) found compatibility or harmony between "temporal sequences of vegetation and spatial sequences along environmental gradients." Forest succession encompassed differential growth, survival, and colonizing ability of plant species with concurrent adaptively to environmental gradients. Successive replacement of plant and plant groups by other groups

(continued)

is partially due to interspecific competition. Drury and Nisbet's model of forest succession was a fundamental shift to more individual properties and correlations in plant size, longevity, and growth.

Connell and Slatyer (1977) identified three models of succession, facilitation, tolerance, and inhibition, with discussions on evidence in nature:

> The first 'facilitation' model suggests that the entry and growth of the later species is dependent upon the earlier species 'preparing the ground' only after this can later species colonize. Evidence in support of this model applies mainly to certain primary successions and in heterotrophic succession.

> A second 'tolerance' model suggests that a predictable sequence is produced by the existence of species that have evolved different strategies for exploiting resources. Later species will be those able to tolerate lower levels of resources than earlier ones. Thus, they can invade and grow to maturity in the presence of those that preceded them. At present there exists little evidence in support of this model.

> A third 'inhibition' model suggests that all species resist invasions of competitors. The first occupants preempt the space and will continue to exclude or inhibit later colonists until the former die or are damaged, thus releasing resources. Only then can later colonists reach maturity. A considerable body of evidence exists in support of this model.

The ideas and theories of succession have merit in certain plant communities. Unfortunately, the processes of plant succession in plant communities are not consistent for many terrestrial plant communities. No one single model or explanation fits all plant communities. Gibson and Brown (1985) state that there are "several broad types of succession which dominate under recognizable circumstances e.g., dune, forest, grasslands, and some old field succession." Whittaker and Levin (1977) recognized the historic failure of successional theories and attempts at unifying a modal concept (generalities). In reality, ecologists recognize the intrinsic complexity of ecological dynamics across the landscape. Plants have evolved in a myriad of different environments and ecosystems in order to compete with other plant species: How can we expect a modal or unifying concept of succession to arise? McIntosh (1981) summed it up with a quote from Whittaker and Levin (1977): "analysis, interpretation, comparison, and modeling of cases rather than on widely applicable generalization" has not been accepted by ecological theorists. Miles (1987) and Glenn-Lewin and van der Maarel (1992) relate that plant species composition and climate associated with the site lead to multiple successional pathways that prevent a general unifying theory.

In ecological site descriptions, state and transition models portray or provide a map of vegetation state change, transitions among the various states, and drivers such as fire, grazing impacts, lack of control of invasive species, and cultivation, but do not explain the mechanisms or processes of plant succession—the interactions between individual plant species, herbivory (if a major factor, see Mack and Thompson 1982), and other environmental factors.

(continued)

As Briske (2017a, b) points out: "The adoption of successional theory—an equilibrium concept [Clemensian succession and the use of rangeland condition classes] was considered a major conceptual advance by rangeland professionals in the early twentieth century [extending on through the century until the early 1990s]. It had a profound influence on the rangeland profession by directly linking it to equilibrium ecology and by indirectly contributing to the steady-state management model of natural resource management. Dyksterhuis (1949) further secured succession in the foundation of rangeland science by operationalizing the range model [or rangeland successional model] on a quantitative basis." The Dyksterhuis (1949) approach to rangeland condition considered how plant species react to grazing pressure. Generally, plant species, usually the most desirable native species, may decrease under heavy grazing pressure, while other plants in the community respond as increasers or invaders. Native species could act as invader species, i.e., juniper and mesquite invasion, and/or introduce exotic species replacing native species. Adaptation of the Dyksterhuis' concept of range condition was eventually adapted into the USDA—Soil Conservation Service where decreaser, increaser, and invasive species were identified within range sites. A similarity index approach was also used to determine range condition, which compared existing vegetation to historic climax vegetation for the site identified 0–25% species similarity as poor condition, 26–50% similarity as fair condition, 51–75% similarity as good condition, and >76% similarity as excellent condition (USDA-SCS 1976). These concepts were pervasive and formed the basis for successional theory in rangeland science and management for 50 years; however, beginning in the 1970s and 1980s, they encountered severe criticism by both Australian and US rangeland scientists (Westoby et al. 1989; Joyce 1993; Briske 2017a, b). The similarity index-based range condition model was solely on comparing existing vegetation composition with a historic climax plant community. Although the similarity index is a useful tool in evaluating plant composition dynamics, it was not entirely accurate in assessing true rangeland health: hydrology, energy dynamics, and soil and surface stability were not directly evaluated. The USDA-NRCS dropped this approach as a health assessment tool (although still used for evaluating plant composition dynamics) and began using the rangeland health model for assessing biotic integrity, soil and surface stability, and hydrologic function (Pellant et al. 2005, 2020).

What caused the shift in successional concepts on rangeland? Rangeland managers and scientists recognized the failures of the preceding rangeland successional model based on Clementsian theory. For years, a dilemma ensued: why was the classic theory of plant succession not a functional model in many western rangeland ecosystems? Many plant communities in fragile arid ecosystems do not exhibit strong ecological resistance and resilience to disturbances when invasive exotic plant species are present in varying

(continued)

amounts. Disturbance from fire, drought, insects, animals, etc. is a natural phenomenon in many plant communities and is necessary to maintain healthy ecosystems.

The geographic spread and the number of invasive plant species have increased significantly over the past 200 years as a result of human activities (di-Castri 1989). On rangelands, exotic annual grass invasion has been especially dramatic and has transformed many native plant community types throughout the United States (Mack 1981). This transformation has been rapid and ubiquitous, and when annual grass dominance occurs, ecosystem function can be compromised (Vitousek et al. 1997). On US rangelands, non-native exotic plants can negatively impact biotic integrity, ecosystem stability, composition and structure, natural fire cycles, diversity, soil biota, vegetation production, forage quality, wildlife habitat, soil physical properties, organic matter dynamics, carbon balance,,nutrient and energy cycles, and hydrology and erosion dynamics in many unique ways (Chapin et al. 2000; Evans et al. 2001; Pierson et al. 2002; Ehrenfeld 2003; Ogle et al. 2003, 2004; Brooks et al. 2004; Norton et al. 2004, Belnap et al. 2005; Hooper et al. 2005; Sommer et al. 2007; Boxell and Drohan 2008; Herrick et al. 2010; Davies 2011).

Inherent resistance is the ability of plant communities to retain vegetation structure, energy and nutrient cycles, and functionality (resist change) with various disturbance factors. Ecological resistance is related to the degree and amount of disruption required to shift the system from one state with mutually reinforcing processes to another state (Peterson et al. 1998; Briske et al. 2008, 2017a, b). Ecological resilience related to how well plants, animals, and organisms in the abiotic environment can regain their vegetation structure, function, and environmental processes with disturbance (fire, drought, land use changes, and management (Peterson et al. 1998).

Are there plant communities where the classic model of secondary succession exists?

Many forest plant communities exhibit seral stages of secondary succession after mature adapted tree species are removed by fire or logging. Alaback (1982) documented forest succession in a Sitka spruce-western hemlock forest (*Picea sitchensis-Tsuga heterophylla*).

Understory vegetation undergoes successional stages during the 1st 300 yr. after logging or fire disturbance . . . Residual shrubs and tree seedlings increase their growth within 5 yr. after overstory removal. Understory biomass peaks at 5 yr. . . Shrubs and herbs are virtually eliminated from the understory after forest canopies close at stand ages of 23–35 yr. Bryophytes and ferns dominate understory biomass during the following century. An understory of deciduous shrubs and herbs is reestablished after 140–160 yr. Thereafter, biomass of the shrubs, herbs, and ferns continues to increase, while bryophyte biomass and tree productivity decline . . . Tree growth also declines after the first century as losses to mortality and fungal attack become more significant . . . The final stages of understory development represent a transition to the dynamic equilibrium of an old-growth forest dominated by *Tsuga heterophylla* .

(continued)

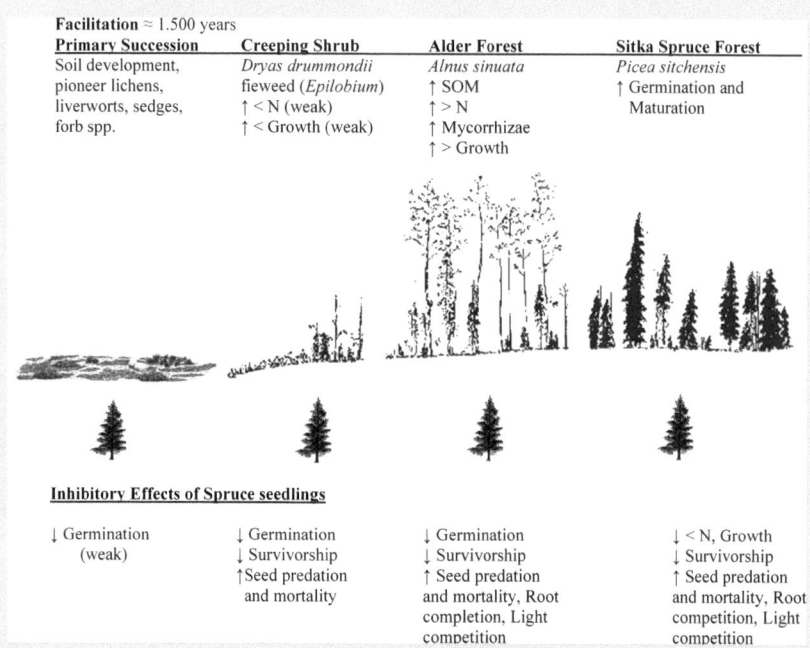

Facilitation ≈ 1.500 years			
Primary Succession	Creeping Shrub	Alder Forest	Sitka Spruce Forest
Soil development, pioneer lichens, liverworts, sedges, forb spp.	*Dryas drummondii* fieweed (*Epilobium*) ↑ < N (weak) ↑ < Growth (weak)	*Alnus sinuata* ↑ SOM ↑ > N ↑ Mycorrhizae ↑ > Growth	*Picea sitchensis* ↑ Germination and Maturation

Inhibitory Effects of Spruce seedlings

| ↓ Germination (weak) | ↓ Germination ↓ Survivorship ↑ Seed predation and mortality | ↓ Germination ↓ Survivorship ↑ Seed predation and mortality, Root completion, Light competition | ↓ < N, Growth ↓ Survivorship ↑ Seed predation and mortality, Root competition, Light competition |

Fig. 2.15 Primary and secondary succession of Sitka spruce (*Picea sitchensis*) forest with some interacting facilitative and inhibitory factors. (Adapted from Chapin et al. 1994)

. . Biomass of tree seedlings . . . increases during the final successional stage. During this period, most tree seedlings are concentrated on well-decayed logs and stumps (Alaback 1982).

Chapin et al. (1994) studied the succession at Glacier Bay, Alaska, and found that multiple factors are associated with primary succession (Fig. 2.15). The author's objective was to investigate facilitation (the process where colonizing species (at the expense of eventually compromising their survival) improve the environment for subsequent establishment of later successional species). Research on plant succession has argued against facilitation in favor of inhibition of later successional species (Drury and Nisbet 1973; Connell and Slatyer 1977; Walker and Chapin 1986). Chapin et al. (1994) stress that generalizations are difficult in studying plant successional mechanisms: "generalizations about the importance of facilitation and other successional mechanisms must be placed in a context of resource availability and environmental severity." Life history traits such as seed size, growth rate, reproduction age, availability of propagules, maximum height, longevity, shade intolerance, root competition and development of soil, organic matter, mycorrhizal populations, and nutrient availability all play an important role as secondary succession advances. The authors summarized that life history traits of individual

(continued)

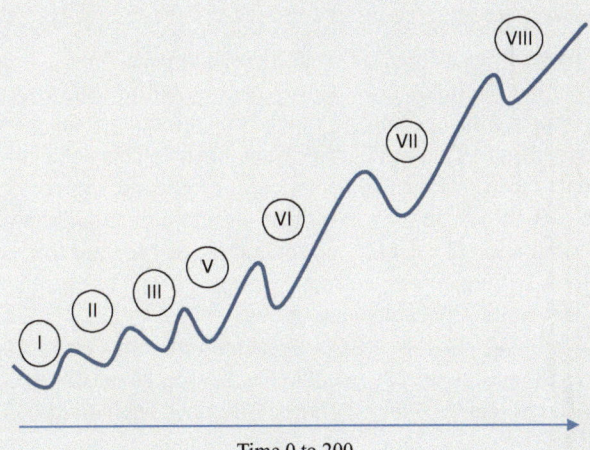

Time 0 to 200

Fig. 2.16 The "ball and cup or trough" model illustrating facilitative shift in successional states (northern Piedmont forest Billings 1978)

plant species, competitive interactions, and successional facilitation (when present) were all interacting factors of succession at Glacier Bay— "the search for a single predominant mechanism seem futile (Chapin et al. 1994)." Start new paragraph Billings (1978) Figure 2.15 should be here Primary and secondary succession ...

Billings (1978) gives an example of old field succession in eastern US forest (North Carolina Piedmont (MLRA 148, Northern Piedmont); ER 45e Northern Inner Piedmont) with abundant rainfall and moderate temperatures. Prior to European settlers clearing the land for crops (corn, tobacco, cotton) in the early eighteenth century, the historic climax vegetation consisted of deciduous (oaks and hickory) and coniferous tree species (loblolly and shortleaf pine). Repeated plantings of these crops soon depleted nitrogen resources, and it was common to abandon the fields and plant new fields. This practice continued through the 1930s, and land abandonment of this type was largely discontinued with the use of better cropping practices and fertilization.

This old field successional pathway (Fig. 2.16) begins with[15]:

- *Stage 1*: 1st year—abandonment of cultivation, invasive weeds establish (crabgrass (*Digitaria sanquinalis*), Canadian horseweed (*Conyza canadensis*). Horseweed overwinters as a rosette.
- *Stage 2*: 2nd summer, horseweed diminishes while white aster (*Aster pilosus*) appears and becomes more dominant.

---

[15] As described by Billings (1978).

(continued)

- *Stage 3*: 3rd summer, broomsedge (*Andropogon virginicus*) appears and gradually replaces aster. Pine seedlings can appear.
- *Stage 4*: Pine seedlings grow slowly at first, but in following years grow rapidly shading the understory, which is unsuitable for the pioneer weeds.
- *Stage 5*: 5–10 yrs. Pine grows rapidly, competition between individual pine trees, many die off. Pine trees compete for light and water.
- *Stage 6*: 10–80 yrs. In the understory, sweet gum (*Liquidambar* spp.) and various oak species establish in understory as they are tolerant of shade conditions. In time, a population of sweet gum, then oak saplings, followed by hickory saplings establish and grow.
- *Stage 7*: 80–140 yrs. Pine trees are reaching maturity, and secondary trees appear in the understory [dogwoods (*Cornus* spp.), sourwood (*Oxydendrum* spp.), and red maple (*Acer rubrum*)]. Old pines begin to die off.
- *Stage 8*: 200 yrs. Deciduous trees now dominant, minor changes in species composition occur.

Many studies have evaluated secondary succession in prairie vegetation (Bruner 1931; Booth 1941; Penfound and Rice 1957a, b; Penfound 1964; Levin 1966; Kaputska and Moleski; 1976; Kirchner 1977; Reichhardt 1982; Collins and Adams 1983; Inouye et al. 1987). Reichhardt (1982) studied successional trends of abandoned fields on shortgrass prairie on the Pawnee National Grassland in Northeastern Colorado. The old field sites since disturbance were 43 years old. Their findings showed that old field species composition was diverse and consisted of perennial grasses, herbaceous forbs, and mixed shrubs. Frequency values of blue grama, the dominant historic perennial grass species, ranged from 2% in old disturbed fields to 40–60% in adjacent undisturbed prairie. The study concluded that "a weak trend toward convergence with the unplowed condition was evident, the successional process is slow, and likely to exceed 50 years if the unplowed condition is ever to be reached (Reichhardt 1982)."

Old field succession in tallgrass prairie theoretically involves four stages of development: pioneer weeds, annual grass, bunch grass, and mature prairie. Collins and Adams (1983) evaluated old field succession in tallgrass prairie in Central Oklahoma. Booth (1941) proposed (although never documented by a field study) that secondary succession in Oklahoma prairie vegetation exhibits four stages of development: pioneer weed establishment, annual grasses, perennial native bunchgrasses, and finally mature prairie. Old field sites were 32 years old, and successive vegetation development was variable and unpredictable. Convergence of historic climax tallgrass vegetation was unpredictable. The sites were largely invaded by shrubs and some tree species.

Summary

Tracing the evolutionary and ecological history of our local crop, pastures, ranges, and garden areas is the first step in understanding site dynamics. These dynamics include soil type, soil erosion potentials, inherent soil fertility, and factors related to historically occurring plant species and productivity. Scientists and land management agencies in the United States classify lands into various hierarchical systems. The purpose of land classification methodologies is to identify and describe similar patterns among biotic and abiotic components, which can be used to understand certain ecological parameters. Different scales of land classification exist and range from biomes to discrete localized ecological sites.

In the United States, various land management agencies such as the USDA Forest Service, USDA Natural Resources Conservation Service, and the Environmental Protection Agency have developed land classifications that meet their mission expectations and goals. In this chapter, two dominant land classification schemes are introduced: ecoregions (I, II, III, IV) and Major Land Resource Areas. Within these broad land classifications, lands can be further dissected into ecological sites and/or habitat types. In farm and ranch planning, ecological sites include details of climate, physiography, soils, vegetation parameters, state and transition models showing potential patterns of change, and specific management information. Two basic land types, forestland and rangeland, are the basis for ecological site classification. Lands cropped or in tame pasture for long periods of time can be tied to either forest or rangeland. Relating existing land use to the historic ecological site identity provides a baseline for how lands respond to management over time. Managing soil health is directly tied to local site potentials at the ecological site level.

References

Alaback, P.B. 1982. Dynamics of Understory Biomass in Sitka Spruce-Western Hemlock Forests of Southeast Alaska: Ecological Archives E063-004. *Ecology* 63: 1932–1948.

Anderson, K.L. 1965. Time of burning as it affects soil moisture in an ordinary upland bluestem prairie in the Flint Hills. *Journal of Range Management* 18: 311–316.

Anderson, K.L., E.F. Smith, and C.E. Owensby. 1970. Burning bluestem range. *Journal of Range Management* 23: 81–92.

Austin, M.P., and T.M. Smith. 1990. A new model for the continuum concept. In *Progress in theoretical vegetation science*, 35–47. Dordrecht: Springer.

Bailey, R.G. 1976. *Ecoregions of the United States (map)*. Ogden: US Department of Agriculture, Forest Service.

———. 1980. *Description of the ecoregions of the United States*. Washington, DC: US Dept. Agriculture, Miscellaneous Publ. 1391.

———. 1983. Delineation of ecosystem regions. *Environmental Management* 7: 365–373.

———. 1995. *Description of the ecoregions of the United States*. 2nd ed. Washington, DC: US Dept. Agriculture, Miscellaneous Publ. 1391.

Barbour, M.G., J.H. Burk, and W.D. Pitts. 1987. Terrestrial Plant Ecology. San Francisco, California: The Benjaminn/Cumming Pub. Co. Inc. 634pp.

Barnes, P.W., and C.J. Nelson. 2003. Forages and grasslands in a changing world. In *Forages: An introduction to grassland agriculture*, ed. R.F. Barnes, C.J. Nelson, M. Collins, and K.J. Moore, vol. 1, 3–24. Ames, Iowa: Iowa State Univ. Press.

Belnap, J., J.R. Welter, N.B. Grimm, N. Barger, and J.A. Ludwig. 2005. Linkages between microbial and hydrologic processes in arid and semiarid watersheds. *Ecology* 86: 298–307.

Bestelmeyer, B.T., J.R. Brown, K.M. Havstad, R. Alexander, G. Chavez, and J.E. Herrick. 2003. Development and use of state-and-transition models for rangelands. *Journal of Range Management* 56: 114–126.

Bestelmeyer, B.T., J.E. Herrick, J.R. Brown, D.A. Trujillo, and K.M. Havstad. 2004. Land management in the American southwest: A state-and-transition approach to ecosystem complexity. *Environmental Management* 34: 38–51.

Bestelmeyer, B.T., A.J. Tugel, G.L. Peacock Jr., D.G. Robinett, P.L. Shaver, J.R. Brown, and K.M. Havstad. 2009. State-and-transition models for heterogeneous landscapes: A strategy for development and application. *Rangeland Ecology & Management* 62: 1–15.

Bestelmeyer, B.T., A. Ash, J.R. Brown, B. Densambuu, M. Fernández-Giménez, J. Johanson, M. Levi, D. Lopez, R. Peinetti, L. Rumpff, and P. Shaver. 2017. State and transition models: Theory, applications, and challenges. In *Rangeland systems processes, management and challenges*, ed. D.D. Briske, 303–346. Cham: Springer Open.

Billings, W.D. 1978. Plants and the ecosystem 3rd ed. Bellmont, California: Wadsworth Publishing Co.

Bisdom, E.B.A., L.W. Dekker, and J.F.T. Schoute. 1993. Water repellency of sieve fractions from sandy soils and relationships with organic material and soil structure. *Geoderma* 56: 105–118.

Bond, R.E., and J.R. Harris. 1964. The influence of the microflora on physical properties of soils I. Effect associated with filamentous algae and fungi. *Australian Journal of Soil Research* 2: 111–122.

Booth, W.E. 1941. Revegetation of abandoned fields in Kansas and Oklahoma. *American Journal Botany* 28: 415–422.

Boxell, J., and P.J. Drohan. 2008. Surface soil physical properties and hydrological characteristics in *Bromus tectorum* L. (cheatgrass) versus *Artemisia tridentata* Nutt. (big sagebrush) habitat. *Geoderma* 149: 305–311.

Bragg, T.B., and L.C. Hulbert. 1976. Woody plant invasion of unburned Kansas bluestem prairie. *Journal of Range Management* 29: 19–24.

Breshears, D.D., O.B. Myers, S.R. Johnson, C.W. Meyer, and S.N. Martens. 1997. Differential use of spatially heterogeneous soil moisture by two semiarid woody species: *Pinus edulis* and *Juniperus monosperma*. *Journal of Ecology* 85: 289–299.

Briske, D.D. 2017a. Rangeland systems: Foundation for a conceptual framework. In *Rangeland systems processes, management and challenges*, ed. D.D. Briske, 1–24. Cham: Springer Open.

———. 2017b. Nonequilibrium ecology and resilience theory. In *Rangeland systems processes, management and challenges*, ed. D.D. Briske, 197–228. Cham: Springer Open.

Briske, D.D., S.D. Fuhlendorf, and F.E. Smeins. 2005. State-and-transition models, thresholds, and rangeland health: A synthesis of ecological concepts and perspectives. *Rangeland Ecology & Management* 58: 1–10.

Briske, D.D., B.T. Bestelmeyer, T.K. Stringham, and P.L. Shaver. 2008. Recommendations for development of resilience-based state-and-transition models. *Rangeland Ecology & Management* 61: 359–367.

Brooks, M.L., C.M. D'Antonio, and D.M. Richardson. 2004. Effects of invasive alien plants on fire regimes. *BioScience* 54: 677–688.

Bruner, W.E. 1931. The vegetation of Oklahoma. *Ecological Monographs* 1: 101–188.

Callaway, R.M. 1997. Positive interactions in plant communities and the individualistic-continuum concept. *Oecologia* 112: 143–149.

Carpenter, S.R., and W.A. Brock. 2006. Rising variance: A leading indicator of ecological transition. *Ecology Letters* 9: 308–315.

Chapin, F.S., L.R. Walker, C.L. Fastie, and L.C. Sharman. 1994. Mechanisms of primary succession following deglaciation at Glacier Bay, Alaska. *Ecological Monographs* 64: 149–175.

Chapin, F.S., E.S. Zavaleta, V.T. Eviner, R.L. Naylor, P.M. Vitousek, H.L. Reynolds, D.U. Hooper, S. Lavorel, O.E. Sala, S.E. Hobbie, and M.C. Mack. 2000. Consequences of changing biodiversity. *Nature* 405: 234.

Chapman, S.S., J.M. Omernik, J.A. Freeouf, D.G. Huggins, J.R. McCauley, C.C. Freeman, G. Steinauer, R.T. Angelo, and R.L. Schlepp. 2001. Ecoregions of Nebraska and Kansas (color poster with map, descriptive text, summary tables, and photographs): Reston, Virginia, U.S. Geological Survey (map scale 1:1,950,000).

Clements, F.E. 1904. *The development and structure of vegetation.* Lincoln: University of Nebraska-Botanical Seminar.

———. 1916. *Plant succession: An analysis of the development of vegetation.* Carnegie: Institution of Washington. Washington, DC, Pub. 242. 512 p.

Collins, S.L., and D.E. Adams. 1983. Succession in grasslands: Thirty-two years of change in a central Oklahoma tallgrass prairie. *Vegetatio* 51: 181–190.

Connell, J.H. 1990. Apparent versus "real" competition in plants. In *Perspectives on plant competition*, ed. J.B. Grace and D. Tilman, 9–26. Sand Diego: Academic Press.

Connell, J.H., and R.O. Slatyer. 1977. Mechanisms of succession in natural communities and their role in community stability and organization. *The American Naturalist* 111: 1119–1144.

Cooper, S.T., K.E. Newman, R. Steele, D.W. Roberts. 1991. Forest habitat types of northern Idaho: A second approximation. USDA For. Serv. Gen. Tech. Rep. INT-236. p 135.

Cowles, H.C. 1899. The ecological relations of the vegetation on the sand dunes of Lake Michigan. Part I.-geographical relations of the Dune Floras. *Botanical Gazette* 27: 95–117.

Crepet, W.L., and G.D. Feldman. 1991. The earliest remains of grasses in the fossil record. *American Journal of Botany* 78: 1010–1014.

Czarnes, S., P.D. Hallett, A.G. Bengough, and I.M. Young. 2000. Root and microbial-derived mucilages affect soil structure and water transport. *European Journal of Soil Science* 51: 435–443.

Dahl, B.E. 1995. Developmental morphology of plants. In *Wildland plants: Physiological ecology and developmental morphology*, ed. D.J. Bedunah and R.E. Sosebee, 22–58. Denver: Society for Range Management.

Daubenmire, R., and J.B. Daubenmire. 1968. *Forest vegetation of eastern Washington and northern Idaho.* Technical Bulletin 60, 104. Pullman: Washington State University, College of Agriculture, Washington Agricultural Experiment Station.

Davies, K.W. 2011. Plant community diversity and native plant abundance decline with increasing abundance of an exotic annual grass. *Oecologia* 167: 481–491.

Dee, R.F., T.W. Box, and E. Robertson Jr. 1966. Influence of grass vegetation on water intake of Pullman silty clay loam. *Journal of Range Management* 19: 77–79.

di Castri, F. 1989. History of biological invasions with emphasis on the Old World. In *Biological invasions: A global perspective*, ed. J. Drake, F. di Castri, R. Groves, F. Kruger, H.A. Mooney, M. Rejmanek, and M. Williamson, 1–30. Wiley USA: New York.

Dice, L.R. 1943. *The biotic provinces of North America.* Ann Arbor: University of Michigan Press.

Doerr, S.H., R.A. Shakesby, and R.P.D. Walsh. 1996. Soil water repellency variation with depth and particle size fraction in burned and unburned *Eucalyptus globulus* and *Pinus pinaster* forest terrain in the Agueda basin, Portugal. *Catena* 27: 25–47.

Doescher, P.S., L.E. Eddleman, and M.R. Vaitkus. 1987. Evaluation of soil nutrients, pH, and organic matter in rangelands dominated by western juniper. *Northwest Science* 61: 97–102.

Drury, W., and I.C.T. Nisbet. 1973. Succession. *Journal of the Arnold Arboretum* 54: 331–368.

Dyksterhuis, E.J. 1949. Condition and management of rangeland based on quantitative ecology. *Journal of Range Management* 2: 104–105.

Eddleman, L.E. 1983. Some ecological attributes of western juniper. In *Research in range management*, 32–34. Washington, DC: USDA ARS, SR-682.

Egler, F.E. 1954. Vegetation science concepts I. Initial floristic composition, a factor in old-field vegetation development. *Vegetatio* 4: 412–417.

Ehrenfeld, J.G. 2003. Effects of exotic plant invasions on soil nutrient cycling processes. *Ecosystems* 6: 503–523.

Ellerbrock, R.H., H.H. Gerke, J. Bachmann, and M.O. Goebel. 2005. Composition of organic matter fractions for explaining wettability of three forest soils. *Soil Science Society of America Journal* 69: 57–66.

Evans, R.D., R. Rimer, and S.P. Belnap. 2001. Exotic plant invasion alters nitrogen dynamics in an arid grassland. *Ecological Applications* 11: 1301–1310.

Eynard, A., T.E. Schumacher, M.J. Lindstrom, D.D. Malo, and R.A. Kohl. 2006. Effects of aggregate structure and organic C on wettability of Ustolls. *Soil and Tillage Research* 88: 205–216.

FAO (Food and Agriculture Organization of the United Nations). 2005. http://www.un.org/esa/sustdev/natlinfo/indicators/methodology_sheets/land/arable_cropland_area.pdf

———. 2018. http://www.fao.org/docrep/005/Y4171E/Y4171E37.htm

Fenneman, N.M. 1928. Physiographic divisions of the United States. *Annals of the Association of American Geographers* 18: 261–263.

———. 1931. *Physiography of the western United States*. New York: McGraw-Hill Co.

Fidanza, M.A. 2003. Combination treatments for fairy ring prove effective. *Turfgrass Trends in GOLFDOM* 59: 62–64.

Forage and Grazing Terminology Committee. 1992. http://forages.oregonstate.edu/fi/topics/pasturesandgrazing/grazingsystemdesign/grazingterminology

Forbis, T.A., A.L. Provencher, L. Frid, and G. Medlyn. 2006. Great basin land management planning using ecological modeling. *Environmental Management* 38: 62–83.

Franco, C.M.M., P.P. Michelsen, and J.M. Oades. 2000. Amelioration of water repellency: Application of slow release fertilisers to stimulate microbial breakdown of waxes. *Journal of Hydrology* 231–232: 342–351.

Friedel, M.H. 1991. Range condition assessment and the concept of thresholds: A viewpoint. *Journal of Range Management* 44: 422–426.

Garten, C.T., Jr., and S.D. Wullschleger. 2000. Soil carbon dynamics beneath switchgrass as indicated by stable isotope analysis. *Journal of Environmental Quality* 29: 645–653.

Gibson, D.J. 2009. *Grasses and grassland ecology*. Oxford, England: Oxford Univ. Press.

Gibson, C.W.D., and V.K. Brown. 1985. Plant succession theory and applications. *Physical Geography: Earth and Environment* 9: 473–493.

Gifford, G.F. 1970. Some water movement patterns over and through pinyon-juniper litter. *Journal of Range Management* 23: 365–366.

Gleason, H.A. 1917. The structure and development of the plant association. *Bulletin of the Torrey Botanical Club* 44: 463–481.

———. 1926. The individualistic concept of the plant association. *Bulletin of the Torrey Botanical Club* 53: 7–26.

———. 1939. The individualistic concept of the plant association. *American Midland Naturalist* 21: 92–110.

Glenn-Lewin, D.C., and E. van der Maarel. 1992. Patterns and processes of vegetation dynamics. In *Plant succession: Theory and prediction*, ed. D.C. Glenn-Lewin, R.K. Peet, and T.T. Veblen, 11–59. London: Chapman & Hall.

Goodall, D.W. 1974. A new method for the analysis of spatial pattern by random pairing of quadrats. *Vegetatio* 29: 135–146.

Hallett, P.D., and I.M. Young. 1999. Changes to water repellence of soil aggregates caused by substrate-induced microbial activity. *European Journal of Soil Science* 50: 35–40.

Harris, G.A. 1977. Root phenology as a factor of competition among grass seedlings. *Journal of Range Management* 30: 172–177.

Herrick, J.E., V.C. Lessard, K.E. Spaeth, P.L. Shaver, R.S. Dayton, D.A. Pyke, L. Jolley, and J.J. Goebel. 2010. National ecosystem assessments supported by scientific and local knowledge. *Frontiers in Ecology and the Environment* 8: 403–408.

Holechek, J.L., R.D. Pieper, and C.H. Herbel. 2004. *Range management principles and management*. Upper Saddle River: Pearson Prentice Hall.

Hooper, D.U., F.S. Chapin, J.J. Ewel, A. Hector, P. Inchausti, S. Lavorel, and B. Schmid. 2005. Effects of biodiversity on ecosystem functioning: A consensus of current knowledge. *Ecological Monographs* 75: 3–35.

Horman, C.S., and V.J. Anderson. 2003. Understory species response to Utah juniper litter. *Journal of Range Management* 56: 68–71.

Horn, H.S. 1974. The ecology of secondary succession. *Annual review of ecology and systematics* 5: 25–37.

Huston, M., and T. Smith. 1987. Plant succession: Life history and competition. *The American Naturalist* 130: 168–198.

Inouye, R.S., N.J. Huntly, D. Tilman, J.R. Tester, M. Stillwell, and K.C. Zinnel. 1987. Old-field succession on a Minnesota sand plain. *Ecology* 68: 12–26.

ISM. 2018. http://www.museum.state.il.us/muslink/forest/htmls/intro_def.html

Jameson, D.A. 1966. Pinyon-juniper litter reduces growth of blue grama. *Journal of Range Management* 19: 214–217.

Joyce, L.A. 1993. The life cycle of the range condition concept. *Journal of Range Management* 1: 132–138.

Kahmen, S., and P. Poschlod. 2004. Plant functional trait responses to grassland succession over 25 years. *Journal of Vegetation Science* 15: 21–32.

Kaputska, L.A., and F.L. Moleski. 1976. Changes in community structure in Oklahoma old field succession. *Botanical Gazette* 137: 7–10.

Keller, D.R., and F.B. Golley, eds. 2000. *The philosophy of ecology: From science to synthesis*. Athens: University of Georgia Press.

King, E.G., and R.J. Hobbs. 2006. Identifying linkages among conceptual models of ecosystem degradation and restoration: Towards an integrative framework. *Restoration Ecology* 14: 369–378.

Kirchner, T.B. 1977. The effects of resource enrichment on the diversity of plants and arthropods in a shortgrass prairie. *Ecology* 58: 1334–1344.

Kostka, S.J. 2000. Amelioration of water repellency in highly managed soils and the enhancement of turfgrass performance through the systematic application of surfactants. *Journal of Hydrology* 231–232: 359–368.

Kucera, C.L. 1970. Ecological effects of fire on tallgrass prairie. In *Proceedings of a symposium on prairie and prairie restoration*, ed. P. Schramm. Galesburg: Knox College.

Kucera, C.L., and J.H. Ehrenreich. 1962. Some effects on annual burning on central Missouri prairie. *Ecology* 43: 334–336.

Kuchler, A.W. 1964. *Potential natural vegetation of the conterminous United States*. New York: American Geographic Society Special Publication 36.

Lawton, J.H. 1999. Are there general laws in ecology? *Oikos* 84: 177–192.

Laycock, W.A. 1991. Stable states and thresholds of range condition on North American rangelands: A viewpoint. *Journal of Range Management* 44: 427–433.

Levin, M.H. 1966. Early stages of secondary succession on the Coastal Plain, New Jersey. *American Midland Naturalist* 75: 101–131.

Lund, H.G. 2006. *Definitions of agroforestry, forest health, sustainable forest management, urban forests, grassland, pasture, rangeland, cropland, agricultural land, shrubland, and wetlands and related terms*. Gainesville: Forest information services. http://home.comcast.net/~gyde/moredef.htm.

MacArthur, R.H. 1972. *Geographical ecology: Patterns in the distribution of species*. Princeton: Princeton University Press.

Mack, R.N. (1981). Invasion of Bromus tectorum L. into western North America: an ecological chronicle. *Agro-ecosystems* 7: 145–165.

Mack, R.N., and J.N. Thompson. 1982. Evolution in steppe with few large, hooved mammals. *The American Naturalist* 119: 757–773.

Major, J. 1974. Kinds and rates of changes in vegetation and chronofunctions. In *Vegetation dynamics*, 7–18. Dordrecht: Springer.

Mazarak, A.P., and E.C. Conrad. 1959. Rates of water entry in three great soil groups after seven years in grasses and small grains. *Agronomy Journal* 51: 264–267.

McIntosh, R.P. 1981. Succession and ecological theory. In *Forest succession*, 10–23. New York: Springer.

———. 1985. *The background of ecology, concept and theory*. Cambridge, UK: Cambridge University Press.

———. 1995. HA Gleason's 'individualistic concept' and theory of animal communities: A continuing controversy. *Biological Reviews* 70: 317–357.

Meeker, D.O., and D.L. Merkel. 1984. Climax theories and a recommendation for vegetation classification: A viewpoint. *Journal of Range Management* 37: 427–430.

Miles, J. 1987. Vegetation succession: Past and present perceptions. In *Colonization, succession and stability*, ed. A.J. Gray, M.J. Crawley, and P.J. Edwards, 1–29. Oxford, UK: Blackwell Scientific Publications.

Miller, R.F., R.F. Angell, and L.E. Eddleman. 1987. Water use by western juniper. In *Proc. Pinyon-juniper conference. Gen. Tech. Rep. INT-215*, ed. R.L. Everett, 418–422. Ogden: USDA Forest Service Intermountain Res. Sta.

Morin, P.J. 2011. *Community ecology*. 2nd ed. Hoboken: Wiley-Blackwell.

Mueller-Dombois, D., and H. Ellenberg. 1974. *Aims and methods of vegetation ecology*. New York: John Wiley and Sons.

Nelson, C.J., ed. 2012. *Conservation outcomes from pastureland and hayland practices: Assessment, recommendations, and knowledge gaps.*. USDA-NRCS. Lawrence: Allen Press.

Norton, J.B., T.A. Monaco, J.M. Norton, D.A. Johnson, and T.A. Jones. 2004. Soil morphology and organic matter dynamics under cheatgrass and sagebrush-steppe plant communities. *Journal of Arid Environments* 57: 445–466.

Ogle, S.M., W.A. Reiners, and K.G. Gerow. 2003. Impacts of exotic annual brome grasses (*Bromus* spp.) on ecosystem properties of northern mixed grass prairie. *American Midland Naturalist* 149: 46–58.

Ogle, S.M., D. Ojima, and W.A. Reiners. 2004. Modeling the impact of exotic annual brome grasses on soil organic carbon storage in a northern mixed-grass prairie. *Biological Invasions* 6: 365–377.

Omernik, J.M. 1987. Ecoregions of the conterminous United States. *Annals of the Association of American Geographers* 77: 118–125.

———. 1995. Ecoregion: A framework for managing ecosystems. *George Wright Society Forum* 12 (1): 35–50.

Owens, M.K., R.K. Lyons, and C.L. Alejandro. 2006. Rainfall partitioning within semiarid juniper communities: Effects of event size and canopy cover. *Hydrological Processes* 20: 3179–3189.

Paul, E.A., and F.E. Clark. 1996. *Soil microbiology and biochemistry*. second ed. San Diego: Academic Press.

Pellant, M., P. Shaver, D.A. Pyke, and J.E. Herrick. 2000. *Interpreting indicators of rangeland health, version 3. Tech. Ref. 1734-6*. Denver: U.S. Dept. of Interior, Bureau of Land Management, National Science and Technology Center Information and Communications Group.

———. 2005. *Interpreting indicators of rangeland health, version 4. Technical Reference 1734-6*, 122. Denver: U.S. Department of the Interior, Bureau of Land Management, National Science and Technology Center.

Pellant, M., P.L. Shaver, D.A. Pyke, J.E. Herrick, F.E. Busby, G. Riegel, N. Lepak, E. Kachergis, B.A. Newingham, and D. Toledo. 2020. *Interpreting indicators of rangeland health, version 5. Tech Ref 1734-6*. Denver: U.S. Department of the Interior, Bureau of Land Management, National Operations 8 Center.

Penfound, W.T. 1964. The relation of grazing to plant succession in the tall grass prairie. *Journal of Range Management* 17: 256–260.

Penfound, W.T., and E.L. Rice. 1957a. Plant population changes in a native prairie plot plowed annually over a five-year period. *Ecology* 38: 148–150.

———. 1957b. Effects of fencing and plowing on plant succession in a revegetating field. *Journal of Range Management* 10: 21–22.

Petersen, S.L., T.K. Stringham, and B.A. Roundy. 1998. A process-based application of state-and-transition models: A case study of western juniper (*Juniperus occidentalis*) encroachment. *Rangeland Ecology & Management* 62: 186–192.

Pickett, S.T.A., S.L. Collins, and J.J. Armesto. 1987. A hierarchical consideration of causes and mechanisms of succession. *Vegetatio* 69: 109–114.

Pierson, F.B., K.E. Spaeth, M.E. Weltz, and D.H. Carlson. 2002. Hydrologic response of diverse western rangelands. *Journal of Range Management* 55: 58–570.

Powell, J.W. 1895. Physiographic regions of the United States. *National Geographic Society Monographs* 1: 65–100.

Reichhardt, K.L. 1982. Succession of abandoned fields on the shortgass prairie, Northeastern Colorado. *The Southwestern Naturalist* 27: 299–304.

Reinhart, K.O. 2012. The organization of plant communities: Negative plant–soil feedbacks and semiarid grasslands. *Ecology* 93: 2377–2385.

Rillig, M.C. 2004. Arbuscular mycorrhizae and terrestrial ecosystem processes. *Ecology Letters* 7: 740–754.

———. 2005. A connection between fungal hydrophobins and soil water repellency. *Pedobiologia* 49: 395–399.

Rillig, M.C., and D.L. Mummey. 2006. Mycorrhizas and soil structure. *New Phytologist* 171: 41–53.

Risser, P.G. 1988. Diversity in and among grasslands. In *Biodiversity*, ed. E.O. Wilson, 176–180. Washington, DC: National Academy Press.

Ritsema, C.J., L.W. Dekker, J.L. Nieber, and T.S. Steenhuis. 1998. Modeling and field evidence of finger formation and finger recurrence in a water repellent sandy soil. *Water Resources Research* 34: 555–567.

Robocker, W.C., and B.J. Miller. 1955. Effects of clipping, burning and competition on establishment and survival of some native grasses in Wisconsin. *Journal of Range Management* 8: 117–120.

Sampson, A.W. 1917. *Important range plants: Their life history and forage value (No. 545)*. Washington, DC: US Department of Agriculture.

———. 1919. *Plant succession in relation to range management*. USDA Bull. 791. Washington, DC: US Department of Agriculture.

Schlossberg, M.J., A.S. McNitt, and M.A. Fidanza. 2005. Development of water repellency in sand-based root zones. *International Turfgrass Society* 10: 1123–1130.

Schott, M.R., and R.D. Pieper. 1985. Influence of canopy characteristics of one- seed juniper on understory grasses. *Journal of Range Management* 38: 328–331.

Shepherd, T.G., S. Saggar, R.H. Newman, C.W. Ross, and J.L. Dando. 2001. Tillage-induced changes to soil structure and organic carbon fractions in New Zealand soils. *Australian Journal of Soil Research* 39: 465–489.

Shipley, B., and P.A. Keddy. 1987a. The individualistic and community-unit concepts as falsifiable hypotheses. In *Theory and models in vegetation science*, 47–55. Dordrecht: Springer.

———. 1987b. The individualistic and community-unit concepts as falsifiable hypotheses. *Vegetatio* 69: 47–55.

Simberloff, D. 2004. Community ecology: Is it time to move on? (An American Society of Naturalists Presidential Address). *The American Naturalist* 163: 787–799.

Smith, E.F., and C.W. Owensby. 1972. Effects of fire on true prairie grasslands. Proc. Annual Tall Timbers Fire Ecology Conf., No. 12, 9–22.

Society for Range Management. 1998. https://globalrangelands.org/glossary/O?term=

Sommer, M.L., R.L. Barboza, R.A. Botta, E.B. Kleinfelter, M.E. Schauss, and J.R. Thompson. 2007. *Habitat guidelines for mule deer: California Woodland Chaparral Ecoregion*. Sacramento: Mule Deer Working Group, Western Association of Fish and Wildlife Agencies.

Spaccini, R., A. Piccolo, P. Conte, G. Haberhauer, and H.H. Gerzabek. 2002. Increases soil organic carbon sequestration through hydrophobic protection by humic substances. *Soil Biology and Biochemistry* 34: 1839–1851.

Spaeth, K.E. 1990. Hydrologic and ecological assessments on a discrete range site. Ph.D. Dissertation, Lubbock, Texas: Texas Tech Univ.

Spaeth, K.E. 2018. National resource inventory analysis, 2018 report. Ecological site investigations. USDA-NRCS Ft. Worth Texas.

Spaeth, K.E., F.B. Pierson, M.A. Weltz, and J.B. Awang. 1996a. Gradient analysis of infiltration and environmental variables as related to rangeland vegetation. *Transactions of ASAE* 39: 67–77.

Spaeth, K.E., F.B. Pierson, M.A. Weltz, and G. Hendricks, eds. 1996b. *Grazingland hydrology issues: Perspectives for the 21st century.* Denver: Society for Range Management.

Stebbins, G.L. 1982. Major trends of evolution in the Poaceae and their possible significance. In *Grasses and grasslands: Systematics and ecology*, ed. J.R. Estes, R.J. Tyrl, and J.N. Brunken, 3–36. Norman, Oklahoma: University of Oklahoma Press.

Stoddart, L.A., A.D. Smith, and T.W. Box. 1975. *Range management.* 3rd ed. New York: McGraw-Hill Book Company.

Stringham, T.K., W.C. Krueger, and P.L. Shaver. 2003. State and transition modeling: An ecological process approach. *Journal of Range Management* 56: 106–113.

Strömberg, C.A. 2011. Evolution of grasses and grassland ecosystems. *Annual Review of Earth and Planetary Sciences* 39: 517–544.

Tiedemann, A.R. 1987. Combustion losses of sulfur from forest foliage and litter. *Forest Science* 33: 216–223.

Tjørve, E. 2003. Shapes and functions of species–area curves: A review of possible models. *Journal of Biogeography* 30: 827–835.

Trimble, D.E. 1980. The geologic story of the Great Plains (No. 1493). Washington, DC: US Government Printing Office.

Turner, B.L., and W.B. Meyer. 1994. Global land use and land cover change: An overview. In *Changes in land use and land cover: A global perspective*, ed. W.B. Meyer and B.L. Turner, 3–10. Cambridge, England: Cambridge University Press.

USA-FED-EPA. 2006. http://www.epa.gov/urban/glossary.htm

USDA-NRCS. 1995. https://www.nrcs.usda.gov/wps/portal/nrcs/detail/national/technical/nra/dma/?cid=nrcs143_014209

———. 2013. Interagency ecological site handbook for rangelands (USDA 2013). Washington, DC.

———. 2006. *Land resource regions and major land resource areas of the United States, the Caribbean, and the Pacific Basin.* Washington, DC: Natural Resources Conservation Service United States Department of Agriculture Handbook 296.

———. 2003. National range and pasture handbook. Washington, DC.

———. 2014. *National ecological site handbook.* 1st ed. Washington, DC: United States Department of Agriculture.

———. 2018. USDA-NRCS 2018 NRI report. Washington, DC.

———. 2019. National resources inventory handbook of instructions for grazing land on-site data collection. Washington, DC. https://www.nrcs.usda.gov/wps/portal/nrcs/main/national/landuse/crops/

USDA-SCS. 1976. National range and pasture handbook. Washington, DC.

USDA-USFS. 2016. https://www.nrs.fs.fed.us/fia/data-tools/state-reports/glossary/default.asp

Vitousek, P.M., C.M. D'antonio, L.L. Loope, M. Rejmanek, and R. Westbrooks. 1997. Introduced species: A significant component of human-caused global change. *New Zealand Journal of Ecology* 21: 1–16.

Walker, L.R., and F.S. Chapin III. 1986. Physiological controls over seedling growth in primary succession on an Alaskan floodplain. *Ecology* 67: 1508–1523.

Weaver, J.E., and F.E. Clements. 1938. *Plant ecology*, 601. New York: McGraw-Hill Book Co.

Westoby, M., B. Walker, and I. Noy-Meir. 1989. Opportunistic management for rangelands not at equilibrium. *Journal of Range Management* 42: 266–274.

White, N.A., P.D. Hallett, D. Feeney, J.W. Palfreyman, and K. Ritz. 2000. Changes to water repellence of soil caused by the growth of white-rot fungi: Studies using a novel microcosm system. *FEMS Microbiology Letters* 184: 73–77.

Whittaker, R.H. 1951. A criticism of the plant association and climatic climax concepts. *Northwest Science* 25: 17–31.

———. 1953. A consideration of climax theory: The climax as a population and pattern. *Ecological Monographs* 23: 1–78.

———. 1962. Classification of natural communities. *The Botanical Review* 28: 1–239.

———. 1965. Dominance and diversity in land plant communities. *Science* 147: 250–260.

———. 1967. Gradient analysis of vegetation. *Biological Reviews* 42: 207–264.

———. 1975. *Communities and ecosystems.* 2nd ed. New York: McMillan Publ. Co.

Whittaker, R.H., and S. Levin. 1977. The role of mosaic phenomena in natural communities. *Theoretical population biology* 12: 117–139.

Wright, H.A., and A.W. Bailey. 1982. *Fire ecology: United States and southern Canada.* New York: John Wiley & Sons.

Wu, X.B., E.J. Redeker, and T.L. Thurow. 2001. Vegetation and water yield dynamics in an Edwards plateau watershed. *Journal of Range Management* 54: 98–105.

Chapter 3
Hydrology and Erosion Processes

Abstract A common theme throughout this book are the environmental and bio-logical components commonly associated with soil health. The soil environment is not an isolated realm; soils develop with specific parent materials, climate, physiography, hydrologic regimes, soil flora, and a particular succession of plant species. In turn, within a particular plant community, the soil environment and associated hydrologic dynamics influence plant growth and often determine individual plant species habitat and adaptability. As plant communities develop and mature, they then impose significant effects on hydrology—a system where hydrologic dynamics and plants assert causal effects throughout the entire local ecosystem. Growing plants are part of soil development and sustainability where aboveground biomass eventually senesces, becomes litter, decomposes, and replenishes organic matter and nutrients. Belowground roots provide microchannels for water infiltration and also supply nutrients and organic matter via root decomposition and soil microbial interactions. In established plant communities (forests, grasslands), where a relative level of equilibrium exists, biotic integrity, hydrologic function, and soil stability all function in concert and erosion levels become sustainable. Disturbances that affect hydrolgy and erosion processes are continual in cropping systems and can be mini-mized in minimal and no-till systems, In natural plant communities, disturbances are also inhernet to the ecological site (climatic events, grazing, insects, rodents, fire—see chapter 2), especially on more fragile sites (shallow soils, steep terrain). Droughts and fire can create an environment where accelerated runoff and erosion may occur. Fires associated with intense heat can increase soil water repellency (hydrophobicity) along with exposed bare ground and subsequent erosion. Soil health and hydrologic processes are closely tied as water is the most limiting factor in terrestrial environments of which a myriad of biotic subsystems depend. Water, the integral constituent of life, together with wind, plays an important part in weath-ering parent materials and eroding the landscape (natural geologic erosion and/or anthropogenically derived). The objective of this chapter is to provide a review of basic hydrologic principles and processes, as they provide a conceptual framework for introducing the following chapters dealing with more detailed information on vegetative influences on hydrology and erosion in cropland and rangeland settings, which ultimately affect soil health. Vegetative cover and subsequent decomposition are recurring themes related to soil health. On all land uses, vegetation and litter/

© Springer Nature Switzerland AG 2020   85
K. E. Spaeth Jr., *Soil Health on the Farm, Ranch, and in the Garden*,
https://doi.org/10.1007/978-3-030-40398-0_3

residue cover are the primary factors in maintaining sustainable runoff and erosion levels, soil surface temperature regulation, and associated soil water evaporation and transpiration from plant tissues.

Hydrology and Erosion Processes

Understanding the hydrologic cycle and the factors that affect infiltration, runoff (hydrologic assessments), and erosion is important for effective and sustained productivity of cropland, grazinglands, and gardens. Hydrology is the science dealing with the occurrence of water on the earth: its physical and chemical properties, transformation, combinations, and movements especially with the course of water movement from the time of precipitation on land and movement to the sea or atmosphere. Ecohydrology focuses more on the functional interrelationships between hydrology and the biotic components of the ecosystem and is considered a new approach to ecosystem function and achieving sustainable management of water (Zalewski 2000; Nuttle 2002). The objectives of watershed management are the management of land for the optimum production of high-quality water, regulation of water yields, and maximum soil stability along with other goods and services from the land.

The hydrologic cycle (Fig. 3.1) represents the cycle of continuous pathways where the various phases of water (rain, snow, hail, sleet) from the atmosphere are

Hydrologic Budget:

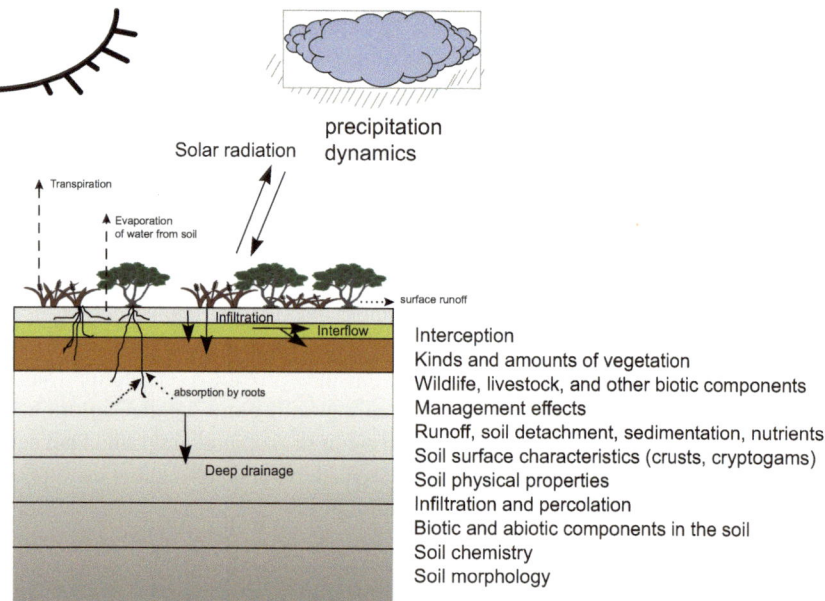

Fig. 3.1 The hydrologic cycle with factors that affect hydrologic processes. (Drawing by author)

transported to land surfaces, then to streams, rivers, ocean, and other storage reservoirs, where evaporation and transpiration by plants return water to the atmosphere. Knowledge of the hydrologic cycle and its components provides the background for understanding the relationships between vegetation-soil-climate-hydrologic interactions. The following chapters deal with specific interactions of plant or crop residues as they are influenced by tillage practices, cover crop influences, and on rangelands the effect of permanent vegetation in the forms of trees, shrubs, grasses, forbs, and biotic crusts on hydrology and erosion.

The basic components of the hydrologic cycle that land users can control to a certain extent are runoff and evaporation of water from the soil. By maintaining and managing vegetative plant cover, whether it is plant residue in crop fields, use of a cover crop, optimum crop stand densities, standing crop biomass, rangeland and pastureland cover, and mulch in the garden, the land user can effect positive actions to reduce soil evaporation, runoff, and subsequent erosion. I mention that runoff and evaporation can be managed to a "certain extent"; we are still at the mercy of nature's cycles and daily climate factors. During extended drought, plant residues may not be replenished at sufficient levels on crop fields and pasture, and over time runoff and erosion risks begin to increase. This creates a tenuous situation as the options to manage runoff and erosion are diminished, thus the need for land users to have contingency management plans for intermittent dry periods and drought. If adequate residue cover is maintained on cropland and vegetative cover on grazinglands, the land user has a temporary buffer for a time to limit soil erosion.

Hydrologic Budget

The hydrologic budget can be expressed as:

$$P - R + Rg - G - E - T - I = S$$

where:

P = Total precipitation
R = Surface runoff
Rg = Groundwater flow that is effluent to a surface stream
G = Groundwater flow
E = Evaporation
T = Transpiration
S = Storage

In Fig. 3.1, the side banner lists many of the interacting factors that influence infiltration and runoff (hydrologic assessments) and subsequently erosivity, both from water and wind. Many of the same factors interact with either protection or advancement of water and wind erosion, e.g., kinds and amounts of vegetation, management effects, soil physical properties, grazing, and biotic components in the

Table 3.1 Common related problems and issues regarding hydrology

Category	Situation
Ecological	Understanding interrelationships: plant/soil complexes, ecology, environmental, and hydrology
	Climatic shifts, vegetation response, and the hydrologic cycle
Management oriented	Tillage effects, cropping rotations, cover crops, application of soil amendments, trampling impacts, and effect of grazing treatments on watersheds
	Conservation practices and their effect on hydrology
	Riparian management and hydrologic implications
Water quantity, quality, and erosion	Enhancement of surface water, groundwater, and aquifer recharge in response to farming practices and vegetation manipulation
	Deficient and unpredictable availability of water supplies
	Flooding
	Polluted surface water, reduced aquatic, fish, and wildlife habitat
	Erosion and sedimentation
	Sludge and animal waste applications
Economic	Economics of conservation, soil health protocols, and watershed restoration

soil. Common problems and issues regarding watersheds can be categorized as ecological, management-oriented, water quality and quantity, erosion, and economic.
Table 3.1 summarizes the most common problems, issues, and factors that affect
watersheds. Table 3.2 shows all the interacting factors that influence the hydrologic cycle.

Precipitation

Precipitation (rain, snow, hail, sleet) is the primary input to the atmosphere-land
interface, which drives the hydrologic cycle. The three major categories of precipitation are convective, orographic, and cyclonic:

- Convective precipitation occurs in the form of light showers and heavy cloud-
  bursts or thunderstorms of extremely high intensity. Often, precipitation intensity varies throughout the storm. Most convective storms are random and last less
  than 1 hour and usually contribute little to overall moisture storage in the soil.
- Orographic precipitation results when moist air is lifted over mountains or other
  natural barriers. Important factors in the orographic process include elevation,
  slope, aspect or orientation of slope, and distance from the moisture source.
- Cyclonic precipitation may be classified as frontal and non-frontal and is related
  to the movement of air masses from high pressure to low pressure regions.

Water originating from other sources such as shallow groundwater or baseflow
reserves may also be utilized by deep-rooted legumes such as alfalfa, shrubs, and
trees on rangelands and phreatophytes (riparian vegetation) along rivers and lakes.

Table 3.2 Interacting factors that affect the hydrologic cycle in watersheds

Soils	Plants	Environmental	Management
• Soil morphology • Soil texture (particle size) • Bulk density • Compaction • Organic matter • Aggregate stability • Nutrient levels • Soil structure • Infiltration rates • Percolation rates • Saturated hydraulic conductivity • Runoff characteristics • Rills and gullies • Porosity • Erosion dynamics • Salinity • Alkalinity • Biotic components • Parent material • Pedogenic processes • Soil chemistry • Soil health attributes	• Types of plants • Rooting morphology • Plant growth form (bunch, sod) • Plant life form (grass, shrub, forb, tree) • Plant biomass, cover, density • Biological soil crust (mosses, lichens, algae) • Plant canopy layers • Plant architecture • Successional dynamics (state and transition model dynamics) • Native vs. introduced plants • Plant competition • Physiological characteristics of plant species • Physiological response to grazing • Biodiversity • Phenological stages	• Climate • Types of storms • Precipitation type • Duration of storm • Intensity of storm • Topography • Geology • Aspect • Slope • Microtopography (soil roughness) • Aquifer usage and recharge dynamics	• Cropping systems • Tillage practices • Cover crops and green manuring • Fertilizer and other soil amendments • Grazing intensity • Timing of grazing • Continuous vs. rotational systems • Chiseling • Herbicides • Seeding • Brush management • Wildfire • Prescribed burning • Past management history • Fencing • Hoof impact • Class of livestock • Disturbance • Stock water location water harvesting • Past disturbance from Farm implements • Recreation • Kinds and types of wildlife

Precipitation in its various forms is measured in mm, cm, and inches over specific periods of time. Land elevation and geography are the principal drivers that affect quantity, timing, and type of precipitation. There are many sources that provide information on local precipitation and temperature regimes.

Raindrop Dynamics and Plant Interception

Vegetation, rocks, or plant litter intercepts raindrops and reduces the kinetic energy of raindrop and droplet impact on the soil surface. Infiltration and, subsequently, runoff are important processes in the hydrologic cycle and are regulated by the kind and amount of vegetation, edaphic, climatic, and topographic influences (Wood and Blackburn 1981). Live plant cover and/or litter are essential to controlling water erosion on all land types. In natural plant communities, the hydrologic condition of a site is the result of complex interactions of soil and vegetation factors. The kind and amount of vegetation and soil characteristics influence many hydrologic processes including interception, infiltration, evaporation, transpiration, percolation, surface runoff, soil water storage, soil erosion, and deposition of sediment. Every plant-soil complex exhibits a characteristic infiltration pattern and hydrologic regime (Gifford 1985). The term plant cover is commonly used in agricultural discussions on cropland, range, and pasture, and horticultural settings and provides many benefits in protection and interception of raindrops on the soil surface to reduce runoff and erosion.

Raindrop sizes vary with storm intensity. This affects soil surface stability and infiltrability. Average drop sizes for various storm intensities are 1.25 mm diameter at 1.27 mm hr^{-1} (0.05 in hr^{-1}), 1.80 mm diameter at 12.7 mm hr^{-1} (0.5 in hr^{-1}), and 2.80 mm diameter at 101.6 mm hr^{-1} (4.0 in hr^{-1}). Generally, an average falling raindrop attains a terminal shape of a hemisphere or oblate shape. On the average, the velocity of raindrops (2.5 mm diameter) is about 2 m sec^{-1}, but terminal velocities vary with storm intensity and raindrop size. At 5 mm, raindrops begin to break up due to the force of falling through the atmosphere. Regarding raindrop velocities, recent studies are indicating that raindrops can often fall at speeds faster than terminal velocity (Montero-Martinez et al. 2009; Thurai et al. 2013; Larsen et al. 2014). Hypotheses related to "superterminal drops" moving 30% faster than normal drop velocity suggest that larger faster-moving raindrops breakup and generate fragments from the "parent raindrops." Raindrops this size will disrupt the soil surface on impact, whereas drops smaller than 1 mm in diameter are less disruptive.

Interception of precipitation by plant canopies and ground cover limits the amount of water that can infiltrate into to soil surface. Raindrop interception is variable and is affected by plant height, leaf area, plant canopy cover, plant architecture, rainfall frequency, rainfall duration, amount of precipitation, type of precipitation, and time of precipitation. Rainfall interception by plant canopies is inversely proportional to intensity of rainfall. Intercepted precipitation can follow two pathways; one is evaporation back to the atmosphere (interception loss); the second is throughfall, where water passes through the plant canopy and/or reaches the ground by stemflow (Branson et al. 1981; Owens et al. 2006; Dunkerley 2008). In prairie vegetation, the ratio of foliage area is 3 to 20 times greater than the soil surface (Clark 1940). During small storms, water intercepted and evaporated without reaching the soil surface may be substantial, especially in shrub, tall grass, mixed grass, and bunchgrass communities. Rainfall interception loss during heavy storms is often a small proportion of the storm's total volume (<10%) (Corbett and Crouse 1968). Raindrops and snow can (1) be intercepted by living vegetation (trees, shrubs, forbs and/or grasses) and stored in the canopy until it evaporates; (2) be intercepted by soil surface plant residues, mulch, or litter; (3) slowly drip off vegetation, droplets, intercepted, and later falling from the canopy of shrubs and trees can form an erosive drip line under the plant; (4) land directly on bare soil; and (5) collect in the plant canopy and run down plant stems. This water is redistributed in a concentrated way and can either infiltrate depending on the volume of water and soil surface conditions or can run off. In a watershed, interception must be considered as a factor in the total water balance and budget (loss of water). Vegetative matter, living or dead, protects mineral soil from sheet and splash erosion, and a lack thereof can lead to rill erosion and in the extreme, ephemeral gully, and gully erosion.

In rangelands, individual shrubs and conifer trees on the average can intercept 50–60% of rainfall from a low-intensity storm, while 5–35% can be intercepted during higher intensity storms (West and Gifford 1976; Branson et al. 1981; Owens et al. 2006; Taucer et al. 2008). For multistorm and annual time scales, gross rainfall interception for individual shrubs is about 5–46% (Hull 1972; Thurow et al. 1987; Domingo et al. 1998; Serrato and Diaz 1998). Annual rainfall interception in shrub and woodland stands averages between 5 and 25% (Thurow et al. 1987; Dunkerley and Booth 1999; Carlyle-Moses 2004).

On an annual basis, tree interception is greater than grass interception; however, at maximum growth, some grasses have as much leaf area per unit area as some trees (Clark 1940) (Fig. 3.2). During the growing season, alfalfa can intercept as much rainfall as a forest (Table 3.3). Grasses in the tallgrass prairie can intercept a substantial amount of rainfall—29–80% during low-intensity storms (Clark 1940). On an annual basis, gross annual rainfall interception by grasslands ranges from 13 to 56%. Water storage by grasses, shrubs, and trees is proportional to average heights and ground cover. Table 3.4 is a summary for average annual interception rates on cropland, forest, woodland, and grassland vegetation types.

Fig. 3.2 Interception rates for various forms of plant cover. POPR, *Poa pratensis* (Kentucky blue-grass); PASP, *Pseudoroegneria spicata* (bluebunch wheatgrass); KOCR, *Koeleria cristata* (prairie June grass); ARTR, *Artemisia tridentata* (big sagebrush); QUVI, *Quercus virginiana* (live oak). (Drawing by author)

Table 3.3 Raindrop size and terminal velocity

Storm type	Raindrop size (mm)	Raindrop terminal velocity (m sec⁻¹)	Raindrop terminal velocity (ft. sec⁻¹)
Light storm intensity			
Small droplets	0.5	2.1	6.9
Large droplets	2.5	6.5	21.3
Moderate storm intensity			
Small droplets	1.0	4.0	13.1
Large droplets	2.6	7.6	24.9
Heavy storm intensity			
Small droplets	1.2	4.6	15.1
Large droplets	4.0	8.8	28.9
Maximum droplet	5.0	9.1	29.9
Hailstone	10.0	10.0	32.8

Data from Gunn and Kinzer (1949), Byers (1959), Rogers (1979), Pruppacher (1981)

Table 3.4 Average annual interception rates for crop, forest, and rangeland vegetation types

Vegetation	Storm amount @ 2.54–5 cm hr⁻¹ (% interception, plant ht. cm)	Annual rainfall interception (%)	Annual rainfall interception by litter layer after passes through canopy (%)
Crop plants			
Wheat field (Leuning et al. 1994)		33	
Wheat (*Triticum aestivum*)	51–60 (16 cm)		
Wheat straw		19	
Oats (*Avena sativa*)	45–72 (12 cm)		
Alfalfa field (*Medicago sativa*) (Clark 1940)		7–36	
Trees and Shrubs			
Brazilian tropical rain forest (Lloyd et al. 1988; Dykes 1997)		8.9–18	
Aspen forest (*Populus tremuloides*) (Dunford and Niederhof 1944)		16	
Ponderosa pine (*Pinus ponderosa*) (Connaughton 1936)		32	
Lodgepole pine (*Pinus contorta*) (Miner and Trappe 1957)		24	

(continued)

Table 3.4 (continued)

Vegetation	Storm amount @ 2.54–5 cm hr^{-1} (% interception, plant ht. cm)	Annual rainfall interception (%)	Annual rainfall interception by litter layer after passes through canopy (%)
Hardwood forest (Beall 1934)		21	
Conifer forest (Tate 1995)		30	
Conifer litter (Tate 1995)		5	
Big sagebrush (*Artemisia tridentata*) (West and Gifford 1976; Hull and Klomp 1974)		4–31	
Redberry juniper (*Juniperus coahuilensis*) (Thurow and Hester 1997)		25.9	40.1
Ashe juniper (*Juniperus ashei*) (Thurow and Hester 1997)		36.7	43
Utah juniper (*Juniperus osteosperma*) (Skau 1964)		17	
Western juniper (*Juniperus occidentalis*) (Young et al. 1984)		2–27	
Shad scale saltbush (*Atriplex confertifolia*) (West and Gifford 1976)		4	
Saltbush (*Atriplex argentea*)			
Chaparral (Rowe and Colman 1951; Tate 1995)		8–11	
Live oak (*Quercus*) (Texas) (Thurow et al. 1987)		46.1	20.7
Rangeland grasses			
Australian grasslands (Dunkerley and Booth 1999)		32	
Prairie grasses, Nebraska (Clark 1940)		29–80	
Big bluestem (*Andropogon gerardii*) (Clark 1940)	47–66 (14 cm)		
Sideoats grama (*Bouteloua curtipendula*) (bunch-type midgrass) (Thurow et al. 1987)		18.1	
Buffalograss (Clark 1940) (*Bouteloua dactyloides*)	17–26 (7 cm)		

(continued)

Table 3.4 (continued)

Vegetation	Storm amount @ 2.54–5 cm hr^{-1} (% interception, plant ht. cm)	Annual rainfall interception (%)	Annual rainfall interception by litter layer after passes through canopy (%)
Curley mesquite (*Hilaria belangeri*) (sod-type shortgrass) (Thurow et al. 1987)		10.8	
Wheatgrass (*Agropyron*) and June grass (*Koeleria*) prairie (Couturier and Ripley 1973)		21–32	
Mixed prairie (Couturier and Ripley 1973)		14–24	
Bluestem prairie (*Andropogon gerardii*) (Clark 1940)		57–84	
Tallgrass prairie (Clark 1940)	43 (9 cm)		
California annual grassland (Kittredge 1948; Tate 1995)		26	
Kentucky bluegrass (*Poa pratensis*) (Haynes 1940)		56	
Western wheatgrass (*Pascopyrum smithii*) (Clark 1940)	46 (12 cm)		
Vegetable crops			
Brussels sprouts (*Brassica oleracea*) (Noble and Morgan 1983)	10–25		

Snow is also intercepted by vegetation canopies, especially trees and shrubs. Snow-retaining capacity is affected by vegetation canopy architecture, density, height, and snowpack conditions (Pomeroy and Gray 1995; Sturm et al. 2001). Accumulation of windblown snow is proportional to the maximum height of the plant canopy, and excess snow deposition is subject to wind and transported off-site or lost to wind-driven sublimation (Sturm et al. 2001). Snow accumulation is greater between tree and shrubs, and snow water equivalent can be greater than 80% (Breshears et al. 1997). A study of four conifer species showed approximately 60% canopy interception of snowfall (Storck et al. 2002).

In the northern and central Great Plains, snow accumulation is a valuable water resource for all land types. In the northern Great Plains, Evans (1975) estimated 24 million acre-feet of water from snowfall. Overwintering crop stubbles can be effective in capturing and holding snow on-site, and snowmelt is an efficient source of water—evaporation is usually low during snowmelt. On undisturbed wheat fields, water storage from total snowfall precipitation can be as high as 53%, and snowmelt water contributes about 45% to fallow-wheat rotations in northeastern Colorado (Greb 1975). On croplands where snow accumulation is substantial, contour strip cropping is effective in rolling topography, and strips of standing stubble and tall-grass barriers act as small windbreaks to capture snow in moisture limited farming areas.

Design Storm Frequencies

Hydrologists, engineers, and climatologists categorize design storm events at specific intervals for estimating runoff volumes and duration at 2, 5, 10, 25, 50, 75, and 100 years. For example, a 25-year design storm event is the intensity and amount that would occur at least once in 25 years or four times in a hundred years. In reality, several storms of equal intensity may occur during the 25-year period or during a single year. The National Oceanic and Atmospheric Administration (NOAA) publishes information on precipitation and storm frequency data (NOAA Atlas 14 Point Precipitation Frequency Estimates). NOAA data tables show average recurrence of storm (1–1000 years) and storm duration (5 minutes to 60 days intervals) (example in Table 3.5).

As an example, in Oklahoma City, Oklahoma, a 2-year design storm could generate 22.3 mm (0.88 in) at 15 minutes, 32.7 mm (1.29 in) at 30 min, and 94.5 mm (3.72 in) at 24 hours. The upper and lower bound probability of rainfall for 2 years and 24 hours is 76.2–117 mm (3–4.62 in) (Table 3.4). For example, an engineer designing a dam and spillway with specifications to safely discharge water from a 50-yr 24-hr storm would have to consider runoff volume for the drainage area based on an average precipitation estimate of 198 mm [7.88 in (6.03–10.4 in)] (Table 3.4). Structural conservation practices on crop fields such as terraces, diversions, and waterways also consider precipitation and runoff based on design storm data. To avoid a disastrous runoff and erosion event, farmers and ranchers should consider the risk of high-intensity storm events and the consequence of heavy rainfall in their management plans. Is residue cover and permanent plant cover adequate to help protect the soil surface from raindrop impact, runoff, and subsequent water erosion?

Infiltration Dynamics

Infiltration of water into the soil is one of the most primary functions in the hydrologic cycle. Infiltration is the process by which water enters the soil surface and is affected by the combined forces of capillarity and gravity (Hillel 1982; Brooks et al. 1991). There are many influencing factors that affect infiltration and water movement through the soil (Fig. 3.3). Soil physical and chemical properties, vegetation characteristics, and soil fauna and flora all interact and affect the process of infiltration (Fig. 3.3). Under dry soil conditions, a higher initial infiltration rate is caused

Table 3.5 National Oceanic and Atmospheric Administration Atlas 14 precipitation frequency estimates for Oklahoma City, Oklahoma

PDS-based precipitation frequency estimates with 90% confidence intervals (in inches)[a]

Duration	Average recurrence interval (years)									
	1	2	5	10	25	50	100	200	500	1000
5-min	0.424 (0.328–0.547)	0.495 (0.383–0.638)	0.614 (0.473–0.794)	0.715 (0.548–0.929)	0.860 (0.640–1.15)	0.975 (0.709–1.32)	1.09 (0.770–1.51)	1.22 (0.823–1.72)	1.38 (0.903–2.00)	1.51 (0.962–2.22)
10-min	0.621 (0.481–0.800)	0.724 (0.560–0.935)	0.899 (0.693–1.16)	1.05 (0.803–1.36)	1.26 (0.937–1.69)	1.43 (1.04–1.93)	1.60 (1.13–2.21)	1.78 (1.21–2.52)	2.02 (1.32–2.93)	2.21 (1.41–3.25)
15-min	0.757 (0.586–0.976)	0.883 (0.683–1.14)	1.10 (0.845–1.42)	1.28 (0.979–1.66)	1.54 (1.14–2.06)	1.74 (1.27–2.36)	1.95 (1.38–2.70)	2.17 (1.47–3.07)	2.47 (1.61–3.58)	2.70 (1.72–3.96)
30-min	1.11 (0.856–1.43)	1.29 (1.00–1.67)	1.61 (1.24–2.08)	1.88 (1.44–2.44)	2.27 (1.69–3.03)	2.57 (1.87–3.48)	2.88 (2.03–3.99)	3.21 (2.17–4.54)	3.65 (2.38–5.29)	4.00 (2.55–5.86)
60-min	1.46 (1.13–1.88)	1.71 (1.32–2.20)	2.14 (1.65–2.76)	2.51 (1.92–3.26)	3.04 (2.27–4.09)	3.48 (2.53–4.72)	3.92 (2.77–5.44)	4.40 (2.98–6.23)	5.05 (3.30–7.32)	5.56(3.54–8.15)
2-hr	1.81 (1.41–2.31)	2.12 (1.66–2.72)	2.66 (2.07–3.41)	3.14 (2.43–4.04)	3.82 (2.88–5.10)	4.38 (3.22–5.91)	4.96 (3.54–6.83)	5.58 (3.82–7.87)	6.44 (4.25–9.29)	7.11 (4.57–10.4)
3-hr	2.02 (1.59–2.57)	2.37 (1.86–3.02)	2.98 (2.33–3.80)	3.52 (2.74–4.51)	4.31 (3.27–5.74)	4.97 (3.68–6.68)	5.66 (4.06–7.77)	6.40 (4.41–8.99)	7.43 (4.94–10.7)	8.26 (5.34–12.0)
6-hr	2.40 (1.91–3.03)	2.80 (2.22–3.54)	3.51 (2.77–4.44)	4.16 (3.26–5.28)	5.14 (3.95–6.82)	5.96 (4.47–7.98)	6.84 (4.96–9.35)	7.80 (5.44–10.9)	9.15 (6.15–13.1)	10.2 (6.69–14.8)
12-hr	2.82 (2.25–3.52)	3.25 (2.60–4.07)	4.05 (3.22–5.07)	4.79 (3.79–6.03)	5.93 (4.61–7.83)	6.91 (5.24–9.20)	7.98 (5.85–10.8)	9.14 (6.45–12.7)	10.8 (7.35–15.4)	12.2 (8.04–17.5)
24-hr	3.25 (2.62–4.02)	3.72 (3.00–4.61)	4.61 (3.70–5.73)	5.45 (4.35–6.80)	6.75 (5.31–8.86)	7.88 (6.03–10.4)	9.11 (6.75–12.3)	10.5 (7.46–14.5)	12.4 (8.54–17.6)	14.0 (9.35–20.0)
2-day	3.70 (3.01–4.54)	4.24 (3.45–5.21)	5.25 (4.25–6.47)	6.20 (5.00–7.66)	7.66 (6.08–9.96)	8.93 (6.89–11.7)	10.3 (7.70–13.8)	11.8 (8.50–16.2)	14.0 (9.70–19.7)	15.8 (10.6–22.3)
3-day	4.01 (3.28–4.90)	4.60 (3.76–5.63)	5.69 (4.63–6.97)	6.69 (5.42–8.23)	8.24 (6.56–10.6)	9.56 (7.42–12.5)	11.0 (8.26–14.6)	12.6 (9.08–17.1)	14.8 (10.3–20.7)	16.6 (11.3–23.4)
4-day	4.28 (3.52–5.21)	4.91 (4.03–5.98)	6.05 (4.95–7.39)	7.10 (5.78–8.70)	8.70 (6.94–11.2)	10.0 (7.82–13.0)	11.5 (8.68–15.3)	13.1 (9.51–17.8)	15.4 (10.8–21.4)	17.2 (11.7–24.2)

7-day	4.97 (4.11–6.00)	5.70 (4.71–6.89)	6.97 (5.74–8.44)	8.10 (6.64–9.86)	9.79 (7.84–12.4)	11.2 (8.75–14.3)	12.7 (9.60–16.6)	14.2 (10.4–19.2)	16.5 (11.6–22.8)	18.2 (12.5–25.5)
10-day	5.58 (4.64–6.71)	6.37 (5.29–7.67)	7.73 (6.40–9.32)	8.91 (7.35–10.8)	10.6 (8.55–13.4)	12.1 (9.47–15.3)	13.5 (10.3–17.6)	15.1 (11.1–20.2)	17.2 (12.2–23.7)	18.9 (13.1–26.4)
20-day	7.32 (6.14–8.72)	8.24 (6.91–9.82)	9.77 (8.17–11.7)	11.1 (9.21–13.3)	12.9 (10.4–16.0)	14.3 (11.3–18.0)	15.8 (12.1–20.3)	17.3 (12.8–22.9)	19.3 (13.8–26.3)	20.9 (14.6–28.9)
30-day	8.72 (7.36–10.3)	9.77 (8.24–11.6)	11.5 (9.66–13.7)	12.9 (10.8–15.4)	14.9 (12.1–18.3)	16.4 (13.1–20.5)	17.9 (13.8–22.9)	19.5 (14.5–25.6)	21.5 (15.4–29.1)	23.1 (16.2–31.8)
45-day	10.4 (8.86–12.3)	11.7 (9.93–13.8)	13.8 (11.6–16.3)	15.4 (13.0–18.3)	17.7 (14.4–21.6)	19.4 (15.5–24.0)	21.1 (16.3–26.8)	22.7 (17.0–29.7)	24.9 (18.0–33.5)	26.5 (18.7–36.4)
60-day	11.9 (10.1–13.9)	13.4 (11.4–15.7)	15.7 (13.4–18.5)	17.7 (14.9–20.9)	20.3 (16.6–24.6)	22.2 (17.8–27.4)	24.1 (18.7–30.4)	25.9 (19.4–33.7)	28.3 (20.4–37.8)	30.0 (21.2–41.0)

Station name, Oklahoma City N DSPL; site ID, 34-6664; latitude, 35.5333°; longitude, −97.4667°; elevation, 32.6 m (1070 ft.). (Courtesy of NOAA National Weather Service https://hdsc.nws.noaa.gov/hdsc/pfds/pfds_map_cont.html)

[a]Precipitation frequency (PF) estimates in this table are based on frequency analysis of partial duration series (PDS). Numbers in parenthesis are PF estimates at lower and upper bounds of the 90% confidence interval. The probability that precipitation frequency estimates (for a given duration and average recurrence interval) will be greater than the upper bound (or less than the lower bound) is 5%. Estimates at upper bounds are not checked against probable maximum precipitation (PMP) estimates and may be higher than currently valid PMP values

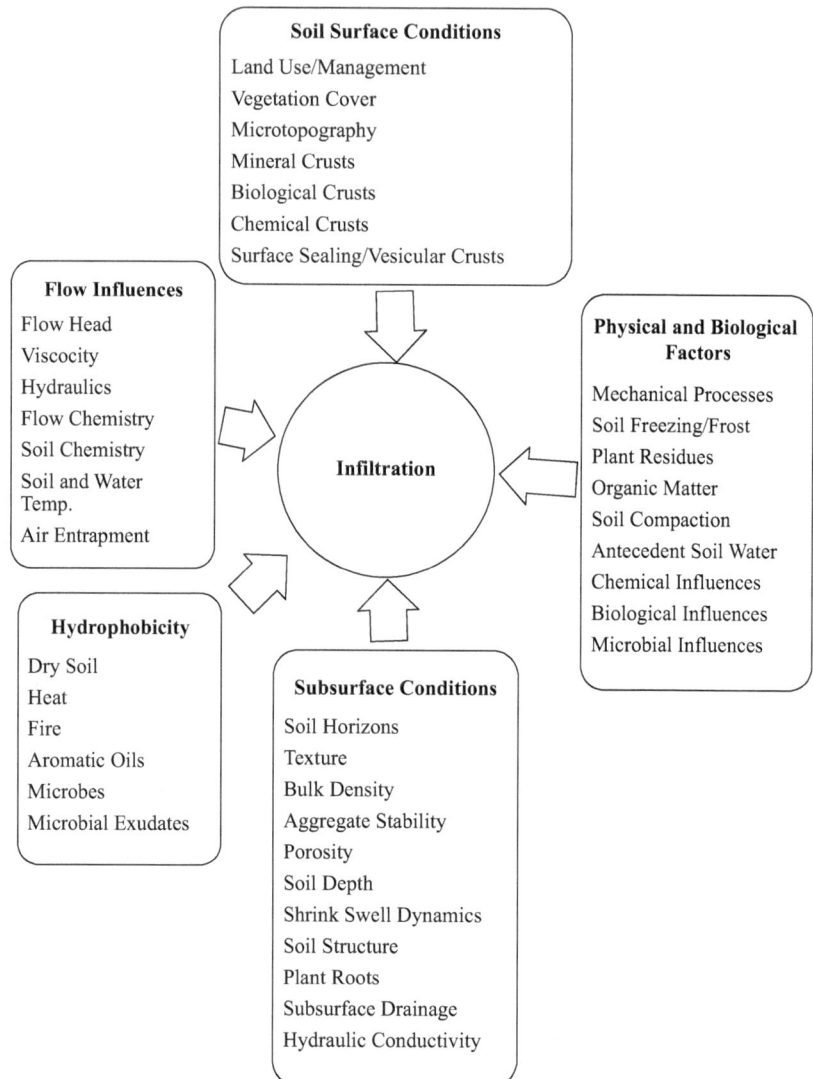

Fig. 3.3 Interrelated factors associated with infiltration. Main broad factors include soil surface conditions, subsurface conditions, flow influences, physical and biological factors, and hydrophobicity. (Adapted from Ward and Elliot 1995)

by the physical attraction of soil particles to water—the matric potential gradient or matric suction gradient that decreases over time until a relatively constant rate is achieved (a curvilinear relationship). As a precipitation event progresses and the soil surface becomes saturated, infiltration rate stabilizes according to a combination of soil and vegetation characteristics—steady-state infiltration. Decreased infiltration over time can be caused by one or more of the following: gradual decreases in the matric suction gradient, deterioration of soil structure, the breakdown of soil aggregate stability, consequential partial sealing of the profile by detachment and migration of pore-blocking particles, and a restricting layer in the soil profile. Organic matter and compounds produced by soil microbia and other soil fauna (earthworms) are associated with the formation, maintenance, and stability of soil aggregates, which is an important diagnostic feature in determining soil health trends on croplands and rangelands. On wet soils, infiltration is associated with saturated hydraulic conductivity, which is highly correlated with soil texture. As the soil profile becomes saturated with water, additional water inputs are then dependent on the physics of downward transmittance or percolation of water.

Text Box 3.1 Physical, Chemical, and Biological Crusts Affect Infiltration

Physical and chemical crusts can be problematic in cropland soils and rangelands where these crusts alter ecosystem functions related to soil health. These crusts can form on the soil surface by raindrop impact on bare soil, evaporative processes forming chemical crusts, and compaction from various sources such as grazing and other related vehicular traffic. Vesicular crusts are common where aggregate stability breaks down and gas bubbles form a crusty layer 1–10 mm thick. As raindrops impact bare ground, whatever remnant of aggregate stability remains is further dispersed to the inherent soil particles. The smaller soil particles wash into and block any existing soil pores and form a crust as the soil dries. As drying occurs, soil surface tension contracts the soil particles into a dense layer. This clogging action can reduce pore continuity and infiltration capacity by 90% (Belnap et al. 2001)—the outcome results in low saturated hydrologic conductivity and increased runoff and soil erosion. As this process continues, poor soil structure, reduced pore space, consistently less and unstable soil aggregates, decreases in organic matter, and less microbial function become more pronounced. In contrast to physical soil crusts reducing hydrologic function, these crusts may have a positive influence on reducing wind erosion—at least temporarily. The presence and perpetuation of soil physical crusts create an environment that is associated with low plant reproductive capacity, seed germination is negligible, and on rangelands, this condition is highly conducive to the beginning of desertification.

Physical crusts can commonly form on silt, clay, and loam soils, and in some soils, salts either in the soil or irrigation water can promote clay dispersion. As aggregates are dispersed, clay particles permeate soil pores, and when soil begins to dry out, the soil surface forms a thin cemented layer, often with cracks, which inhibits water infiltration exacerbating runoff and potential erosion (Fig. 3.4). Soils with low organic matter are also more prone to developing physical crusts and soil sealing on tilled cropland. The formation of physical soil crusts is common after a rain event on freshly tilled and planted

(continued)

Fig. 3.4 Physical soil crusts on cropland have reduced infiltration capacity and porosity, have higher density, and are an indication of high evaporation rates in the presence of sunlight. (Courtesy of USDA-NRCS, Soil Health Division)

fields and often results in poor seedling emergence and the need to reseed. In the case of soil crust formation in newly seeded and/or plant emergence, light chiseling or the use of a rotary hoe is a remedial measure. Soil physical crusts can be minimized by less intense tillage practices and maintaining plant residues. Increasing organic matter and incorporating certain soil amendments, where applicable, can help alleviate clay dispersion and formation of soil crusts. The tendency of some soils such as weathered acid soils, low salinity, and high-sodium soils in semiarid areas can be improved by gypsum (calcium sulfate).

Biological Soil Crusts

Biological crusts (also referred to as microphytic, mirobiotic, crytogamic, and cryptobiotic) where prevalent, are comprised of various microorganism complexes (green algae, cyanobacteria, lichens, mosses, microfungi, and bacteria). Belnap et al. (2001) provide an extensive treatise on biological crusts. The hyphae and filaments of cyanobacteria and microfungi permeate several millimeters of the surface soil, and their exudates bind soil particles together, forming a crust that stabilizes and protects soil surfaces from wind and water erosion. Biological crusts inhabit soil surfaces and are especially obvious in more arid environments [Arches National Park near Moab, Utah is an excellent example (Fig. 3.5)], although they exist in most rangeland environments: for example, the Great Basin, Colorado Plateau, Sonoran Desert, Columbia Basin, and in grassland environments. Microbiotic crusts can also be found throughout the world in varied environments such as deserts,

Fig. 3.5 Biological soil crusts comprised of algae, cyanobacteria, and nonvascular plants (mosses, lichens) have evolved in many rangeland ecosystems. Canyonlands and Arches National Parks are hosts to good examples of biological soil crusts that help stabilize soil interspaces around vegetation patches. (Courtesy of USGS Canyonlands Research Station)

semideserts, woodlands, shrub steppes, evergreen shrub and forests, and alpine and tundra ecosystems and may comprise up to 70% of living ground cover in some rangeland plant communities.

The ecological roles of biological soil crusts can be unique in different rangeland ecosystems and include carbon and nitrogen fixation (Skujins 1984), modification of water balance, hydrology, and erosion of surface soil (Loope and Gifford 1972; Blackburn 1975; Danin 1978; Brotherson and Rushforth 1982; Williams et al. 1995; Eldridge and Kinnell 1997); regulation of albedo (solar energy reflected from the soil surface), effects on vascular plant germination and establishment (Harper and Clair 1985; Belnap 1994; Zaady et al. 1997); and provide refugia for soil invertebrates involved with decomposition and mineralization and nutrient transfer to surrounding plants (Whitford 1996). Elevation, ecological gradients, soils and topography, disturbance factors, timing and amount of precipitation, endemic plants and growth habits, plant community structure, and microhabitats all influence the ecologic aspects of biological crusts. Different species, successional traits, and species recovery mechanisms from disturbance exist among different rangeland plant community types, thus making biological crusts unique. For example, in Wyoming big sagebrush communities, various lichen species can be used as indicators of late succession. Belnap et al. (2001) explain that "some lichens are only present in late-successional communities because they grow upon other lichens or mosses . . .

a)   b)

c)   d)

Tortula ruralis

Fig. 3.6 (**a**) Low sagebrush (*Artemisia arbuscular*) community at Reynolds Creek, Idaho, showing shrub coppice and interspace zones. This low sagebrush site occurs on very shallow lithic soils.
Native grasses, forbs, and some biocrusts inhabit the interspace area. (**b**) Close-up of low sagebrush plant community where soil in interspace areas has eroded by overland water flow and wind
erosion. Note the plant pedestals (elevated plants) in the interspace areas—this is an indicator of
erosion on this site. (**c**) Sagebrush coppice with moss (*Tortula ruralis*) in non-hydrated state. (**d**)
Close-up of *Tortula ruralis*. (Photos by author)

Acarospora schleicheri grows upon *Diploschistes muscorum*, which in turn parasit-
izes the lichen genus *Cladonia* . . . *Massalongia carnosa* primarily grows on mosses
and is not present until mosses become well-established within a site. *Texosporium
sancti-jacobi* is restricted to old-growth sagebrush communities and occurs only on
decaying organic matter."

In order to create the setting where biotic crusts proliferate, envision rangeland
environments in semiarid and arid regions as heterogenous in appearance with
patches of vegetation such as sagebrush, juniper, or other shrub clusters with open
interspaces (Fig. 3.6a, b). These open interspace areas exhibit higher runoff, sediments, and nutrients which are transported during more intense storm events to
other vegetation patches or sink areas downslope or off-site completely (Blackburn
1975; Schlesinger et al. 1990; Pierson et al. 2001; Ludwig et al. 2005; Cánton et al.
2011). Biological crusts are an integral component of the interspace areas between

shrubs and are sources of runoff, sediment, and nutrients that are transported to depositional areas or sinks (zero slope transitions and other shrub coppice zones). Biotic crust components also inhabit coppice zones under shrubs: twisted moss (*Tortula ruralis*) is a common moss found in the understory of sagebrush coppice dunes. For example, in sagebrush plant communities, if the native grass component and biotic crusts are vacant in the interspace areas, hydrologic function from a rangeland health perspective has been significantly decreased—at least a moderate departure from expected conditions based on the preponderance of evidence evaluation (see chapter? on rangeland health).

Biotic crusts are delicate, and when disturbed or destroyed by trampling, the hydrologic function of the site is significantly impacted where water and nutrient fluxes change on site and downslope (Barger et al. 2006). Biological crusts are often sacrificed in grazed areas; however, there are management strategies that can be used to minimize damage. Several recommendations can be used for minimizing grazing impacts on biological crusts: grazing when moist or wet on sandy soils and grazing during dry conditions on clayey soils (Marble and Harper 1989; Memmott et al. 1998); light to moderate stocking in early- to mid-wet season; at low-to-mid elevation sites, winter grazing is less destructive (Miller et al. 1994; Burkhardt 1996); remove livestock in advance before drought to allow regrowth of crustal organisms; avoid very wet and muddy soil conditions as crusts can be buried by hoof impact (Kaltenecker et al. 1999); use of rest-rotation grazing systems can minimize the frequency of hoof action disturbance; and use of animal distribution tools on areas with no or low crust development with salt placement and strategic livestock water developments.

What is the effect of biological soil crusts on infiltration? Various studies indicate no effect, increases, or decreases in infiltration capacity. Biological crusts affect hydrologic and erosion dynamics because they inhabit and modify the soil surface (Fletcher and Martin 1948; Kleiner and Harper 1972, 1977; Warren 2001). Many studies have documented biological crusts with increased infiltration and decrease runoff (Loope and Gifford 1972; Brotherson and Rushforth 1982; Harper and Clair 1985; Greene and Tongway 1989; Belnap and Gardner 1993; Eldridge 1993; Chamizo et al. 2012); whereas, some studies find no significant effect (Eldridge et al. 1997; Williams et al. 1995, 1999). In contrast, other studies suggest that both physical and biological crusts decrease infiltration and increase runoff (Solé-Benet et al. 1997; Eldridge et al. 2000; Cánton et al. 2011; Chamizo et al. 2012). There are several hypotheses that attempt an explanation for the variance of biological crust effects on infiltration. Eldridge and Rosentreter (1999) point out that biological crust morphology of many different component organisms has a major impact on hydrologic function. Belnap (2006) discusses possible reasons for conflicting results:

1. Lack of detailed analysis of crust characteristics that have varying effects on surface and subsurface soil properties.
2. Some studies use different procedures for disturbing crusts—this has an effect on the structure of the original surface and subsoil soil structure.

3. The use of different water or rainfall application methods—infiltrometers vs. rainfall simulators—makes it difficult to compare studies.
4. Lack of attention in defining crust successional gradients and identification of associated species.

In addition, varying chemical exudates (polymeric polysaccharides) associated with biological crust species create stable soil aggregates, and these chemical dynamics are not well understood as they relate to water repellency and hydrophobicity on dry and wet soils. In summary, biological crust species characteristics (early and late-successional species) can have a significant effect on infiltration rates. Chamizo et al. (2012) found that infiltration does not predictably increase linearly with crust development: infiltration increased with cyanobacterial biomass increases and was greatest with late-successional moss-dominated biological soil crusts. Late-successional crustose and squamulose lichen dominated soil crusts were associated with lower infiltration rates (Table 3.6).

Table 3.6 Summary of biotic crusts on infiltration and runoff

Study	Location	Crust type	Results
Eldridge et al. (2000)	Loess hillslopes, Israel, Central-Western Negev	Cyanobacterial dominant crusts	Removal resulted in three–fivefold increase in sorptivity and steady-state infiltration
Brotherson and Rushforth (1982)	Colorado Plateau, pinyon-juniper	Lichen and algal	Infiltration rate significantly faster with biological crusts
Loope and Gifford (1972)	Colorado Plateau, pinyon-juniper	Lichen	Infiltration rate from rainfall simulations was significantly faster with biological crusts
Chamizo et al. (2012)	Mediterranean semidesert, Tabernas desert	Cyanobacterial and moss	Infiltration increases as cyanobacterial biomass increases and was highest in moss crusts. Late-successional crustose and squamulose lichen crusts showed very low infiltration rates
Williams et al. (1999)	Colorado Plateau, South Central Utah	Cyanobacteria dominated	Cyanobacteria-dominated microbiotic crusts, living or dead, did not significantly affect saturated hydraulic conductivity on sandy loam soils in South Central Utah
Williams et al. (1995)	Colorado Plateau, South Central Utah	Cyanobacterium (*Microcoleus vaginatus*)	Microphytic crusts did not significantly influence the infiltration capacity. Interrill erosion was greatest in the chemically killed treatment and lowest in the control treatment

There is variability in effects, which are dependent on soil texture, species comprising the crust, wet-and-dry conditions, successional status of crusts and surrounding vegetation, and chemical exudates associated with different species

Infiltration Capacity

When rainfall rates exceed infiltration capacity of the soil, surface runoff and/or ponding on the soil surface occurs. The infiltration capacity of the soil is dependent on factors such as soil texture, porosity of the soil, soil structure, soil surface conditions, the nature of the soil colloids, organic matter content, soil depth or the presence of impervious layers, the presence of macropores, antecedent soil water content, soil frost, and temperature of the soil.

Infiltration Rate

Infiltration rate is the volume flux of water moving into the soil profile per unit area of surface area.

Infiltration Curve

Infiltration rate curves graphically show infiltration rate over time (Fig. 3.7a). Initially, dry soils infiltrate water more readily than moist soil, and for wet or moist soil, suction gradients are small at the onset of a rainfall event and diminish as time progresses. Figure 3.7b shows a relative comparison of infiltration curves for sand, loam, and clay.

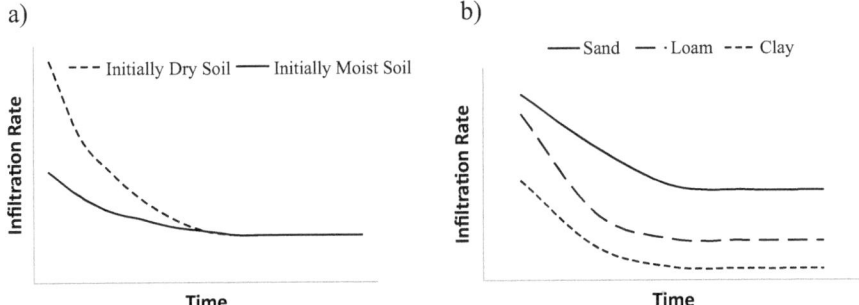

Fig. 3.7 Infiltration as a function of time for an initially dry (**a**) and wet soil. (**b**) Respective infiltration curves for sand, loam, and clay

Infiltrability

Infiltrability denotes the infiltration flux resulting when water, at atmospheric pressure, is freely available at the soil surface (Hillel 1982). Soil infiltrability depends upon the initial wetness, suction, texture, structure, soil layering, its uniformity, aggregate stability, and bulk density. Infiltrability may be high in clay soils as wetting or precipitation begins due to macropores and cracks in the soil surface; however, as these cracks swell, infiltrability decreases. On clayey soils, infiltrability gradually decreases over time because as clay particles expand, air pockets become entrapped, and the bulk compression of soil air is prevented from escaping as it is displaced by water.

Hydraulic Conductivity

Hydraulic conductivity is the ratio of the volume of water passing through a cross-sectional unit area per unit time (flux) to the hydraulic gradient (the driving force acting on the liquid). Hydraulic saturated conductivity is often symbolized as Ks and differs between unsaturated and saturated soil conditions. In saturated soil, there is a positive pressure potential; however, in unsaturated soil, there is a subatmospheric pressure, or suction, which is analogous to a negative pressure potential. The higher the saturated hydraulic conductivity of the soil, the higher its infiltrability. Percentages of sand, silt, and clay are highly correlated with hydraulic conductivity along with bulk density and organic matter (Wösten and Van Genuchten 1988) (Fig. 3.8, Table 3.7).

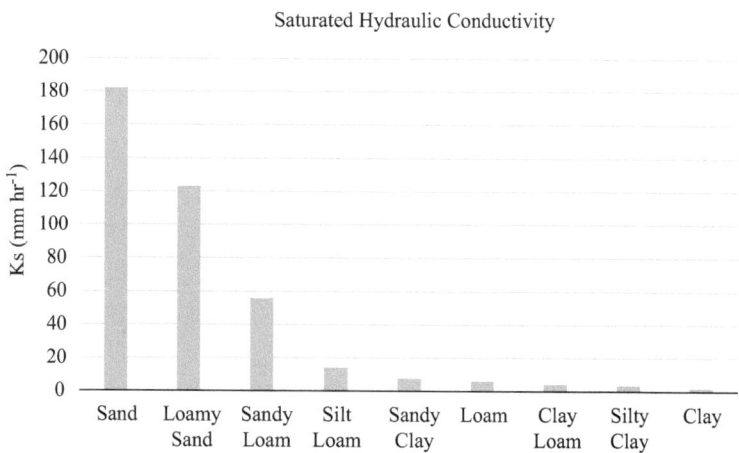

Fig. 3.8 Saturated hydraulic conductivity comparisons with soil texture. (Spaeth data from USDA-NRCS data)

Table 3.7 Approximate relationships between soil texture, water storage, and water intake rates under irrigation conditions (Spaeth USDA-NRCS data from many sources)

Texture	Water mm @ 0.3 m (in ft. of soil)	Max. rate of water intake in hr^{-1} (mm hr^{-1}) (bare soil conditions)
Sand	12.7–17.8 (0.5–0.7)	0.75 (19)
Fine sand	17.8–22.9 (0.7–0.9)	0.60 (15.2)
Loamy sand	17.8–27.9 (0.7–1.1)	0.50 (12.7)
Loamy fine sand	20.3–30.5 (0.8–1.2)	0.45 (11.4)
Sandy loam	20.3–35.6 (0.8–1.4)	0.40 (10.2)
Loam	25.4–45.7 (1.0–1.8)	0.35 (8.9)
Silt loam	30.5–45.7 (1.2–1.8)	0.30 (7.6)
Clay loam	33.0–53.3 (1.3–2.1)	0.25 (6.4)
Silty clay	35.6–63.5 (1.4–2.5)	0.20 (5.1)
Clay	35.6–63.5 (1.4–2.5)	0.15 (3.8)

Percolation

Infiltration is only as rapid as the rate at which water moves through the soil macropores and flows downward by the effect of gravity. Downward movement of water through the soil profile is percolation, and downward movement of soil water is deep drainage. The amount of water lost to deep drainage depends upon soil infiltrability, evapotranspirational demands, substrate and geological conditions, and rooting dynamics of plants.

Soil Water Characteristics

Soil water loss can occur by surface runoff, evaporation, transpiration, leaching, ratios of air and water in soil pores, soil temperature dynamics, and ultimately the biotic component of soils. Soil water dynamics affect most of the physical and chemical characteristics of soil, which are tied to soil health principles. Total soil water potential (Ψ_t) is affected by various forces: gravitational (Ψ_g), matric (Ψ_m), hydrostatic (Ψ_h), and osmotic (Ψ_o)—all having singular potentials.

$$\Psi_t = \Psi_g + \Psi_m + \Psi_o + \Psi_h + \ldots \left(\text{other possible potentials}\right)$$

The physics of soil water potentials and dynamics are quite complex but are ultimately related to how water moves in the soil—between wet and dry soils. Regardless of the complexity of soil water dynamics, there is one important point to remember about the behavior of water in soils: water moves from areas with high water potential to areas with lower water potential. In farming and ranching enterprises, soil water content and storage capacity are of primary importance to growing a crop or forage plants. There is a curvilinear relationship between soil water

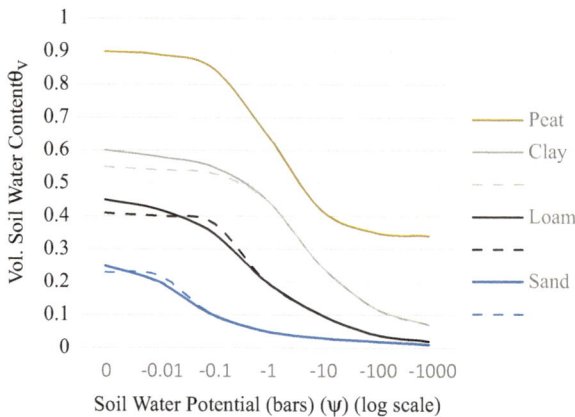

Fig. 3.9 Soil water potential curves for peat and three mineral soil textures. Curves represent saturated soil condition and progressive drying; dashed lines are effects of soil compaction or poor soil aggregate stability for soil textures, respectively. Units for volume soil water content = m^3 H_2O/m^3 soil. (Data from Rawls et al. 1982, 2004; Schwärzel et al. 2002; Weil and Brady 2017). The volume of soil water content is affected by pore sizes and decreases with soil water potential. Clays sustain more water at given potentials than loam and sand. Clay content in the soil determines the amount of soil micropores, as soil water potential increases, water is held more tightly in the micropores—about half of the soil water can be held so strongly that plants cannot utilize it

potential (Ψ_t) and moisture content in the soil (θ) (Fig. 3.9). Several soil physical properties influence soil water content: soil texture, soil structure, soil aggregate stability, and bulk density as imposed by compaction from tillage implements or grazing animals.

During infiltration, a moisture profile is produced and is comprised of the saturation and transition zone, transmission zone, wetting zone, and wetting front (Fig. 3.10). The saturation and transition zones are fully saturated; the transmission zone is the ever-lengthening unsaturated zone of uniform water content. The wetting zone is the area where the transmission zone joins the wetting front. The wetting front is the line of delineation where the soil changes from wet to dry. Depth of the wetting front is an important factor for sustained plant growth. Plants with laterally extending fibrous roots as well as a deep tap roots can utilize precipitation from low precipitation events as well as subsurface water.

Plant Available Water

Wilting point, field capacity, and plant available water are important concepts for agricultural producers and vary significantly with soil texture (Fig. 3.11). These concepts are especially important in determining irrigation schedules for crops. Plant growth and yield are reduced at 40–60% of the plant available water (PAW) content (Fig. 3.12) (Ward and Elliot 1995).

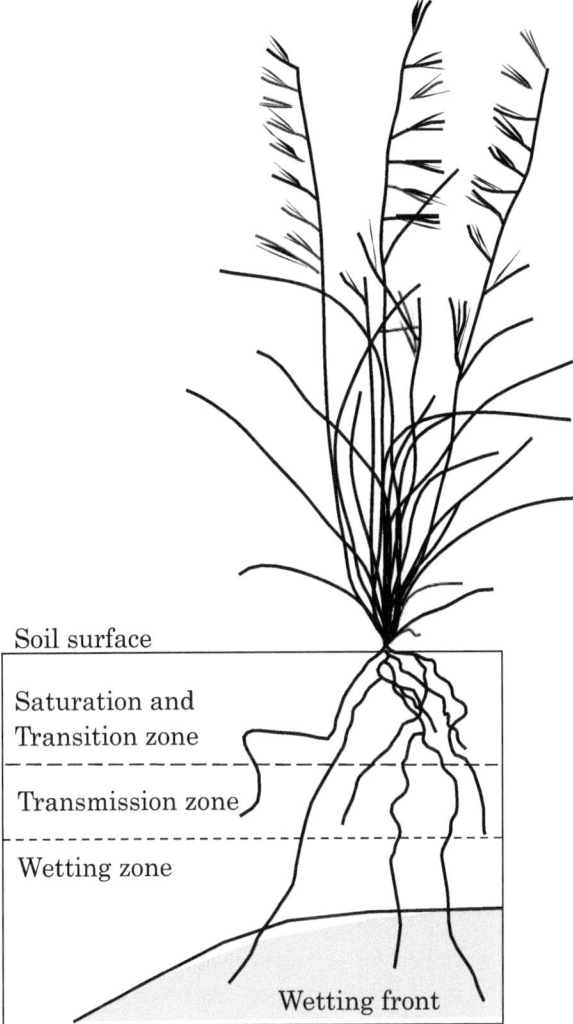

Fig. 3.10 Moisture profile during infiltration

For example, to determine the water needed to increase a loam soil from a critical water deficit (~50%) to field capacity, obtain upper and lower bounds for loam (English units, wilting point = 1.1 in; field capacity = 3.2 in).θ paw = (3.2 in – 1.1 in) = 2.1

The crop root zone is 2 ft. and PAW = (2.1) (2) = 4.2 in (106.6 mm).

If the critical deficit is 50% of PAW, then (0.5) (4.2) = 2.1 in of water is needed to raise the soil water content from the critical value to field capacity.

Wilting point is the moisture content of the soil (oven dry basis) where plants wilt and fail to recover a turgid state in a dark environment. Wilting point is typically at 10–15 bars.

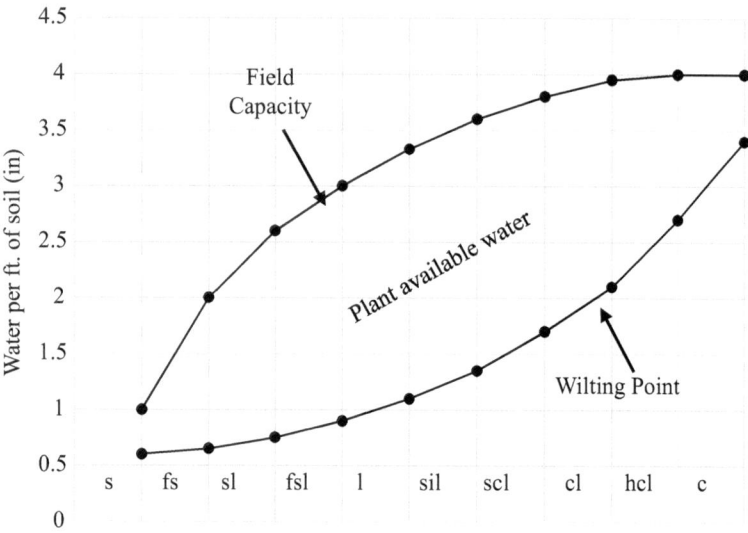

Fig. 3.11 Plant available water capacities of different soil textural groups

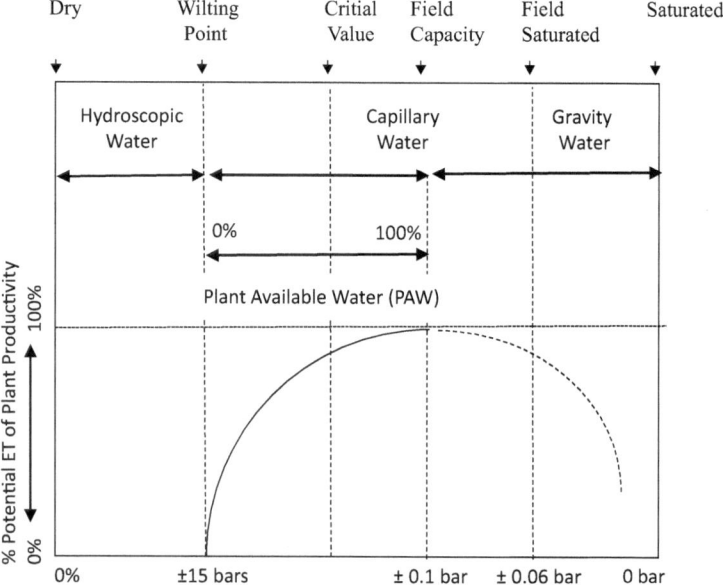

Fig. 3.12 Expected plant yield response with soil water content at 0 to 15 bars (wilting point). The critical value (40–60%) where plant production falls off due to waning available plant water. (After Ward and Elliot 1995)

Field capacity is the percentage of water remaining in the soil after 2–3 days after saturation and drainage has stopped.

In situations where significant erosion has occurred and topsoil with inherent levels of organic matter lost, soil moisture holding capacity is compromised. Soil water retention is a function of soil physical properties and organic matter (Rawls et al. 1991; Wösten et al. 2001); from a soil health perspective, building up organic matter is the only practical alternative to enhancing soil water retention. The effects of organic matter on soil water retention cannot be overemphasized. Rawls et al. (2003) reviews the literature on the effects of organic matter on soil water retention, some research in using organic matter in regression models is contradictory, but there is agreement that organic matter "is an important factor when water contents at field capacity and wilting point are measured directly" (Rawls et al. 2003).

Runoff

As precipitation rate exceeds infiltration rate, soil depressions fill (surface storage) and ponding occurs, then runoff is initiated. Runoff is the antithesis of infiltration; in rainfall simulation experiments, infiltration is determined by subtracting runoff from rainfall rate. As water begins to move downslope, this process is called surface runoff or overland flow. In healthy stable range and forestland, water flow channels follow natural microtopography of the landscape and are usually stabilized by vegetation. Often, rills and water flow patterns are difficult to distinguish. Rills usually follow a linear pattern, are more numerous, are deeper than wide, and appear to be newly formed compared to water flow channels. When rangelands begin to degrade, water flow patterns will change as evidenced by the number, width, length, and connectivity of the channels (Pellant et al. 2020). As hydrologic function and soil surface stability change due to a variety of disturbances, soil deposition may also be evident. In determining rangeland health hydrologic function and soil surface stability, water flow paths, rills, and gullies are evaluated on-site (see chapter? on assessing rangeland health.

There are several mechanisms related to surface runoff generation:

1. Hortonian overland flow (or infiltration excess flow)—when the input of water on the land surface exceeds infiltration capacity. Hortonian flow is common during intense rainfall events.
2. Saturation overland flow—when ponded or saturated soil surfaces and profiles shed water; precipitation falling directly into stream channels.
3. Groundwater reappearing on the surface as springs and returns to stream channels (Horton 1933; Dingman 2002).

Subsurface flow water percolates down into the soil; it may be intercepted by a rock layer or other restricted layer in the soil (hardpan, *fragipan*, *claypan*, plow pan, traffic pan, or root pan). Water then may move laterally (interflow) along the layer and eventually discharge to a water body (creeks, streams, lakes). When overland flows begin, sheetflow (interrill), rill erosion, ephemeral gullies, and extreme event gullies may form. As water enters stream channels, it becomes streamflow. Baseflow is that portion of precipitation which percolates into the soil profile and is released slowly and sustains streamflow between periods of rainfall and snowmelt. Baseflow does not respond quickly to rainfall.

The rate and distribution of runoff from a watershed are determined by a combination of physiographic, land use, and climatic factors. Runoff is closely linked to nutrient cycling, erosion, and contaminant transport. Runoff can be a sensitive indicator of ecosystem change. Factors influencing runoff include:

- Precipitation form (rain, snow, hail)
- Type of precipitation (convective, orographic, cyclonic)
- Seasonal distribution of precipitation
- Intensity, duration, and distribution of precipitation
- Plant community types and the characteristics of vegetative cover
- Kind of vegetation as well as quantity of vegetation
- Plant cover and biomass
- Watershed topography and geology
- Physical and chemical soil characteristics
- Evapotranspiration
- Antecedent soil moisture
- Degree of compaction from land use practices

High-intensity convective thunderstorms are typically associated with runoff, as rainfall intensity and amount are greater than infiltration capacity. These types of storms, especially high return period storms >5, 10. 25, 50, 100 frequencies, can cause rills, gullies, and irreparable soil loss, especially when low cover and production and soil compaction are present.

Watershed Hydrograph

Various environmental processes (climate-soils-plants management) determine how excess water becomes streamflow. Hydrograph analysis is the most widely used method of analyzing the dynamics of surface runoff in subwatersheds and entire watersheds. A hydrograph is a continuous graph showing the properties of streamflow (discharge)—the rise and decline of the rainfall event, streamflow, runoff, subsurface flow, and baseflow over time (Fig. 3.13).

Baseflow

Baseflow is that portion of precipitation which percolates into the soil profile and is released slowly and sustains streamflow between periods of rainfall and snowmelt. Baseflow does not respond immediately with rainfall.

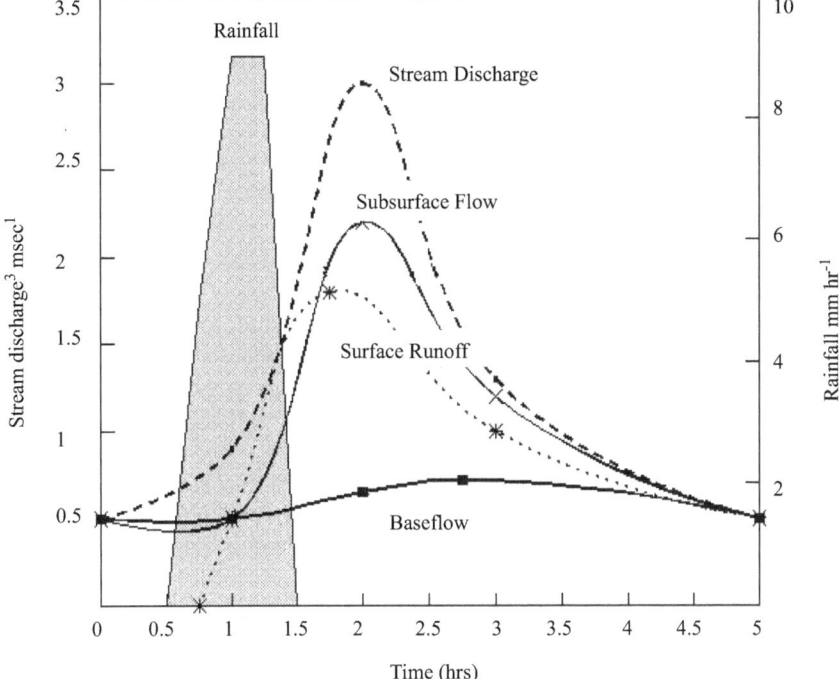

Fig. 3.13 Example hydrograph of a watershed showing the relationship of water flow pathways

Subsurface Flow

Subsurface flow is infiltrated water that is impeded by a restrictive layer in the soil (e.g., hardpan, caliche layer, bedrock). Subsurface water is diverted laterally and flows through the soil until it arrives at a stream channel where it is considered part of the storm hydrograph.

Evaporation, Transpiration, and Evapotranspiration

Evaporation

Evaporation, transpiration, and the additive effects of both are evapotranspiration (ET), which accounts for the greatest loss of water from the environment. In evaluating water balance, runoff and deep percolation of water also are processes associated with water losses but to a much lesser degree. Evaporation is the process of liquid water changing to water vapor as a result of lower atmospheric vapor pressure of the water surface. Evaporation is an integral feature of the hydrologic cycle and one variable that can be partially managed in terrestrial environments to minimize evaporative water loss. This can be accomplished by managing vegetation and plant residue cover. The principle driver of evaporation is the sun (solar energy). On crop fields, plant residues insulate the soil surface against direct solar radiation, thus buffering evaporative water loss. On rangeland, providing adequate cover and maintaining forage plant height can minimize soil water evaporation losses. As soil surface temperatures rise, so does the vapor pressure gradient. Soil surface evaporation is a factor that can be managed by prescribed grazing and maintaining minimum plant heights. Minimum grazing height by species is not only important for insuring plant recovery and continued photosynthesis, but adequate plant cover and height also help buffer soil surface temperature and help regulate soil water evaporation.

There are three main components of evaporation: atmospheric pressure (percent humidity), heat, and movement of air. In order for water to evaporate, water molecules must be located near the surface interface between air and water and must have adequate movement and sufficient kinetic energy to exceed intermolecular forces. Soil texture is highly correlated with evaporation rates; water evaporates more readily in lighter textured soils and more slowly in clay soils (Fig. 3.14). However, soils with higher clay content are more prone to developing physical crusts, especially in newly tilled situations after a rainstorm where subsequent sunshine rapidly evaporates surface moisture. As a point of reference, soil water evaporation and ET were measured on corn stover, wheat stubble, and bare ground at the Kansas State University Research-Extension Center near Garden City, Kansas (silt loam, 1.5% SOM). The 3-year average of evaporation and ET was as follows: bare ground 1.6 mm day^{-1}, 5.2 mm day^{-1}; corn stover 0.82 mm day^{-1}, 6.7 mm day^{-1}; and wheat stubble 0.9 mm day^{-1}, 5.5 mm day^{-1} (Klocke et al. 2009).

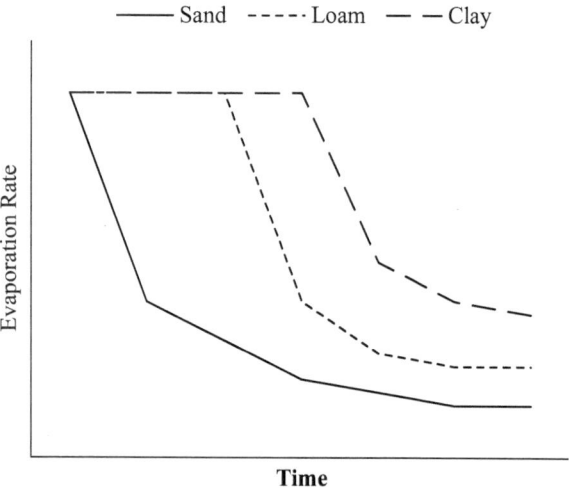

Fig. 3.14 Rates of evaporation over time in sand, loam, and clay soils. (Drawn by author)

Transpiration

Transpiration is a biological process of water vapor exchange to the atmosphere. Transpiration rate is dependent upon the water vapor pressure gradient between plant intercellular spaces and the atmosphere. As water vapor fills plant intercellular spaces, it then diffuses out to the atmosphere through stomata, lenticels (horizontal slit like areas in the bark of a woody stems or roots), or other plant openings that may be present. Any part of the plant anatomy can transpire water; however, increasing temperature, bright sunlight, low humidity, high air pressure, and wind are associated with increased transpiration rates. Leaf size is an indication of transpiration potential as large leaves transpire more water compared to smaller leaves. Transpiration increases about 20–30% for every 18° F rise in temperature. More than 90% of the soil water from plant uptake is transpired and is lost to the atmosphere. Water, the basic constituent of metabolic processes in the plant, is used in photosynthesis and synthesis of hormones, chlorophyll, and other plant pigments. The exchange of gases in photosynthesis requires moist cell surfaces. When the balance of water absorption by roots falls below transpiration rates, leaves may wilt, and stomata close. Stomata remain closed at night when plants are restoring water balance. Unlike soil water evaporation, transpiration is a factor beyond the control of land management.

Evapotranspiration

Evapotranspiration (ET) includes evaporation from soil and water surfaces and transpiration from plants. Evapotranspiration affects water yield and largely determines what proportion of precipitation input to a watershed becomes streamflow. Evaporation of soil water from cropland and rangelands is loss that could contribute to crop or forage production. Vegetation cover, by shading and reduction of wind velocity can reduce soil evaporation rates. The greater the vegetation cover, the greater the interception and transpiration loss which usually offsets the benefits of reduced evaporation. Evapotranspiration is a critical component in ecohydrologic processes and influences local climate and weather (Hill et al. 2008; Betts et al. 1999). Environmental and biological factors such as precipitation and soil moisture affect the upper limits of ET, while net radiation, wind speed, and relative humidity affect daily variations. As a point of reference, peak daily ET in a corn/sunflower crop was 1.5–4.5 mm day^{-1} (Huizhi and Jianwu 2012). In rangeland plant communities, ET in Pinyon-Juniper communities is 63-97% of annual precipitation (AP); ET in honey mesquite, Texas is 95% of AP; ET is 80-83% of AP in California chaparral; and 1.78 mm day in Wyoming big sagebrush (30.5 cm, 12 in. AP) (Branson et al. 1981).

Erosion

Erosion is a natural process; approximatively 80% of the world's land surface is vulnerable to geologic erosion (Thurow 1991). An extreme example: in the United States, the Badlands National Park, geologic erosion has been active for the last 500,000 years (Stetler et al. 2011) and current average soil loss in the park is estimated at ~2.5 cm yr^{-1} (Stoffer 2003), approximately equivalent to 325 Mg ha-1 yr^{-1} (144.9 t ac^{-1} yr^{-1}). Water and wind erosion, fluvial activity, climatic cycles associated with weathering (freeze, thaw, drying, wetting), stream migration, and mass wasting consisting of soil creep, slumps, and slides are the forces behind the development of landscapes (valleys, stream channels, and canyons). In Chap. 2, state and transition diagrams are discussed and can document hydrologic and erosion progression (Williams et al. 2016). Soil loss associated with a state change is usually irreversible. For example, Bakker et al. (2004) evaluated several studies and found that crop productivity decreases 4.3–10.9% per 10 cm of soil loss. Soil erodibility is multivariate in nature, and many factors are involved such as intensity, duration, and amount of precipitation; topography—land slope and shape (linear, convex, concave); types and amounts of vegetation; organic matter content and associated aggregate stability; soil particle size (texture); and nature of parent material (Butzer 1974).

In many natural terrestrial plant communities, a relative balance exists between weathering and erosion:

$$E = W + S$$

where E = erosion by runoff and mass movement, W = rate of weathering and soil formation, and S = soil wash and other incoming upslope colluvium (Bunting 1965). The soil building process continues where $E < W + S$. Where ecological balance is disrupted, $E > W + S$. Erosion continues until a new equilibrium is reached or erosion continues to bare rock.

Human intervention, including grazing and cultivation, has been associated with accelerated erosion since civilizations began. The story of soil erosion is repetitive since the dawn of cultivation and agriculture. Throughout history, civilizations have suffered the consequences of soil erosion from wind and water, resulting in hunger, starvation, impoverishment, and sometimes migration, devastation, or destruction of civilizations. In Chap. 1, I discuss the history of agriculture and how erosion affected whole civilizations to individual pioneers breaking the prairie sod in America. In Chap. 6, the effect of erosion on organic matter is discussed. Ignorance plays a big part in the story of agriculture and erosion; however, it is interesting to note that various observant and inspired individuals throughout history recognized the degrading patterns of erosion saw solutions to the problem: Plato 437–327 BC, Xenophon 434–355 BC, Pliny 23–79 AD, Varro 116 BC–27 BC, Jared Eliot 1685–1763, Samuel Deane (1733–1814), and Hugh Hammond Bennet (1881–1960). One of the most alarming episodes in America's history is the Dust Bowl, where

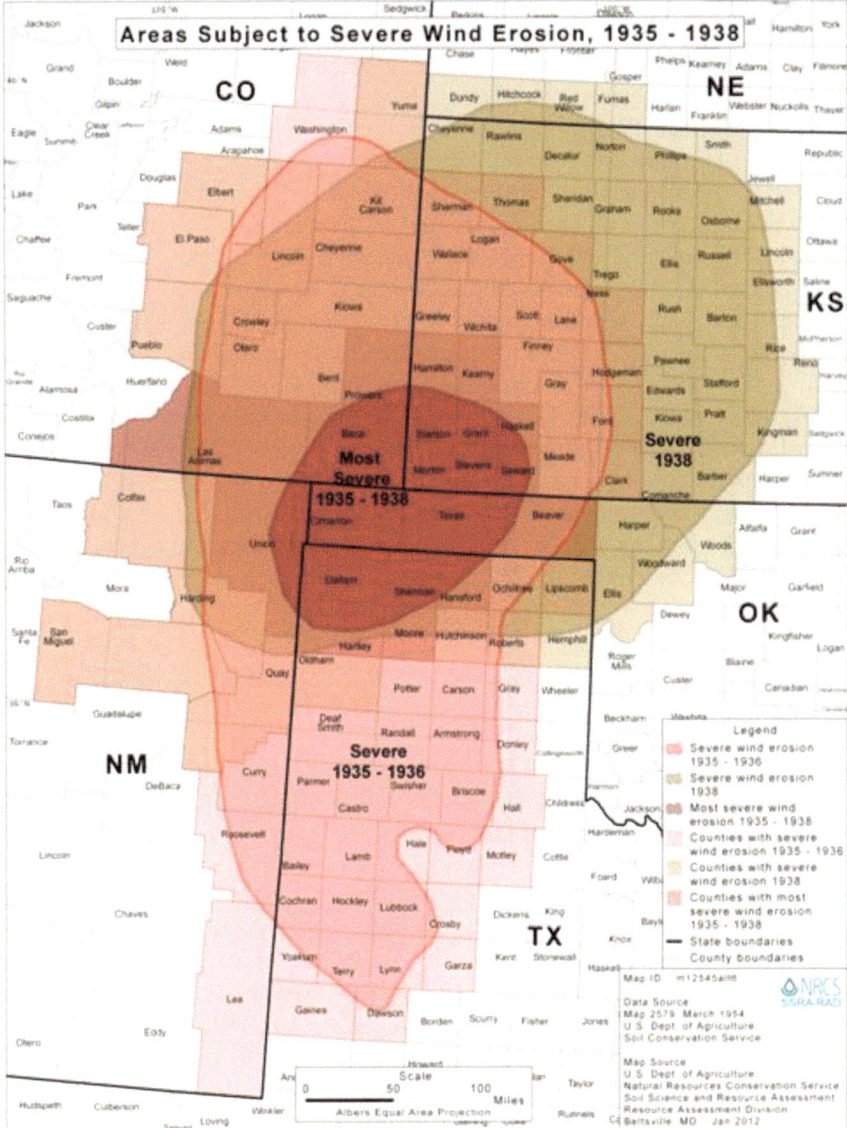

Fig. 3.15 Primary region of the Great Plains and the effects of the Dust Bowl (Boundaries from Lavin et al. (2011). Courtesy of USDA-NRCS (2012))

extensive areas of the North American Great Plains experienced severe drought that precipitated large dust storms and soil erosion (Fig. 3.15). In Chap. 1, some of the farming practices used at that time are discussed. The almost decade-long drought instigated and exacerbated the erosion process. Most likely, most emergency actions such as cover crops would have been short-lived or difficult to establish as the drought continued. The practice and consequence of plowing the soil and exposing open fallow fields during the winter set the stage for disaster.[1] There were many reasons for the lack of immediate action, the main factor being economic: farmers lived from day-to-day and year-to-year, poverty was common, and life was tenuous. In addition, the impacts of the Great Depression greatly affected farm commodity prices and farm incomes, and unemployment was rampant. Legislators were also slow to react for a number of reasons: lack of knowledge, political inaction, and the more pressing problems of the Great Depression. Hugh Hammond Bennet's story and the creation of the Soil Conservation Service is an interesting tale (see Chap. 1). Plainly, the Dust Bowl was a combination of an environmental and economic catastrophe where banks went bankrupt, and farms were abandoned. By the end of the Civil War and the era of the 1930s, about 30% of the Great Plain prairies[2] had been converted to cropland; the remaining lands were grazed with livestock (Cunfer 2005). Even in the 1930s, the United States was a modern society with adequate knowledge of natural resource management, yet it was not until 1935 that the USDA established the Soil Conservation Service. As discussed in Chap. 1, it took a crisis to make progress toward natural resource management. Unfortunately, by the end of the Dust Bowl, many areas in the Midwestern Great Plains had cumulatively lost more than 75% of its topsoil from wind and water erosion (Hornbeck 2009).

Water Erosion Dynamics (Figs. 3.16 and 3.17)

Water erosion is classified into for basic categories: rill erosion, interrill erosion, ephemeral gully erosion, and gully erosion. The main features that influence soil erosion are slope steepness, slope length, and slope shape. Once runoff begins on a steep slope, water accelerates as overland flow resulting in increased incidence of soil erosion. Rills consist of small channels formed by runoff; the spacing between rills is interrills. Interrill erosion includes soil loss by raindrop splash and erosion from shallow overland flow. Soil splashed into the air can be 50 to 90 times greater than runoff losses (Ward and Elliott 1995). On a level surface, raindrop soil splash can be 0.6 m (2 ft.) in height and spread 1.5 m (5 ft.) laterally (Ellison 1947). Ephemeral gullies are deeper than rills and are small erosion channels caused by concentrated overland flow from runoff between two adjacent slopes (a natural drainage) after a single rainstorm event (Casali et al. 2000). Casali et al. (2000) group ephemeral gullies into three categories: (1) classic, formed by concentrated runoff and flow within a field; (2) drainage, formed by concentrated runoff and flow from adjacent upland fields; and (3) discontinuity, areas where farming practices

---

[1] The author worked in Lane County, western Kansas, helping farmers and ranchers apply conversation. Level terraces, development of irrigation, cover crops, and reduced and no-till practices have had significant impacts on reducing and controlling soil erosion.

[2] In Fig. 3.15, the mixed-grass and shortgrass prairies are host to shorter statured grasses, mostly warm season species, with few cool season species that are drought tolerant to a point. This area was more prone to the effects of the drought compared to the tallgrass prairies further east.

Fig. 3.16 This photo depicts severe rill, interrill, ephemeral gully, and the beginning of actual gullies in northwest Iowa after heavy rains (1/1/1999). (Photo by Lynn Betts. (Courtesy of USDA-NRCS))

Fig. 3.17 On a level surface, raindrop soil splash can be 0.6 m (2 ft.) in height and spread 1.5 m (5 ft.). (Courtesy of USDA-ARS (unknown photographer))

have altered the slope, e.g., field boundaries near roads and edges of fields. Ephemeral gullies can appear on cropland each year but can be filled in with tillage operations. Gullies can form from rills and/or ephemeral gullies; they consist of deeper channels that by definition cannot be crossed during tillage operations. Each of these erosion types can form on cropland, pastureland, and rangelands. If rills occur in garden situations, some remedial action is needed, such as use of mulch and/or a terracing system. Soil factors and climate events including rainfall events, melting snow, topography, and land use are all factors that influence water erosion.

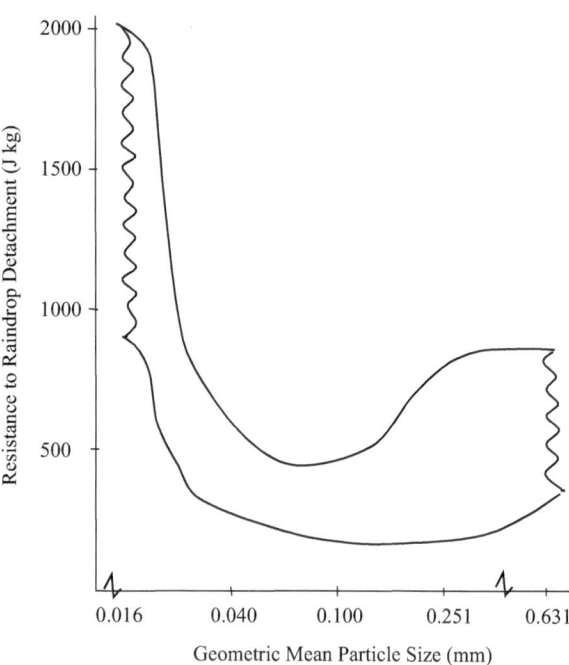

Fig. 3.18 Corresponding geometric soil particle size with raindrop detachment. (Adapted from Poesen 1985; Parsons and Abrahams 1993)

The dominant climatic variables related to soil erosion are rainfall storm intensity, energy of raindrop impact, and amount of rainfall (Nearing et al. 2005). Pruski and Nearing (2002) estimate that for every one percent increase in total rainfall amount, erosion correspondingly increases 0.85% holding rainfall intensity constant. However, if rainfall intensity and amount change in tandem, predicted erosion increases by 1.7% or every 1% increase in total rainfall. At the plant-soil interface, several variables are highly associated with erosion potentials: (1) amount of bare ground or soil exposed, (2) foliar plant cover, (3) ground cover (percent litter or residue, basal plant cover[3]), (4) slope of the field or site, and (5) physical properties of the soil surface. The pertinent soil physical properties that play an important role in soil erosion potential are soil texture (particle size), bulk density of the soil, soil aggregate stability, organic matter content, inherent porosity, and soil structure.

Raindrop impact and runoff are complementary processes; each can operate separately or in combination (Salles et al. 2002). Many studies of splash erosion have produced equations for explaining splash detachment to rain erosivity indices (Bisal 1960; Gilley and Finkner 1985; Nearing and Bradford 1985; Salles et al. 2002). The physics can be quite complex, but the main point is when soil detachment occurs, a threshold of raindrop erosivity has exceeded the weight and binding forces of soil particles and aggregates. Minimal kinetic energy is needed for soils with a mean geometric mean particle size of 0.125 mm, whereas particle sizes between 0.016 and 0.063 mm are most susceptible to detachment (Fig. 3.18) (Poesen 1985; Parsons and Abrahams.1993; Morgan 2009; Zhao et al. 2011).

---

[3] On range and pastureland, other ground cover variables such as rock cover and biological crusts such as mosses and lichen cover can be prominent.

Detachment of soil particles occurs more freely where a thin film of water exists as opposed to dry soil and increases as the depth of water increases. Once the depth of water on the soil surface equals the diameter of the raindrop (~2.5 mm), the water layer provides a barrier to raindrop splash and detachment (Dickey et al. 1986). Raindrop splash can disrupt soil aggregates, and the force of the impact can partially seal the surface soil and reduce infiltration. Runoff occurs when rain intensity exceeds infiltration capacity and then forces of overland flow are the second stage of erosion. The factors associated with overland flow transport are the amount and velocity of the water flowing in combination with slope. Little to no slope has significantly less runoff compared to increasing slope. During the runoff process, as water accelerates across non-vegetated or residue limiting soil surfaces, additional soil particles are displaced. In addition to live plants, plant residue, and rock cover, soil physical properties such as particle size or texture, soil structure, and aggregate stability influence erosion. Organic matter is linked closely with aggregate stability and is a stabilizing factor along with exudates from microorganisms.

In cropland settings, surface crusts often form after rainfall events, usually a few millimeters thick, which inhibit water infiltration by the clogging and infilling of soil pores spaces by fine soil particles that detach from more stable aggregates upon rainfall impact. Soil moisture content plays a role in the structural state of soils and rainfall intensity. Morgan (2009) made these generalizations:

1. On dry soil with high rainfall intensity, soil aggregates are easily dispersed and break down quickly by slaking. Breakdown of soil particles is by compression of air. On smooth soil surfaces with little microtopography, infiltration capacity is quickly reduced; however, on rougher soil surfaces, there is greater storage of water in depressions, which slows runoff.
2. For soil aggregates that are partially wetted with low rainfall intensity, micro-cracks can occur where larger aggregates are reduced to smaller aggregates. Infiltration capacity can be maintained because of larger pore spaces between soil aggregates.
3. For saturated aggregates, infiltration capacity is dependent upon saturated hydraulic conductivity associated with soil texture. Higher amounts of rain are needed to seal soil pores; soils with less than 15% clay are vulnerable to sealing with high-intensity rainfall.

Vegetative cover is indispensable in helping to reduce runoff and erosion on any land use type; however, cultivated land poses the highest runoff and erosion risks (Fig. 3.19). Agricultural nonpoint source pollution (NPS) is an important concern in the United States (EPA 2005) as it impacts streams, rivers, lakes, estuaries, and wetlands. Nonpoint source pollution is so termed, because unlike pollution from known or point sources (e.g., industrial, sewage treatment plants), the contaminants come from many dispersed sources. The sources of NPS pollution are sediment from cultivation in cropping systems, animal feeding operations, application of pesticides, runoff and nutrient leaching from rainfed, irrigated crop fields, and over-grazed pastures (EPA 2005). Sediment that finds its way to water courses is the number one pollutant from agricultural fields (NRC 1993; EPA 2005; Zuazo and

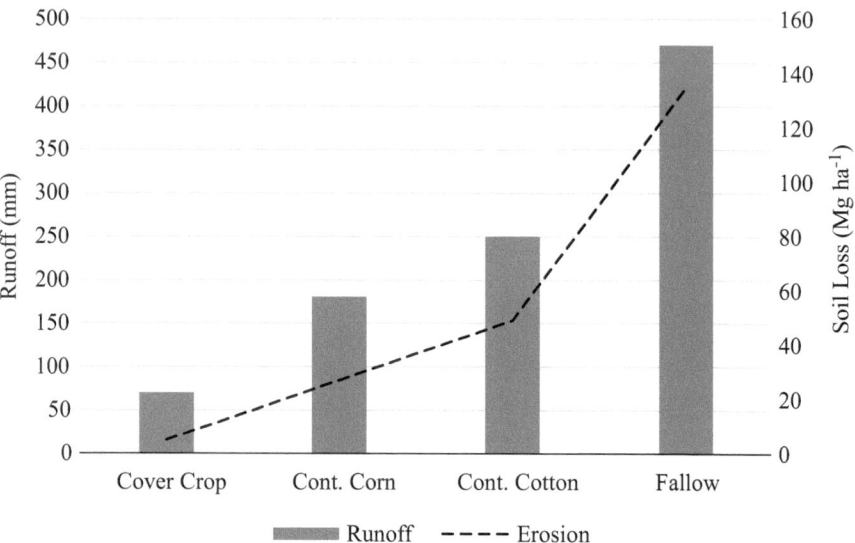

Fig. 3.19 Runoff and soil loss comparisons cover crop (Bermuda grass and crimson clover after corn), continuous corn, continuous cotton, and fallow (cultivated with no vegetative cover). (After Carreker et al. 1978)

Pleguezuelo 2008). Soil erosion and degradation have been a plague since the beginning of agriculture; however, there are remedial measures that can be taken to alleviate the impacts of accelerated runoff and erosion and wind erosion. Erosion removes mineral soil, organic matter, and nutrients, which result in reducing agricultural productivity. In Chap. 2, an example of an ecological site was given with its inherent state and transition pathways. The converted cropland state was a result of cultivation of native prairie vegetation. If excessive erosion has occurred over time and a critical threshold has been crossed, the site will no longer support a profitable cropping enterprise. The most viable option may be reseeding to native grasses or adapted introduced pasture grasses. An important point regarding soil erosion: soil forms slowly and on the average, soil is lost 13–40 times faster than soil building renewable processes (Zuazo and Pleguezuelo 2008).

Hydrology and Erosion Models: The Importance of Plant Litter and Foliar Cover

Plant and litter cover are important parameters in hydrology and erosion models such as the Revised Universal Soil Loss Equation (RUSLE), Water Erosion Prediction Project (WEPP), Erosion Risk Management Tool (ERMiT), and the Rangeland Hydrology and Erosion Model (RHEM). Soil scientists often assign a soil loss tolerance rate (T value) to soils, which relates to long-term sustainability.

The concept of soil loss tolerance is widely discussed in the literature (Wight and Siddoway 1982; Schertz 1983; Pierce et al. 1983; Johnson 1987; Pretorius and Cooks 1989; Renard et al. 1997; Nearing 2002; Li et al. 2009) and is the basis for evaluating potential erosion risk and soil sustainability. The use of the T value in the United States started in 1956 and included ten influencing factors:

1. Soil formation rate from parent material
2. Rate of development of topsoil from subsoil
3. Soil depth
4. Crop yield reduction by erosion
5. The transition of soil properties associated with plant growth
6. Plant nutrient loss from erosion
7. Incidence of rill and gully erosion
8. Deposition of sediment
9. Sediment delivery off of field
10. Application of sustainable soil conservation practices

The USDA-NRCS (2018) defines the T-factor as: "the maximum rate of annual soil loss that will permit crop productivity to be sustained economically and indefinitely on a given soil". Soil loss tolerance or permissible soil loss/sustainability factors are assigned to most soils by USDA Natural Resources Conservation Service [T commonly ranges from 2.2 to 11.2 Mg ha^{-1} yr^{-1} (1 to 5 ton ac^{-1} yr^{-1})] to evaluate management effects vs. soil loss on all land types. In conservation planning, if cropping or grazing management is associated with T-factors less than the assigned value, then erosion is considered to be at sustainable limits. Soil loss tolerance factors can also be used as a proxy indicator of soil quality (Johnson 2005). Controversy surrounds the T value concept: Nearing (2002) contends that T values for the United States and soils worldwide are inadequate for two reasons: the original science is outdated and environmental issues have changed. New research is needed and a more scientific approach to the concept is needed. Li et al. (2009) propose that three criteria be considered in developing or revising the concept: (1) soil formation should be considered in determining T values; (2) determine long-term relationships between erosion and productivity; and (3) examine the relationship between soil loss and deterioration of the soil and water quality both on-site and off-site. From a statistical viewpoint, the following variables related to a realistic T value equation should consider climate and precipitation, depth of A horizion, soil parent material and texture, annual vegetative productivity in cropped systems and natural productivity and litter production in undistrubed natural plant communities.

Predicting soil loss has been used since the 1940s (Wischmeier and Smith 1978; Schertz 1983; Pierce et al. 1983; Johnson 1987; Pretorius and Cooks 1989; Spaeth et al. 2003). Since the mid-1940s, the United States Department of Agriculture (USDA) began using erosion prediction equations and models as a guide in conservation planning to select suitable structural and field management practices. After several years of testing (1952–1962), the USDA-NRCS first officially applied the Universal Soil Loss Equation (USLE) on cropland, pastureland, and rangeland in 1965 to predict sheet and rill erosion (Renard et al. 1997). USLE technology was based on two handbook publications (Wischmeier and Smith 1965, 1978). Wischmeier (1976) stated: "the USLE was designed to predict soil loss from sheet and rill erosion" and predicted soil loss was defined as "that soil moved off the particular slope segment represented by the selected topographic factor." In 1985, an update of USLE began, and the Revised Universal Soil Loss Equation (RUSLE) subsequently replaced USLE in USDA-NRCS around 1997 (Renard et al. 1997). Since then, the USDA-NRCS has been using RUSLE and updated variants (RUSLE2) in conservation planning. Also, in 1985, the USDA Agricultural Research Service and other cooperating agencies initiated the Water Erosion Prediction Project (WEPP). The official release of the WEPP model in 1995 provided a "state-of-the-art" process-based erosion model for cropland and was based on the physical dynamics of infiltration, hydrology, plant science, hydraulics, and erosion mechanics (Nearing et al. 1989; Flanagan and Nearing 1995; Flanagan et al. 2007). The advantage of WEPP over the empirically based USLE and RUSLE was the ability to estimate spatial and temporal distributions of net soil loss under a broad range of conditions (Nearing et al. 1990). However, WEPP was ahead of its time, and practical implementation within USDA-NRCS was difficult since the model only existed on CD's, there was no Internet, and the computer interface was cumbersome. The USDA-NRCS (as of 2020) is initiating the transition to WEPP on cropland to estimate soil erosion risks. The cropland WEPP model outputs runoff and erosion estimates, with impressive graphics and statistical outputs (see example in Chap. 4) (Fig. 3.20).

On rangeland, USDA initiated rainfall simulation experiments in the late 1980s and early 1990s because it was becoming more apparent that USLE technology was

Fig. 3.20   Rainfall simulation experiment at the USDA Deep Loess Research Station near Treynor, Iowa, to evaluate residue cover. (Photo by John Gilley (Courtesy of USDA-ARS))

Fig. 3.21 Rainfall simulation experiment on rangeland with rotating boom rainfall simulator to obtain data on infiltration, runoff, and soil loss (c. 1987 to 1992 experiments for developing RHEM). (Photo by Scott Bauer (Courtesy of USDA-ARS))

not applicable to rangeland (Weltz et al. 1998; Spaeth et al. 2003; Nearing et al. 2011) (Fig. 3.21). A critique of USLE by Hawkins (1985) stated: the USLE "does not lead directly to erosion but produces the intermediate product of storm runoff . . . the complications of time and spatial variations in site properties are usually not considered, even when of known consequence." Weltz et al. (1998) discussed several limitations regarding the USLE on rangeland: "USLE is a lumped empirical model that does not separate factors that influence soil erosion, such as plant growth, decomposition, infiltration, runoff, soil detachment, or soil transport. The USLE was designed to estimate sheet and rill erosion from hillslope areas. It was not designed to address soil deposition and channel or gully erosion within watersheds." Renard et al. (1991) also summarized: "the fundamental erosion processes and their interactions are not represented, explicitly in the USLE." In the USDA-NRCS, the accepted state of the art hydrology and erosion model on range and pastureland is the Rangeland Hydrology and Erosion Model (RHEM). Originally the WEPP developmental effort in USDA started in 1985 combined both cropland and rangeland; however, the disciplines gradually digressed as WEPP on cropland and RHEM on range and pastureland. The Rangeland Hydrology and Erosion Model is a coordinated project between three USDA agencies: Agricultural Research Service (ARS), Natural Resource Conservation Service (NRCS), and the US Forest Service (USFS) (Wei et al. 2009). The RHEM model is designed for government agencies, land managers, and conservationists who need sound and science-based technology to model and predict runoff and erosion rates on rangelands and to assist in assessing rangeland conservation practice effects. RHEM is a process-based erosion prediction tool specific for rangeland application and is based on fundamentals of infiltration, hydrology, plant science, hydraulics, and erosion mechanics. RHEM can be used to evaluate runoff and erosion as a consequence of plant species and growth form changes from disturbances such as fire, brush management, and climate change. RHEM will also evaluate the statistical risk from various storm events (2, 5, 10, 25, 50, 75, and 100 yr.). Outputs of RHEM include tables and figures of average precipitation, number of storms producing runoff, runoff, soil loss, and hydrology and erosion risks for the design storm events (see example in Chap. 5).

Water erosion indictors are included in the rangeland health model, an assessment tool where 17 indicators are evaluated from a reference state, which is usually the historic plant community (Pellant et al. 2020). Two assessments in rangeland health, soil and surface stability, and hydrologic function evaluate sheet and rill erosion, gully erosion, water flow patterns, litter movement, plant pedestalling to determine these assessments (see chapter?).

Text Box 3.2 1 mm of Soil Loss?
Several publications have provided estimates of average soil loss:

- 0.1 mm yr^{-1} for our most recent geologic epoch (Wilkinson and McElroy 2007)
- 0.021 mm yr^{-1} the average erosion rate at 21 meters per million years (m/m.y.) (Summerfield and Hulton 1994)
- Loess and glacial till areas of Iowa, United States, ~0.8–1.9 Mg ha^{-1} yr^{-1} (Ruhe and Daniels 1965; Walker 1966)
- On cultivated US croplands, average estimated erosion rates are about 600 m/m.y. (0.6 mm yr^{-1}) (USDA NRCS 2000)

If 1 mm of soil erodes on a silt loam with a bulk density of 1.33 Mg m^3, what is the equivalent soil loss in Mg ha^{-1} and US tons ac^{-1}?

For 1 mm soil depth = (1 m) (1 m) (0.001 m) = 0.001 m^3
Weight of soil 1 m^2 at 1 mm depth = (0.001 m^3) (Bulk density 1.33 Mg m^3) = 0.0013 Mg m^3
Weight of soil for 1 hectare at 1 mm depth = (100 m) (100 m) (0.0013 Mg m^3) = 13 Mg ha^{-1} (5.79 t ac^{-1})

Wind-Induced Erosion (Fig. 3.22)

Wind erosion is a major erosion factor in dry semiarid and arid cropping regions where it can exceed water erosion. It is important to keep in mind that many areas are prone to both water and wind erosion as wet and dry periods may occur simultaneously. Rangelands, especially fragile plant community types in the Great Basin, Desert Southwest, and the western fringe of the shortgrass prairie regions, are highly susceptible to wind erosion. It is estimated that wind erosion displaces and

Fig. 3.22 Wind erosion from dust storm in cotton field near Lubbock, Texas (June 2010). (Photo by Scott Van Pelt (Courtesy of USDA-ARS))

transports about 2/3 more soil than water erosion. In six states in the Great Plains, annual soil losses from wind erosion exceed water erosion, average wind erosion is about 4 Mg ha⁻¹ (1.8 t ac⁻¹) in Nebraska to 29 Mg ha⁻¹ (12.9 t ac⁻¹) in New Mexico (Weil and Brady 2017).

Wind erosion is most active when the soil surface is relatively dry. All soil types can erode, but silty and sandy soils are most prone. As with water erosion, there are three basic processes: (1) detachment, (2) transportation, and (3) deposition. During transport of windblown soil particles, saltation occurs where soil particles exhibit a series of short bounces along the soil surface. Saltation accounts for about 50–90% of soil movement by wind. As saltation begins, soil creep can occur where soil particles are rolled across the surface after contact with soil particles. Soil creep generally moves soil particles 0.5 to 1 mm in diameter accounting for 5–25% of windblown erosion. Suspension, the last mode of the wind erosion process, produces the most noticeable effect and occurs when fine particles are lifted into the atmosphere. Suspended soil particles can contain over three times more organic matter and nitrogen compared to the parent material remaining (Bennett 1939). Suspended soil particles can be carried high into the atmosphere and transported great distances (see Chap. 1, wind erosion photos during the dust bowl). Generally, most suspended soil and movement occurs below 1 meter.

The abrasive power of wind erosion is greatly increased when moving air is full of soil particles. Rapidly moving soil particles act as a force (sand blasting effect) to dislodge soil clods and aggregates and remove soil around the base of plants (plant pedestalling). As with water erosion on cropland systems, cover crops and crop residue are part of the "best management" approach to control wind erosion soil loss. By maintaining standing crop residues in wind-prone areas, wind velocity can be reduced. In addition, standing crop residues can aid in trapping snow. On rangelands, wind erosion is insidious as it becomes more difficult to stop after the process starts. Wind erosion is one of the ten indicators for assessing soil and site stability in the rangeland health model (Pellant et al. 2019).

Vegetative Influences and Manipulation on Soil Physical Properties Related to Hydrology

Soil physical properties affect hydrology (infiltration, runoff, and water flow) and subsequent soil detachment and finally sediment transport across the landscape. Soil health and intrinsic soil physical properties are inextricably linked and are key to the soil production capability and sustainability potential. Soil physical properties such as texture, soil structure, bulk density, and aggregate stability define the parameters of infiltration and movement of air in the soil, which affect chemical and biological processes in the soil. Soil health on cropland and in gardens must consider how tillage practices and management, residue management, and cover crops affect soil aggregation, which are key factors associated with infiltration dynamics. On grazinglands, the intrinsic parameters of soil bulk density and aggregate stability are directly affected by grazing intensity and management. In the garden, a loose friable soil is maintained by additions of compost, manure, and mulches.

Many hydrologically important soil physical properties such as aggregate stability, porosity, soil hydrologic conductivity and infiltration, hydraulic conductivity, soil structure, soil temperature, soil moisture, and bulk density can be influenced by vegetation cover and decomposition and subsequent additions and maintenance of organic matter. In the adjoining chapters, the effects of vegetation on cropland, rangeland, and gardens are covered in more detail. Higher infiltration capacity, hydraulic conductivity, and lower erosion rates are common in no-till/minimum tillage cropland systems where plant residues remain on the soil surface. The effects of increased organic matter in conservation tillage systems are associated with greater porosity, microaggregate formation and soil aggregate stability, and decreased bulk density (Shaver et al. 2002; Chan et al. 2003). Burning crop residues can have adverse effects on soil porosity due to blockage of soil macropores by the burnt ash (Valzano et al. 1997). Studies have shown that continual cultivation and incorporation of crop residues tend to decrease macro- and meso-porosity, increase compaction, increase bulk density, and alter soil structure and aggregates (Mapa et al. 1986). There are many advantages to increasing soil organic matter; however, the accumulation of significant amounts of organic carbon can induce water repellency at the soil surface (Harper et al. 2000). In garden systems, an abundance of fine decomposed organic matter from plant residues and high manure applications can create a fine surface texture similar to peat moss (has water repellent properties until saturated). This situation is usually short-lived as the fine organic material becomes more incorporated into the upper profile of the soil. Cultivation will decrease water repellency by mixing and increasing the mineralization of organic matter. Soil water repellency is also a characteristic feature on burned and unburned range and forest-lands (see Chap. 5).

Conclusions

The hydrologic cycle and its function are integral to the health of the soil. Soil health is dependent on proper hydrologic function; organic matter cycling cannot proceed without water and associated hydrologic dynamics. Every soil has intrinsic physical and chemical characteristics that define infiltration, runoff, erosion, and sediment yield parameters. As a soil matures, soil physical properties reach a level that is associated with the original soil parent material (and any deposited material), climate, geomorphic characteristics, and plants that inhabit the site. Primary and secondary plant succession patterns associated with the development of soil and the ecological site are in continuous transition until some relative level of equilibrium is reached. There are many views on whether natural plant communities and land-scapes reach equilibrium or are perpetually in a state of nonequilibrium. Briske et al. (2017) contend that qualification of equilibrium ecology with resilience theory provides a more realistic interpretation of plant-herbivore interactions and vegeta-tion responses compared to nonequilibrium models. However, for the purposes of understanding the soils we farm and raise livestock on, we need a reference so that we know what the intrinsic characteristics were historically—many of the upland grassland and forest soils that were converted for intensive agricultural purposes were for all practical purposes stable with respect to a given particle size (texture), bulk density, soil structure, organic matter content, and the soil flora and fauna. Plant succession, net primary productivity, and use by native fauna in natural plant communities also help define the intrinsic characteristics of a soil. Understanding basic hydrology and its functionality at a particular location is the first important step in addressing issues of soil health on the farm, ranch, and in the garden.

References

Bakker, M.M., G. Grovers, and D.A. Rounsevell. 2004. The crop productivity-erosion relation-ship: An analysis based on experimental work. *Catena* 57: 55–76.

Barger, N.N., J.E. Herrick, J. Van Zee, and J. Belnap. 2006. Impacts of biological soil crust dis-turbance and composition on C and N loss from water erosion. *Biogeochemistry* 77: 247–263.

Beall, H.W. 1934. The Penetration of Rainfall through Hardwood and Softwood Forest Canopy. *Ecology* 15: 412–415.

Belnap, J. 1994. Potential role of cryptobiotic soil crust in semiarid rangelands. In *Proceedings of the ecology and management of annual rangelands*, ed. S.B. Monsen and S.G. Kitchen General

Technical Report INT-GTR-313, 179–185. Ogden: USDA Forest Service, Intermountain Research Station.

――――. 2006. The potential roles of biological soil crusts in dryland hydrologic cycles. *Hydrological Processes* 20: 3159–3178.

Belnap, J., and J.S. Gardner. 1993. Soil microstructure in soils of the Colorado Plateau: the role of the cyanobacterium Microcoleus vaginatus. *The Great Basin Naturalist* 1: 40–47.

Belnap, J., R. Rosentreter, S. Leonard, J. Kaltenencker, J. Williams, and D. Eldridge. 2001. Biological crusts: Ecology and management. In *USDI-BLM technical reference 1730–2*. Denver: Colorado.

Bennett, H.H. 1939. *Soil conservation*. New York: McGraw Hill.

Betts, A.K., J.H. Ball, and P. Viterbo. 1999. Basin-scale surface water and energy budgets for the Mississippi from the ECMWF reanalysis. *Journal Geophysical Research* 104.

Bisal, F. 1960. The effect of raindrop size and impact velocity on sand splash. *Canadian Journal Soil Science* 40: 242–245.

Blackburn, W.H. 1975. Factors influencing infiltration and sediment production of semiarid rangelands in Nevada. *Water Resources Research* 11: 929–937.

Branson, F.A., G.F. Gifford, K.G. Renard, and R.F. Hadley. 1981. *Rangeland Hydrology*. Dubuque, Iowa: Kendall Hunt Publishing Company.

Breshears, D.D., P.M. Rich, F.J. Barnes, and K. Campbell. 1997. Overstory-imposed heterogeneity in solar radiation and soil moisture in a semiarid woodland. *Ecological Applications* 7: 1201–1215.

Briske, D.D., A.W. Illius, and J.M. Anderies. 2017. Nonequilibrium ecology and resilience theory. *Rangeland Systems*: 197–227.

Brotherson, J.D., and S.R. Rushforth. 1982. Influence of cryptogamic crusts on moisture relationships of soils in Navajo National Monument, Arizona. *Great Basin Naturalist* 43: 73–78.

Brooks, K.N., P.F. Ffolliott, H.M. Gregersen, and J.L. Thames. 1991. *Hydrology and the management of watersheds*. Ames, Iowa: Iowa State University.

Bunting, B.T. 1965. *The geography of soil*. Chicago: Aldine.

Burkhardt, J.W. 1996. Herbivory in the Intermountain West. An overview of evolutionary history, historic cultural impacts and lessons from the past. Station Bulletin 58, Idaho Forest, Wildlife and Range Experiment Station, University of Idaho, Moscow.

Butzer, K.W. 1974. Accelerated soil erosion: A problem of man-land relationships. In *Perspectives on environment*, ed. R. Manners and M.W. Mikesell. Washington D.C.: Association of American Geographers, Pub No. 13.

Cánton, Y., A. Solé-Benet, J. de Vente, C. Boix-Fayos, A. Calvo-Cases, C. Asensio, and J. Puigdefabregas. 2011. A review of runoff generation and soil erosion across scales in semiarid southeastern Spain. *Journal of Arid Environments* 75: 1254–1261.

Carlyle-Moses, D.E. 2004. Throughfall, stemflow, and canopy interception loss fluxes in a semiarid Sierra Madre Oriental matorral community. *Journal of Arid Environments* 58: 181–202.

Carreker, J.R., S.R. Wilkinson, A.P. Barnett, and J.E. Box. 1978. Soil and water systems for sloping lands. USDA-ARS-160.

Casali, J., S.J. Bennett, and K.M. Robinson. 2000. Processes of ephemeral gully erosion. *International Journal of Sediment Research* 15: 31–41.

Chamizo, S., Y. Cantón, R. Lázaro, A. Solé-Benet, and F. Domingo. 2012. Crust composition and disturbance drive infiltration through biological soil crusts in semiarid ecosystems. *Ecosystems* 15: 148–161.

Chan, K.Y., D.P. Heenan, and H.B. So. 2003. Sequestration of carbon and changes in soil quality under conservation tillage on light-textured soils in Australia: A review. *Australian Journal Experimental Agriculture* 43: 325–334.

Clark, O.R. 1940. Interception of rainfall by prairie grasses, weeds, and certain crop plants. *Ecological Monographs* 10: 43–277.

Connaughton, C. 1936. Fire damage in the ponderosa pine type in Idaho. *Journal of Forestry* 34: 46–51.

Corbett, E.S., and R.P. Crouse. 1968. *Rainfall interception by annual grass and chaparral... losses compared. Res. Paper PSW-RP-48*, 12. Berkeley: Pacific Southwest Forest & Range Experiment Station, Forest Service, US Department of Agriculture.

Couturier, D.E., and E.A. Ripley. 1973. Rainfall interception in mixed grass prairie. *Canadian Journal of Plant Science* 53: 659–663.

Cunfer, G. 2005. *On the Great Plains agriculture and environment.* College Station: Texas A&M University Press.

Danin, A. 1978. Plant species diversity and plant succession in a sandy area in the Northern Negev. *Flora* 167: 409–422.

Dickey, E.C., P.J. Jasa, and D.P. Shelton. 1986. *Estimating residue cover.* Neb. Guide G86-793. Lincoln, Neb.: University of Nebraska Cooperative Extension Service.

Dingman, L. 2002. *Physical hydrology.* 2nd ed. Upper Saddle River: Prentice Hall. 600 p.

Domingo, F., G. Sánchez, M.J. Moro, A.J. Brenner, and J. Puigdefábregas. 1998. Measurement and modeling of rainfall interception by three semi-arid canopies. *Agricultural and Forest Meteorology* 91: 275–292.

Dunford, E.G., and H.C. Niederhof. 1944. Influence of aspen, lodgepole pine, and open grassland types upon factors affecting water yield. *Journal of Forestry* 42: 673–677.

Dunkerley, D.L. 2008. Intra-storm evaporation as a component of canopy interception loss in dryland shrubs. Observations from Fowlers Gap, Australia. *Hydrological Processes* 22: 1985–1995.

Dunkerley, D.L., and T.L. Booth. 1999. Plant canopy interception of rainfall and its significance in a banded landscape, arid western New South Wales, Australia. *Water Resources Research* 35: 1581–1586.

Dykes, A.P. 1997. Rainfall interception from a lowland tropical rainforest in Brunei. *Journal of Hydrology* 200 (1): 260–279.

Eldridge, D.J. 1993. Cryptogam cover and soil surface condition: Effects on hydrology on a semi-arid woodland soil. *Arid Soil Research and Rehabilitation* 7: 203–217.

Eldridge, D.J., and P.I.A. Kinnell. 1997. Assessment of erosion rates from microphyte-dominated monoliths under rain-impacted flow. *Australian Journal of Soil Research* 35: 475–489.

Eldridge, D.J., and R. Rosentreter. 1999. Morphological groups: A framework for monitoring microphytic crusts in arid landscapes. *Journal of Arid Environments* 41: 11–25.

Eldridge, D.J., E. Zaady, and M. Shachak. 2000. Infiltration through three contrasting biological soil crusts in patterned landscapes in the Negev, Israel. *Catena* 40: 323–336.

Ellison, W.D. 1947. Soil erosion studies, part I. *Agricultural Engineering* 1947: 145–146.

EPA (Environmental Protection Agency). 2005. *Protecting water quality from agricultural runoff. EPA 841-F-05-001.* Washington, D.C.

Evans, C.E. 1975. Snow management in the Great Plains. Nebraska. Agricultural Experiment Station, & Great Plains Agricultural Council. University of Nebraska, Agricultural Experiment Station. Proc. Great Plains Council Pub. No. 73, 186 pp.

Flanagan, D.C., and M.A. Nearing, eds. 1995. *NSERL report no. 10 USDA-ARS.* West Lafayette: National Soil Erosion Research Laboratory.

Flanagan, D.C., J.E. Gilley, and T.G. Franti. 2007. Water Erosion Prediction Project (WEPP): Development history, model capabilities, and future enhancements. *Transactions of the ASABE* 50: 1603–1612.

Fletcher, J.E., and W.P. Martin. 1948. Some effects of algae and molds in the raincrust of desert soils. *Ecology* 29: 95–100.

Gifford, C.F. 1985. Cover allocation in rangeland watershed management: a review. In *Watershed Management in the Eighties.* Proceedings of the American Society of Civil Engineers Symposium, New York; 23–31.

Gilley, J.E., and S.C. Finkner. 1985. Estimating soil detachment caused by raindrop impact. *Transactions of the ASAE* 28: 140–146.

Greb, B.W. 1975. Snowfall characteristics and snowmelt storage at Akron, Colorado. In *Proc. snow management in the Great Plains. Great Plains Council Pub. No. 73*, 45–64.

Greene, R.S.B., and D.J. Tongway. 1989. The significance of (surface) physical and chemical properties in determining soil surface condition of red earths in rangelands. *Australian Journal Soil Research* 27: 213–225.

Gunn, R., and G.D. Kinzer. 1949. The terminal velocity of fall for water drops in stagnant air. *Journal of Meteorology* 6: 243–248.

Harper, K.T., and L.L. St. Clair. 1985. Cryptogamic soil crusts on arid and semiarid rangelands in Utah: effects of seedling establishment and soil stability. Final report, Bureau of Land Management, Utah State Office, Salt Lake City.

Harper, R.J., I. McKissock, R.J. Gilkes, D.J. Carter, and P.S. Blackwell. 2000. A multivariate framework for interpreting the effects of soil properties, soil management and land use on water repellency. *Journal of Hydrology* 231/232: 371–383.

Hawkins, R.H. 1985. *Hydrology and the universal soil loss equation: Application to rangelands. BLM-YA-PT-85-001-4340.* Denver: BLM.

Haynes, J.L. 1940. Ground rainfall under vegetation canopy of crops. *Journal of the American Society Agronomy* 32: 176–184.

Hill, T.C., M. Williams, and J.B. Moncrieff. 2008. Modeling feedbacks between a boreal forest and the planetary boundary layer. *Journal Geophysical Research* 113.

Hillel, D. 1982. *Introduction to soil physics.* Cambridge, Massachusetts: Academic Press Inc.

Hornbeck, R. 2009. *The enduring impact of the American dust bowl: Short and long-run adjustments to environmental catastrophe. Working paper 15605.* Cambridge: National Bureau of Economic Research.

Horton, R.E. 1933. The role of infiltration in the hydrologic cycle. *Transactions of the American Geophysical Union* 14: 446–460.

Huizhi, L., and F. Jianwu. 2012. Seasonal and interannual variations of evapotranspiration and energy exchange over different land surfaces in a semiarid area of China. *American Meteorological Society* 51: 1875–1888.

Hull, A.C. 1972. Rainfall and snowfall interception of big sagebrush. *Utah Academy of Science, Arts, and Letters* 49: 64.

Hull, A.C., and G.J. Klomp. 1974. *Yield of crested wheatgrass under four densities of big sagebrush in southern Idaho. No. 1483.* Washington D.C. US Department of Agriculture.

Johnson, L.C. 1987. Soil loss tolerance: fact or myth? *Journal of Soil and Water Conservation* 42: 155–160.

Johnson, L.C. 2005. Soil loss tolerance--fact or myth? *Journal of Soil and Water Conservation* 60: 52A–52A.

Kaltenecker, J.H., M. Wicklow-Howard, and R. Rosentreter. 1999. Biological soil crusts in three sagebrush communities recovering from a century of livestock trampling. In *Proceedings shrubland ecotones*, ed. E.D. McArthur, W.K. Ostler, and C.L. Wambolt, 222–226. RMRS-P-11. USDA Forest Service, Rocky Mountain Research Station. Ogden, Utah.

Kittredge, J. 1948. *Forest influences: The effects of woody vegetation on climate, water, and soil, with applications to the conservation of water and the control of floods and erosion.* New York: McGraw-Hill.

Kleiner, E.F., and K.T. Harper. 1972. Environment and community organization in grasslands of Canyonlands National Park. *Ecology* 53: 229–309.

———. 1977. Soil properties in relation to cryptogamic ground cover in Canyonlands National Park. *Journal of Range Management* 30: 202–205.

Klocke, N.L., R.S. Currie, and R.M. Aiken. 2009. Soil water evaporation and crop residues. *Transactions of the American Society of Agricultural and Biological Engineers* 52: 103–110.

Larsen, M.L., A.B. Kostinski, and A.R. Jameson. 2014. Further evidence for superterminal raindrops. *Geophysical Research Letters* 4: 6914–6918.

Lavin, S.J., F.M. Shelley, and J.C. Archer, eds. 2011. *Atlas of the Great Plains.* Lincoln: University of Nebraska Press.

Leuning, R., A.G. Condon, F.X. Dunin, S. Zegelin, and O.T. Denmead. 1994. Rainfall interception and evaporation from soil below a wheat canopy. *Agricultural and Forest Meteorology* 67: 221–238.

Li, L., S. Du, L. Wu, and G. Liu. 2009. An overview of soil loss tolerance. *Catena* 78: 93–99.

Lloyd, C.R., J.H. Gash, J.H., and W.J. Shuttleworth. 1988. The measurement and modelling of rainfall interception by Amazonian rain forest. *Agricultural and Forest Meteorology* 43: 277–294.

Loope, W.L., and G.F. Gifford. 1972. Influence of a soil microfloral crust on select properties of soils under pinyon-juniper in southeastern Utah. *Journal of Soil Water and Conservation* 27: 164–167.

Ludwig, J.A., B.P. Wilcox, D.D. Breshears, D.J. Tongway, and A.C. Imeson. 2005. Vegetation patches and runoff-erosion as interacting ecohydrological processes in semi-arid landscape. *Ecology* 86: 288–297.

Mapa, R.B., R.E. Green, and L. Santo. 1986. Temporal variability of soil hydraulic properties with wetting and drying subsequent to tillage. *Soil Science Society America Journal* 50: 1133–1138.

Marble, J.R., and K.T. Harper. 1989. Effect of timing of grazing on soil-surface cryptogamic communities in Great Basin low-shrub desert: A preliminary report. *Great Basin Naturalist* 49: 104–107.

Memmott, K.L., V.J. Anderson, and S.B. Monsen. 1998. Seasonal grazing impact on cryptogamic crusts in a cold desert ecosystem. *Journal of Range Management* 51: 547–550.

Miller, R.F., T.J. Svejcar, and N.E. West. 1994. Implications of livestock grazing in the intermountain sagebrush region: Plant composition. In *Ecological implications of livestock Herbivory in the West*, ed. M. Vavra, W.A. Laycock, and R.D. Pieper, 101–146. Denver: Society for Range Management.

Miner, N.H., and J.M. Trappe. 1957. *Snow interception, accumulation, and melt in lodgepole pine forests in the Blue Mountains of eastern Oregon*. Pacific Northwest Forest and Range Experiment Station, Forest Service, US Department of Agriculture. Portland, Oregon.

Montero-Martinez, G., A.B. Kostinski, R.A. Shaw, and F. Garcia-Garcia. 2009. Do all raindrops fall at terminal speed? *Geophysical. Research Letters* 36 (11).

Morgan, R.P.C. 2009. *Soil erosion and conservation*. 3rd ed. Hoboken, New Jersey: Wiley-Blackwell.

National Research Council (NRC). 1993. *Vétiver gGrass: A thin green line against erosion. Board on Science and Technology for International Development*. Washington D.C.: National Academy Press.

Nearing, M.A. 2002. *Toward a new definition of soil loss tolerance for the United States*. International Soil Conservation Organization Conference, Conference abstracts.

Nearing, M.A., and J.M. Bradford. 1985. Single waterdrop splash detachment and mechanical properties of soils. *Soil Science Society America Journal* 49: 547–552.

Nearing, M.A., G.R. Foster, L.J. Lane, and S.C. Finkner. 1989. A process-based soil erosion model for USDA water erosion prediction project technology. *Transactions of the ASAE* 32: 1587–1593.

Nearing, M.A., L. Deer-Ascough, and J.M. Laflen. 1990. Sensitivity analysis of the WEPP hillslope profile erosion model. *Transactions of the ASAE* 33: 839–0849.

Nearing, M.A., V. Jetten, C. Baffaut, O. Cerdan, A. Couturier, M. Hernandez, and V. Souchère. 2005. Modeling response of soil erosion and runoff to changes in precipitation and cover. *Catena* 61: 131–154.

Nearing, M.A., H. Wei, J.J. Stone, F.B. Pierson, K.E. Spaeth, M.A. Weltz, and M. Hernandez. 2011. A rangeland hydrology and erosion model. *Transactions of the ASABE* 54: 901–908.

Noble, C.A., and R.P.C. Morgan. 1983. Rainfall interception and splash detachment with a Brussels sprouts plant: A laboratory simulation. *Earth Surface Processes and Landforms* 8: 569–577.

Nuttle, W.K. 2002. Eco-hydrology's past and future in focus. *Earth Science and Space News* 83: 205.

Owens, M.K., R.K. Lyons, and C.L. Alejandro. 2006. Rainfall partitioning within semiarid juniper communities: Effects of event size and canopy cover. *Hydrological Processes* 20: 3179–3189.

Parsons, A.J., and A.D. Abrahams. 1993. Mechanisms of overland flow generation and sediment production on loamy and sandy soils with and without rock fragments. In *Overland flow:*

Hydraulics and erosion mechanics, ed. A.J. Parsons and A.D. Abrahams, 275–305. New York: Chapman and Hall.

Pellant, M., P.L. Shaver, D.A. Pyke, J.E. Herrick, F.E. Busby, G. Riegel, N. Lepak, E. Kachergis, B.A. Newingham, and D. Toledo. 2020. *Interpreting indicators of rangeland health, version 5. Tech 7 ref 1734–6*. Denver: U.S. Department of the Interior, Bureau of Land Management.

Pierce, F.J., W.E. Larson, R.H. Dowdy, and W. Graham. 1983. Productivity of soils: Assessing long-term changes 610 due to erosion. *Journal of Soil and Water Conservation* 38: 39–44.

Pierson, F.B., P.R. Robichaud, and K.E. Spaeth. 2001. Spatial and temporal effects of wildfire on the hydrology of a steep rangeland watershed. *Hydrological Processes* 15: 2905–2916.

Poesen, J. 1985. An improved splash transport model. *Zeitschrift für Geomorphologie* 29: 193–211.

Pomeroy, J.W., and S.M. Gray. 1995. *Snowcover: Accumulation, relocation and management. National Hydrology Research Institute Science report no. 7*, 144. Saskatoon: National Hydrology Research Institute.

Pretorius, J.R., and J. Cooks. 1989. Soil loss tolerance limits: An environmental management tool. *Geo Journal* 19: 67–75.

Pruppacher, H.R. 1981. Microstructure of Atmospheric Clouds and Precipitation. *In clouds: their formation, optical properties and effects*, eds. P. Hobbs and A. Deepak. Cambridge: Massachusetts, Academic Press.

Pruski, F.F., and M.A. Nearing. 2002. Climate-induced changes in erosion during the 21st century for eight US locations. *Water Resources Research* 38: 34–31.

Rawls, W.J., A. Nemes, and Y.A. Pachepsky. 2004. Effect of soil organic carbon on soil hydraulic properties. *Developments in Soil Science 30*: 95–114.

Rawls, W.J., D.L. Brakensiek, and K.E. Saxton. 1982. Estimation of soil water properties. *Transactions of the ASAE* 25: 1316–1320.

Rawls, W.J., T.J. Gish, and D.L. Brakensiek. 1991. Estimating soil water retention from soil physical properties and characteristics. *Advances in Soil Science* 16: 213–234.

Rawls, W.J., Y.A. Pachepsky, J.C. Ritchie, T.M. Sobecki, and H. Bloodworth. 2003. Effect of soil organic carbon on soil water retention. *Geoderma* 116: 61–76.

Renard, K.G., G.R. Foster, G.A. Weesies, and J.P. Porter. 1991. RUSLE: Revised universal soil loss equation. *Journal of Soil and Water Conservation* 46: 30–33.

Renard, K.G., G.R. Foster, G.A. Weesies, D.K. McCool, and D.C. Yoder. 1997. Predicting soil erosion by water: a guide to conservation planning with the revised universal soil loss equation (RUSLE). In *USDA agriculture handbook 703*. United States Government Printing, Washington, D.C.

Rogers, R.R. 1979. *A short course in cloud physics*, 2 ed. Oxford, United Kingdom: Pergamon Press.

Rowe, P.B., and E.A. Colman. 1951. *Disposition of rainfall in two mountain areas of California. No. 156498*. United States Department of Agriculture, Economic Research Service. Washington, D.C.

Ruhe, R.V., and R.B. Daniels. 1965. Landscape erosion-geologic and historic. *Journal of Soil and Water Conservation* 20: 52–57.

Salles, C., J. Poesen, and D. Sempere-Torres. 2002. Kinetic energy of rain and its functional relationship with intensity. *Journal of Hydrology* 257: 256–270.

Schertz, D.L. 1983. The basis for soil loss tolerances. *Journal of Soil and Water Conservation* 38: 10–14.

Schlesinger, W.H., J.F. Reynolds, G.L. Cunningham, L.F. Huenneke, W.M. Jarrell, R.A. Virginia, and W.G. Whiteford. 1990. Biological feedbacks in global desertification. *Science* 247: 1043–1048.

Schwärzel, K., M. Renger, R. Sauerbrey, and G. Wessolek. 2002. Soil physical characteristics of peat soils. *Journal of Plant Nutrition and Soil Science* 165: 479–486.

Serrato, F.B., and A.R. Diaz. 1998. A simple technique for measuring rainfall interception by small shrub: "Interception flow collection box". *Hydrological Processes* 12: 471–481.

Shaver, T.M., G.A. Peterson, L.R. Ahuja, D.G. Westfall, L.A. Sherrod, and G. Dunn. 2002. Surface soil physical properties after twelve years of dryland no-till management. *Soil Science Society America Journal* 66: 1296–1303.

Skau, C.M. 1964. Interception, throughfall, and stemflow in Utah and alligator juniper cover types of northern Arizona. *Forest Science* 10: 283–287.

Skujins, J. 1984. Microbial ecology of desert soils. *Advances in Microbial Ecology* 7: 47–91.

Solé-Benet, A., A. Calvo, A. Cerdà, R. Làzaro, R. Pini, and J. Barbero. 1997. Influence of micro-relief patterns and plant cover on runoff related processes in badlands from Tabernas (SE Spain). *Catena* 31: 23–38.

Spaeth, K.E., F.B. Pierson, M.A. Weltz, and W.H. Blackburn. 2003. Evaluation of USLE and RUSLE estimated soil loss on rangeland. *Journal of Range Management* 1: 234–246.

Stetler, L.D., R. Benton, and M. Weiler. 2011. Erosion rates from Badlands National Park. In *International Symposium on Erosion and Landscape Evolution (ISELE), 18–21 September 2011*, 26. Anchorage, Alaska: American Society of Agricultural and Biological Engineers.

Stoffer, P.W. 2003. *Geology of Badlands National Park: A preliminary report. USGS open-file report 03–35*. Menlo Park: USGS.

Storck, P., D.P. Lettenmaier, and S.M. Bolton. 2002. Measurement of snow interception and canopy effects on snow accumulation and melt in a mountainous maritime climate, Oregon, United States. *Water Resources Research* 38: 5-1–5-16.

Sturm, M., J.P. McFadden, G.E. Liston, F.S. Chapin, C.H. Racine, and J. Holmgren. 2001. Snow-shrub interactions in arctic tundra: A hypothesis with climatic implications. *Journal of Climate* 14: 336–344.

Summerfield, M.A., and N.J. Hulton. 1994. Natural controls of fluvial denudation rates in major world drainage basins. *Journal of Geophysical Research* 99: 13871–13883.

Tate, K.W. 1995. *Interception on rangeland watersheds. Rangeland watershed program fact sheet no. 36., U.C. Coop. Extension and NRCS*. Davis: University of California.

Taucer, P.I., C.L. Munster, B.P. Wilcox, M.K. Owens, and B.P. Mohanty. 2008. Large-scale rainfall simulation experiments of juniper rangelands. *Transactions of the American Society of Agricultural and Biological Engineers* 51: 1951–1961.

Thurai, M., V.N. Bringi, W.A. Peterson, and P.N. Gatlin. 2013. Drop shapes and fall speeds in rain: Two contrasting examples. *Journal of Applied Meteorology and Climatology* 52: 2567–2581.

Thurow, T.L. 1991. Hydrology and erosion. In *Grazing management: An ecological perspective*, ed. R.K. Heitschmidt and J.W. Stuth, 141–159. Portland: Timber Press, Inc.

Thurow, T.L., and J.W. Hester. 1997. How an increase or reduction in juniper cover alters rangeland hydrology. In *Juniper symposium proceedings*, 9–22. San Angelo: Texas A&M University.

Thurow, T.L., W.H. Blackburn, S.D. Warren, and C.A. Taylor Jr. 1987. Rainfall interception by midgrass, shortgrass, and live oak mottes. *Journal of Range Management* 40: 455–460.

USDA-NRCS. 2012. United States Department of Agriculutre-Natural Resources Conservation Service. http://www.nrcs.usda.gov/Internet/FSE_MEDIA/stelprdb1049472.png
———. 2018. *Updated T and K factors*. Washington, D.C.

Valzano, F.P., R.S.B. Greene, and B.W. Murphy. 1997. Direct effects of stubble burning on soil hydraulic and physical properties in a direct drill tillage system. *Soil Tillage Research* 42: 209–219.

Walker, P.H. 1966. Postglacial environments in relation to landscape and soils on the Cary Drift, Iowa. *Research Bulletin (Iowa Agriculture and Home Economics Experiment Station)* 35 (549): 1.

Ward, A.D., and W.J. Elliot. 1995. *Environmental hydrology*. New York: Lewis Publishers.

Warren, S.D. 2001. Biological soil crusts and hydrology in North American Deserts. In *Biological soil crusts: Structure, function, and management*, Ecological studies (analysis and synthesis), ed. J. Belnap and O.L. Lange, vol. 150. Berlin/Heidelberg: Springer.

Wei, H., M.A. Nearing, J.J. Stone, D.P. Guertin, K.E. Spaeth, F.B. Pierson, M.H. Nichols, and C.A. Moffett. 2009. A new splash and sheet erosion equation for rangelands. *Soil Science Society America Journal* 73: 1386–1392.

Weil, R.R., and N.C. Brady. 2017. *The nature and properties of soils. New York.* New York: Pearson.

Weltz, M.A., M.R. Kidwell, and H. Dale Fox. 1998. Influence of abiotic and biotic factors in measuring and modeling soil erosion on rangelands: State of knowledge. Soil Erosion on Rangeland. *Journal of Range Management* 51: 482–495.

West, N.E., and G.F. Gifford. 1976. Rainfall interception by cool-desert shrubs. *Journal of Range Management* 29: 171–172.

Wight, J.R., and F.H. Siddoway. 1982. Determinants of soil loss tolerance for rangelands. In: *Amer. Society Agronomy Special Publ.* 4567–74.

Whitford, W.G. 1996. The importance of the biodiversity of soil biota in arid ecosystems. *Biodiversity and Conservation* 5: 185–195.

Wilkinson, B.H., and B.J. McElroy. 2007. The impact of humans on continental erosion and sedimentation. *Geological Society of America Bulletin* 119: 140–156.

Williams, J.D., J.P. Dobrowolski, and N. West. 1999. Microbiotic crust influence on unsaturated hydraulic conductivity. *Arid Soil Research and Rehabilitation* 13: 145–154.

Williams, J.D., J.P. Dobrowolski, N.E. West, and D.A. Gillette. 1995. Microphytic crust influence on wind erosion. *Transactions of the ASAE* 38: 131–137.

Williams, C.J., F.B. Pierson, K.E. Spaeth, J.R. Brown, O.Z. Al-Hamdan, M.A. Weltz OZ, M.A. Nearing, J.E. Herrick, J. Boll, P.R. Robichaud, D.C. Goodrich, P. Heilman, D.P. Guertin, M. Hernandez, H. Wei, S.P. Hardegree, E.K. Strand, J.D. Bates, L.J. Metz, and M.H. Nichols. 2016. Incorporating hydrologic data and ecohydrologic relationships into ecological site descriptions. *Rangeland Ecology and Management* 69: 4–19.

Wischmeier, W.H. 1976. Use and misuse of the universal soil loss equation. *Journal of Soil and Water Conservation* 31: 5–9.

Wischmeier, W.H., and D.D. Smith. 1965. Predicting rainfall erosion losses from cropland east of the Rocky Mountains. In *USDA handbook 282.* Washington, D.C.: U.S. Gov. Print. Office.

———. 1978. Predicting rainfall erosion losses, a guide to conservation planning. In *USDA Handbook 537.* Washington, D.C.: U.S. Gov. Print. Office.

Wood, M.K., and W.H. Blackburn. 1981. Grazing systems: Their influence on infiltration rates in the Rolling Plains of Texas. *Journal of Range Management* 34: 331–335.

Wösten, J.H.M., and M.T. Van Genuchten. 1988. Using texture and other soil properties to predict the unsaturated soil hydraulic functions. *Soil Science Society of America Journal* 52: 1762–1770.

Wösten, J.H.M., Ya.A. Pachepsky, and W.J. Rawls. 2001. Pedotransfer functions: Bridging the gap between available basic soil data and missing soil hydraulic characteristics. *Journal of Hydrology* 251: 123–150.

Young, J.A., R.A. Evans, and D.A. Easi. 1984. Stemflow on western juniper trees. *Weed Science* 32: 320–327.

Zaady, E., Y. Gutterman, and B. Boeken. 1997. The germination of mucilaginous seeds of *Plantago coronopus, Reboudia pinnata, and Carrichtera annua* on cyanobacterial soil crust from the Negev Desert. *Plant and Soil* 190: 247–252.

Zalewski, M. 2000. Ecohydrology—The scientific background to use ecosystem properties as management tools toward sustainability of water resources. *Ecological Engineering* 16: 1–8.

Zhao, P., M.A. Shao, W. Omran, and A. monem Mohamed Amer. 2011. Effects of erosion and deposition on particle size distribution of deposited farmland soils on the Chinese loess plateau. *Revista Brasileira de Ciência do Solo* 35: 2135–2144.

Zuazo, V.H.D., and C.R.R. Pleguezuelo. 2008. Soil-erosion and runoff prevention by plant covers: A review. *Agronomic Sustainable Development* 28: 65–86.

Chapter 4
Cover Crop Dynamics on Hydrology and Erosion

Abstract The number of acres of cover crops in the United States has increased by about 50% in the last 5 years. The primary reasons for planting cover crops are to provide protective vegetative cover during dormant crop periods, suppress weeds, add and scavenge plant nutrients, and improve soil health via additional organic matter. There are many benefits to planting cover crops, but there are also some negative aspects that must be considered, especially on local levels. Not all agronomic practices function the same way across the U.S. landscape. Cover crop species have special niches and attributes relative to both crop production and soil health. Thorough knowledge of cover crop dynamics in specific locales is important to maximize benefits as well as minimize negative outcomes. The objective of this chapter is to explore the relationships between cover crops (above and below ground) on hydrology and erosion dynamics. The focus on hydrology is important as it is the main ecological driver and starting point for understanding and initiating management practices that enhance soil health.

History and Agronomic Impacts of Cover Crops

Cover crops have been an important practice in agriculture for millennia (Groff 2015), primarily for prevention and degradation from wind and water erosion and improving soil fertility. The common practice of cover cropping involves planting a stand of annual plants, usually cereal grains, legumes, and/or mixtures of grains, legumes, mustards, or other herbaceous plants. Ancient farmers realized and observed that cover crops improved subsequent plant vigor and productivity. The Egyptians discovered that legumes such as pea, lentil, and clover improved or helped maintain soil fertility; around 300 BC, the Greeks realized the value of using broad beans aka fava beans (*Vicia faba*) cover crop; and the Romans planted lupines and beans (*Phaseolus* spp.) for soil improvement (Pieters and McKee 1938). In 1679, Malpighi (1628–1694, Italian biologist and physician) recognized nodules on legume roots but thought they were insect galls (Hirsch 2009). It was not until 1888 that Hermann Hellriegel in collaboration with Hermann Wilfarth discovered the true nature of legumes' ability to fix atmospheric nitrogen via organisms living in the root nodules. Also, in 1888, Martinus Beijerinck in Holland isolated and cultured these microorganisms from different legume species and discovered the nodule

© Springer Nature Switzerland AG 2020
K. E. Spaeth Jr., *Soil Health on the Farm, Ranch, and in the Garden*,
https://doi.org/10.1007/978-3-030-40398-0_4

bacteria and classified them into the genus *Rhizobium* (*rhiza* = root; *bios* = life) (Hirsch 2009).

In the eighteenth century, George Washington and Thomas Jefferson were early advocates of cover crops, and farmers in Maryland and Virginia used partridge pea (*Chamaecrista fasciculata*) as green manure. The dynamics and science behind the advantages of cover cropping and green manuring were not precisely known at that time, but the anecdotal evidence of minimizing soil loss during the dormant growing period and enhancement of subsequent crops was recognizable. The use of cover crops and green manure often appears in the literature; however, some agronomists stress that these terms are not identical by definition. Pieters and McKee (1938) state that the purpose of a cover crop is to provide protection from wind and water erosion during the winter without reference to "turning under" the crop (i.e., the crop could be grown to maturity), whereas green manuring specifically refers to cover crops that are turned into the soil to enhance nutrient exchange and adding organic matter to the soil. Before the introduction of synthetic fertilizers (primarily beginning in 1903[1]), the use of cover crops (and/or green manuring) was prevalent during the nineteenth century; however, after World War II, the use of cover crops declined and was virtually a "distant memory" by the mid-1960s (Groff 2015). Renewed interest in using cover crops began to gain momentum again in the 1990s to present. The 2017 US Agricultural Census reported that about 6 million ha (15 million ac) was planted to cover crops in the United States. The idea of growing cover crops in vegetable gardens has the same benefits as crop fields in production agriculture and has gained momentum, especially with organic growers (Rodale 2017).

Cover crops can include annual cereals (annual grasses), annual and perennial legumes, and perennial grasses and forbs (Fig. 4.1). Figure 4.2 shows most of the common cover crops, categorized as cool season/warm season crops, legume and non-legume, annual/perennial, plant architecture characteristics, and relative water use.

Advantages and Disadvantages of Cover Crops

With the resurgence of soil quality and health in popular and scientific literature, cover cropping is a prominent issue. The Sustainable Agriculture Research and Education (SARE) organization states that in the last 5 years, there is more enthusiasm about the use of cover crops. They found that cover crops are normally planted during dormant periods to provide ground cover and weed suppression, add and

---

[1] The German chemists Fritz Haber and Carl Bosch work culminated in what is commonly referred to as the Haber-Bosch process (work done between 1904 and 1913) where atmospheric nitrogen (N_2) is converted to ammonia (NH_3) in a reaction with hydrogen (H_2) using a metal catalyst under high temperatures and pressures (Russel and Williams 1977).

Text Box 4.1 Fact sheet for three common cover crops

Brassica (Raphanus sativus)

- Cool season plant, but winter-kills
- Rapid root growth 2 mm day^{-1} °C^{-1} (Thorup-Kristensen 2001)
- Deep tap root (2m depth @ 4 mo) (Thorup-Kristensen et al. 2009)
- Brassicas penetrate compacted soil more readily than grasses (Chen and Weil 2010)
- Greater air permeability than cereal rye and no cover crops (Chen et al. 2014)
- Macroporosity increases, large tap roots leave many large soil macropores for water infiltration in spring. Less runoff (Weil et al. 2009; White and Weil 2011; Lounsbury and Weil 2015)
- Capture nitrate deeper in soil profile, transport nutrients to soil surface <C >N higher ratio of N to C (SARE 2012)
- Nitrate N (NO$_3$-N) > than grasses (Thorup-Kristensen 2001)
- Study: Less NO$_3$ conc. in groundwater 1.5 µg nitrate L^{-1} (Thorup-Kristensen 2001)
- Biofumigation potential (Chew 1988; Haramoto and Gallandt 2004; Wick et al. 2017)
- Nonmychorrhizal, increase P uptake by reducing rhizosphere pH with organic acid extrusion from roots (Hedley et al. 1982; Hoffland et al. 1989)
- Mycorrhizal colonization of corn roots can be retarded after forage radish, but equivalent to no cover crop in most years (White and Weil 2010)
- Greater potential to scavenge residual soil NO$_3$-N than legume cover crops (Nielsen and Jensen 1985; Groffman et al. 1987; Wagger et al. 1998; Isse et al. 1999). Cause may be due to faster root growth (Sainju et al. 1998)
- Soybean yields were significantly greater following forge radish/ winter rye cover crop (Williams and Weil 2004)

Winter wheat (Triticum aestivum)

- 2× rooting depth than spring wheat growth in fall and throughout winter. Increased N uptake and leaching losses (Thorup et al. 2009)
- < Bulk density, > aggregate stability, > infiltration < runoff (Villamil et al. 2006; Steele et al. 2012; Wick et al. 2017)
- Effective at reducing compaction, biodrilling (Cresswell and Kirkegaard 1995)
- >AMF potential for subsequent crops (Boswell et al. 1998; Kabir and Koide 2002)
- Reduce wind and water erosion (Dabney et al. 2001)

Cereal rye (Secale cereale)

- Winter hardy, one of the most cold-tolerant plants, easy to establish, productive, reaches inflorescence faster than other cereal cover crops in temperate climates. Best where fertility is low (Pieters and McKee 1938; Dabney et al. 2001)
- Widely used cover crop (Singer et al. 2007)
- Rapid growth, fibrous root sys., good root penetration in compacted soils, > macropores, soil moisture potential (Villamil et al. 2006)
- >Propagule formation and biomass of AMF benefitting corn crop than no cover crop (Boswell et al. 1998; Kabir and Koide 2002; White and Weil 2010; Lehman et al. 2012)
- Reduce wind and water erosion (Dabney et al. 2001)
- < Bulk density (7% dec Villamil et al. 2006), > aggregate stability, > infil. < runoff (Villamil et al. 2006; Steele et al. 2012; Wick et al. 2017)

Cereal rye cont.

Caution: Broad potential of increasing pathogen densities in soil (Fusarium graminearum, F. oxysporum, Pythium sylvaticum, P. torulosum) (Bakker et al. 2016). Residue associated with wetter and cooler soil temperatures, host to root-rot pathogens could infect young corn seedlings. Short intervals between killing rye and planting corn increase risk. Longer intervals of terminating rye (10–14 days) before planting corn reduce chances of yield loss. The 2-week interval can allow for soils to warm up and reduce susceptibility of corn seedlings

Fig. 4.1 (**a**) Rye cover crop. (Courtesy of USDA-ARS, Steven Ausmus); (**b**) Angus calves in cover crop of pearl millet. (Courtesy of USDA-ARS, Steven Knapp); (**c**) cover crop mixture that includes oat, proso millet, canola, sunflower, dry pea, soybean, and Pasja turnip. (Courtesy of USDA-ARS, Mark Liebig); (**d**) wheat cover crop (foreground) planted around young apple trees (background). (Courtesy of USDA-ARS, Brian Prechtel); and (**e**) cover crop of turnips and radishes in a garden setting (on right, photo by author)

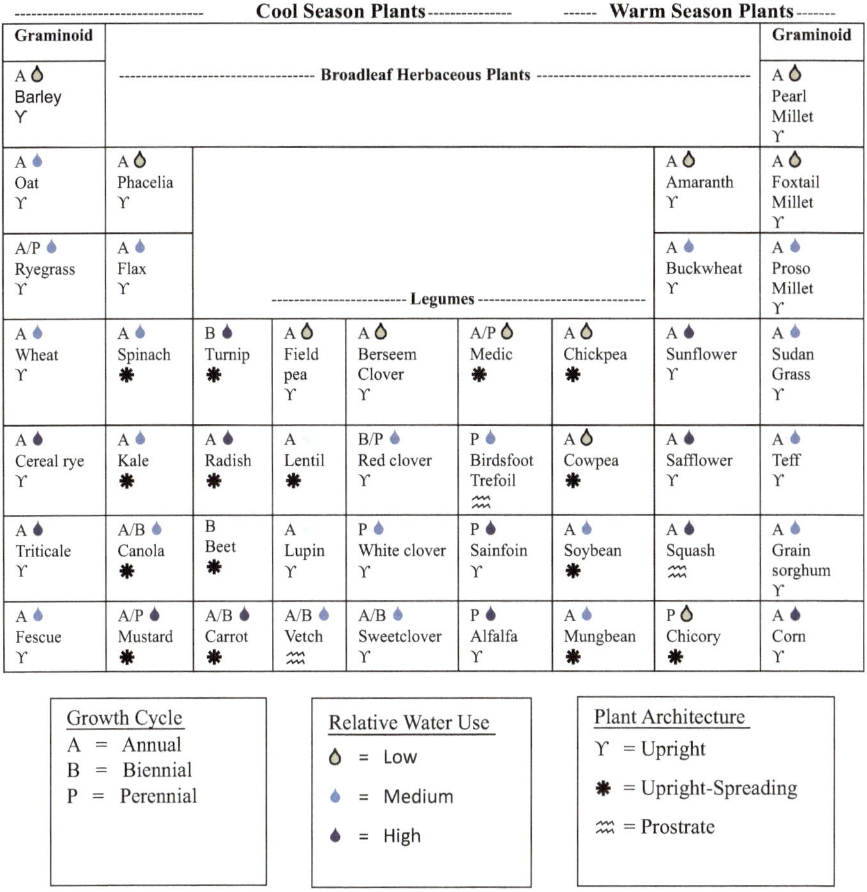

Graminoid	Cool Season Plants							Graminoid
A ◊ Barley Υ	Broadleaf Herbaceous Plants							A ◊ Pearl Millet Υ
A ● Oat Υ	A ◊ Phacelia Υ						A ◊ Amaranth Υ	A ◊ Foxtail Millet Υ
A/P ● Ryegrass Υ	A ● Flax Υ			Legumes			A ● Buckwheat Υ	A ● Proso Millet Υ
A ● Wheat Υ	A ● Spinach ✱	B ● Turnip ✱	A ◊ Field pea Υ	A ◊ Berseem Clover Υ	A/P ◊ Medic ✱	A ◊ Chickpea ✱	A ● Sunflower Υ	A ● Sudan Grass Υ
A ● Cereal rye Υ	A ● Kale ✱	A ● Radish ✱	A Lentil ✱	B/P ● Red clover Υ	P ● Birdsfoot Trefoil 〰	A ◊ Cowpea ✱	A ● Safflower Υ	A ● Teff Υ
A ● Triticale Υ	A/B ● Canola ✱	B Beet ✱	A Lupin Υ	P ● White clover Υ	P ● Sainfoin Υ	A ● Soybean ✱	A ● Squash 〰	A ● Grain sorghum Υ
A ● Fescue Υ	A/P ● Mustard ✱	A/B ● Carrot ✱	A/B ● Vetch 〰	A/B ● Sweetclover Υ	P ● Alfalfa Υ	A ● Mungbean ✱	P ◊ Chicory ✱	A ● Corn Υ

Growth Cycle	Relative Water Use	Plant Architecture
A = Annual	◊ = Low	Υ = Upright
B = Biennial	◐ = Medium	✱ = Upright-Spreading
P = Perennial	● = High	〰 = Prostrate

Fig. 4.2 Cover crop plant choices and inherent characteristics. (Adapted from USDA-ARS Northern Great Plains Research Laboratory 2018)

scavenge nutrients, and improve soil health. The 2017 USDA Agricultural Census reported that the number of non-Conservation Reserve Program (CRP) acres had increased significantly over the past 5 years. In 2012, about 10 million acres on ~133,500 farms used cover crops; in 2017, the number of acres increased to over 15 million acres on ~153,400 farms—a 50% increase in 5 years.

There are many benefits to cover cropping, but as in life and nature, there are always issues that challenge us. The use of cover crops has many advantages, but there are negatives associated as well. Using cover crops in cropping systems requires pre-planning. The advantages and disadvantages need to be considered: contributions to soil health; agronomic dynamics—before and after planting the main crop, interactions with various cropping systems (conventional, reduced tillage, and no-till methods); economics; soil building, pros and cons associated with pests; and environmental factors (soils, climatic factors).

The benefits of cover crops are numerous:

- Vegetative protective cover—reduce and manage excess runoff and soil loss during dormant crop periods.
- Help protect water quality to nearby watercourses.
- Reduce wind erosion.
- Improve air quality.
- Enhance surface soil physical properties such as aggregate stability.
- Suppress and control weeds.
- Enhance and help maintain soil organic matter in conjunction with other practices such as residue management, additions of manure, and compost.
- Reduce losses from applied nutrients and pesticides.
- Offset soil organic carbon losses from residue and stover removal from field.
- Initiate the slow process of sequestering organic carbon lost through cultivation over time.
- Legumes can supply N to the soil.
- Enhance microbial activity in the soil.
- In some cases, suppress soil-borne pathogens and nematodes in the soil.
- Increase arbuscular mycorrhizal fungi (AMF) with the following benefits: enhance nutrient uptake by plants (especially phosphorus), provide protection for the host plant against pathogenic organisms, increase some environmental tolerance to drought, and provide an organic matter sink from plant to soil.
- Offset fertilizer (nutrient) applications to the field during dormant seasons where nutrient losses due to runoff may occur.
- Dead stubble can act as a "nurse crop" or cover for perennial seedlings such as native grasses.
- Wildlife enhancement.

There can be disadvantages to growing cover crops; some farmers ask: "If cover crops are so great, why isn't everyone planting them"? The reason may not be related to soil health improvement but may be due to extra financial outlay, additional equipment needed, climatic inconsistencies with establishment and termination of the crop, concerns over water usage by the cover crop, and fostering potential diseases which may affect production of the succeeding crop. Hoorman and Islam (2009) points out timing of terminating a cover crop can be critical; however, in the long term, cover crops can increase soil organic matter and improve infiltration capacity and water storage—the initial decline in available water is offset by enhanced water balance, especially if residue management is practiced (Hoorman and Islam 2009).

Several authors provide specific caveats to growing cover crops, e.g., where soils remain cool and wet in spring, and especially in semiarid dryland farming where moisture balance may be diminished for subsequent crops (Unger and Vigil 1998;

Lu et al. 2000; Dabney et al. 2001; Bakker et al. 2016). Some negative aspects[2] reported by farmers are cost and time involved in establishing and terminating the cover crop; lack of species winter hardiness; in some cases increased risk of some disease proliferation by host organisms, insect, and other pests; and allelopathy[3] (Lu et al. 2000; Dabney et al. 2001). Various studies have demonstrated instances of lower production yields in soybeans, corn, and cotton following cover crop establishment (Zhu et al. 1989; Johnson et al. 1998; Dinnes et al. 2002; Kaspar and Bakker 2015; Pantoja et al. 2015; Bakker et al. 2016; Acharya et al. 2017). In Iowa corn plantings, Acharya et al. (2017) found that short intervals between winter rye cover crop termination and corn seedling emergence resulted in increased corn seedling diseases (*Pythium* spp.), reduced seedling density, and shoot growth. To minimize corn yield production, they recommend a 10-to-14 day interval between winter rye termination and corn planting. Variances in grain crop yield following legumes can be greater compared to no cover crops (Franklin et al. 1989; Ott and Hargrove 1989); however, risks can be offset by using an adapted species and establishing the legume crop early to improve winter survival (Bowen et al. 1991). Several important factors are recommended to reduce seedling stand loss, poor growth, and yield: (1) terminate and desiccate cover crop 2–3 weeks prior to planting; (2) obtain good seed-to-soil contact; (3) ensure proper depth placement of seed; and (4) ensure that crop residue does not interfere with planting seed and inhibit seedling germination and establishment. In summary, Hoorman and Islam (2009) stress that: "every cover crop species has its own niche and attributes for agricultural production. A wrong combination of cover crops may exert negative attributes, so a thorough understanding of cover crops selection and management is needed to minimize negative outcomes."

Cover Crop Benefits to Hydrology and Erosion

One of the fundamental advantages of planting cover crops is providing vegetative cover to control excess runoff and soil erosion during dormant or fallow periods. The scientific literature is replete with examples of how cover crops can reduce runoff, enhance infiltration capacity, and reduce soil water erosion (Brill and Neal 1950; Hall et al. 1984; Zhu et al. 1989; Reeves 1994; Meyer et al. 1997; Dabney 1998; Unger and Vigil 1998; Dabney et al. 2001; Hartwig and Ammon 2002; Nearing et al. 2005; Zuazo and Pleguezuelo 2008, De Baets et al. 2011; Kaspar and Singer 2011). Often, runoff and erosion are correlated; however, in some situations,

---

[2] Often listed as disadvantages by farmers and in the literature, many of these are more precisely inconveniences or economic considerations.

[3] Allelopathy, the chemical inhibition by plants, is where a compound is released or exuded into the soil environment, thereby affecting the growth of different neighboring competing plants and/or plants of the same species.

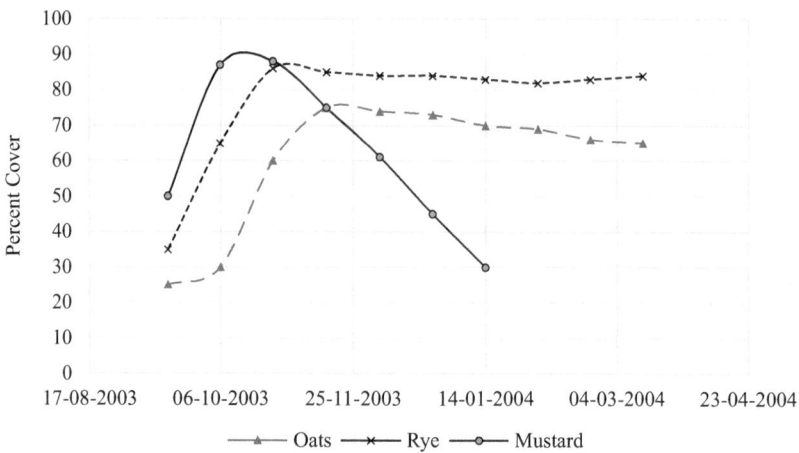

Fig. 4.3 Progression of plant cover crops during the growing season. (Adapted from MESAM 2007; De Baets et al. 2011)

runoff occurs without significant erosion, e.g., where permanent grass cover, or well-established legume crops. In cultivated fields, above average storms with more intensive rainfall (either short duration or sustained rainfall for several days) can result in substantial runoff, accelerated overland flow, and erosion, especially on fields with steeper slopes that are not terraced.

De Baets et al. (2011) evaluated the erosion reduction potential of various cover crops [white mustard (*Sinapis alba*), ryegrass (*Lolium perenne*), oats (*Avena sativa*), cereal rye (*Secale cereale*), and fodder radish (*Raphanus sativus* subsp. *oleiferus*)] at sites in Belgium. White mustard provided 90% cover after seeding (1.5 months); however, winter-kill occurred, and cover values decreased throughout the winter months. Oats and rye developed cover more slowly than mustard; however, cover remained steady throughout the winter months (Fig. 4.3).

Figure 4.4a, b show the relationship of three variables that are related to erosion protection: stem density, vegetation cover, and erosion-reducing potential of plant roots. Aboveground biomass of oats, cereal rye, and ryegrass developed slower that herbaceous cover crops like phacelia and mustard, but they provide denser cover (cover ~80%) during the critical winter period (Fig.4.4) (De Baets et al. 2011). Mustard is susceptible to frost and dies off during winter. Oats also winter-kill at temperatures lower than (−15 °C, 5 °F). Fibrous root species such as ryegrass, cereal rye, and oats have high potential to control concentrated flow erosion, while white mustard had good potential, and radish demonstrated the lowest potential to protect against erosion (Fig. 4.4a, b) (De Baets et al. 2011). Herbaceous forb cover crops such as white mustard (*Sinapis alba*), radish (*Raphanus sativus*) cv. Defender, and rape (*Brassica napus*) cv. Dwarf Essex may in certain crops have other advantages, such as reductions in certain nematodes and disease (Opperman et al. 1988;

Abawi and Widmer 2000; Wang et al. 2004). Again, there is no "magic bullet," as some studies show that cover cropping may have little effect on plant-parasitic nematode populations (Geary et al. 2008; Grabau and Chen 2016) or may increase nematode populations as the plants may be a host for nematodes that are already present (legume winter cover crops where fields are infested with *Meloidogyne incognita*) (Mercer and Miller 1997; Timper et al. 2006).

In a study by Zhu et al. (1989), four different winter cover crop treatments were evaluated with subsequent spring no-till soybean plantings in Missouri on a Mexico silt loam: no planted cover crop—volunteer weedy species, chickweed (*Stellaria media*), Canadian bluegrass (*Poa compressa*), and downy brome (*Bromus tectorum*). The volunteer no-planted cover crop and planted cover crops provided 71, 92, 99, and 99% cover during the critical erosion period. Canada bluegrass, a perennial grass species, and downy brome, a winter annual grass, provided the greatest winter cover. The planted winter cover crops provided significantly greater soil cover (30–50%) during the critical erosion period and late spring to early summer. Soil loss and runoff were reduced with the three cover crops compared to the control treatment of volunteer weeds (Fig. 4.5a, b). Percent reductions of soil loss and

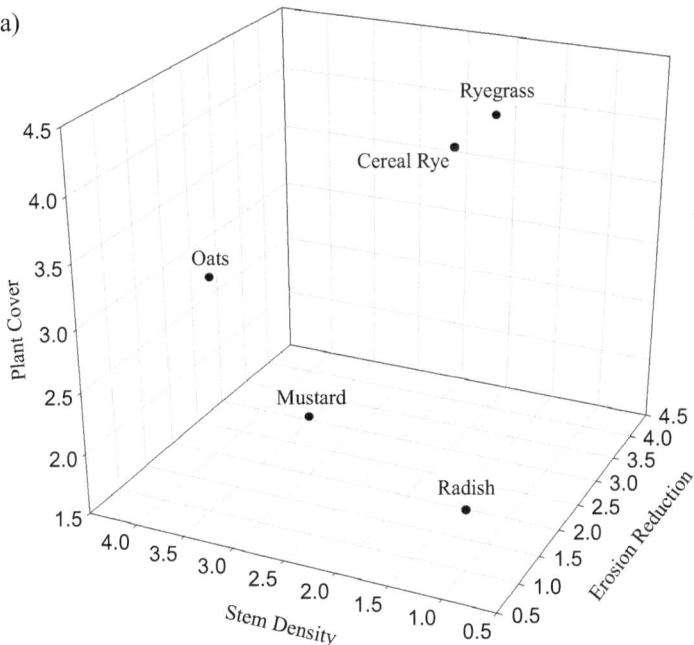

Fig. 4.4 (a) Suitability of five cover crops for controlling concentrated flow erosion. Units are scores (0–4 are dimensionless, 4 having the highest benefit) for the indicators: SD stem density, ER erosion-reducing potential of plant roots, and C plant cover. (b) Pyramid diagram showing three-dimensional relationship between SD, ER, and C. The higher the volume in the pyramids, the greater the concentrated erosion reduction potential of the various cover crops. (Adapted data from De Baets et al. 2011, MESAM. 2007)

b)

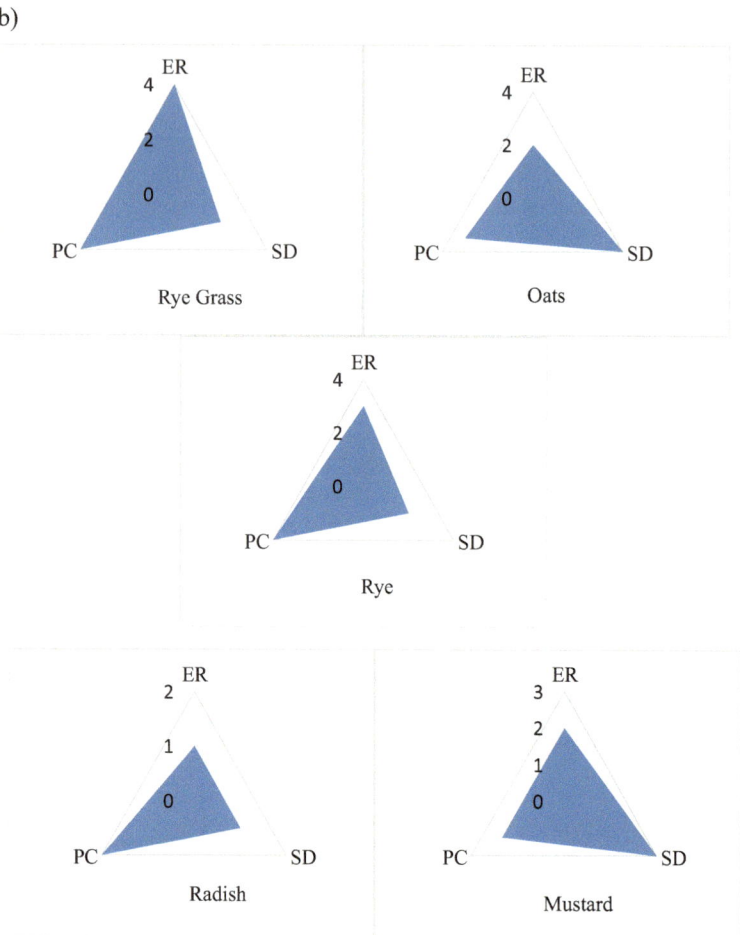

Fig. 4.4 (continued)

runoff were chickweed (87%, 44%), downy brome (95%, 53%), and Canada blue-grass (96%, 45%), respectively. Nutrients were more concentrated in the cover crop runoff (1.6–7.7×) due to less runoff volume; however, total nutrient lost was higher in the volunteer weed cover (Table 4.1). The overall average reduction in nutrient losses was 7–77% in the planted cover crop treatments. The downside to these cover crop treatments was reduced no-till soybean yield with the cover crops (Zhu et al. 1989).

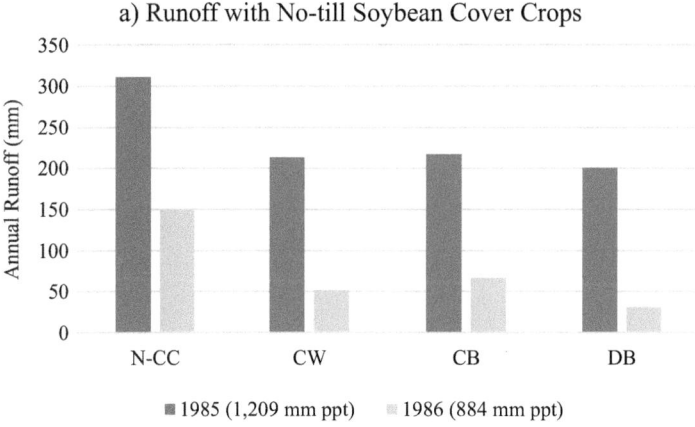

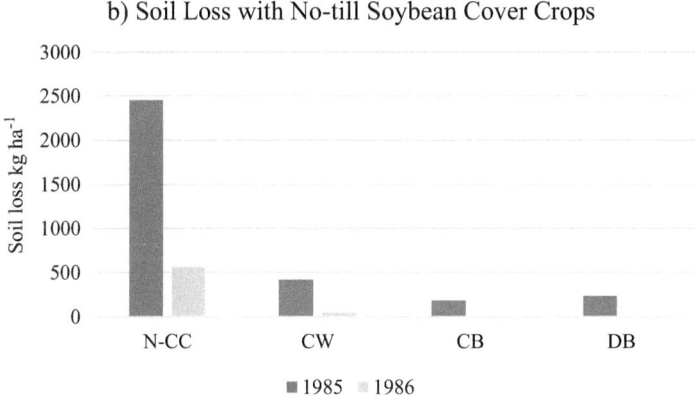

Fig. 4.5 (**a**) Runoff losses from no-till soybeans with cover crop treatments, (**b**) soil losses associated with cover crop treatments. N-CC volunteer no cover crop-weedy species; CW chickweed (*Stellaria media*), CB Canadian bluegrass (*Poa compressa*), and DB downy brome (*Bromus tectorum*). (Data from Zhu et al. 1989)

Table 4.1 Dissolved nutrients in runoff

Cover type	$NO_3^- $ -N	$NH_4^+ $ -N	$PO_4^{3-} $ -P
kg ha yr^{-1}			
Chickweed	0.77	0.0.11	0.17
Downy brome	0.84	0.11	0.27
Canada bluegrass	0.88	0.10	0.43
Volunteer weeds	3.36	0.17	0.46

Data from Zhu et al. 1989

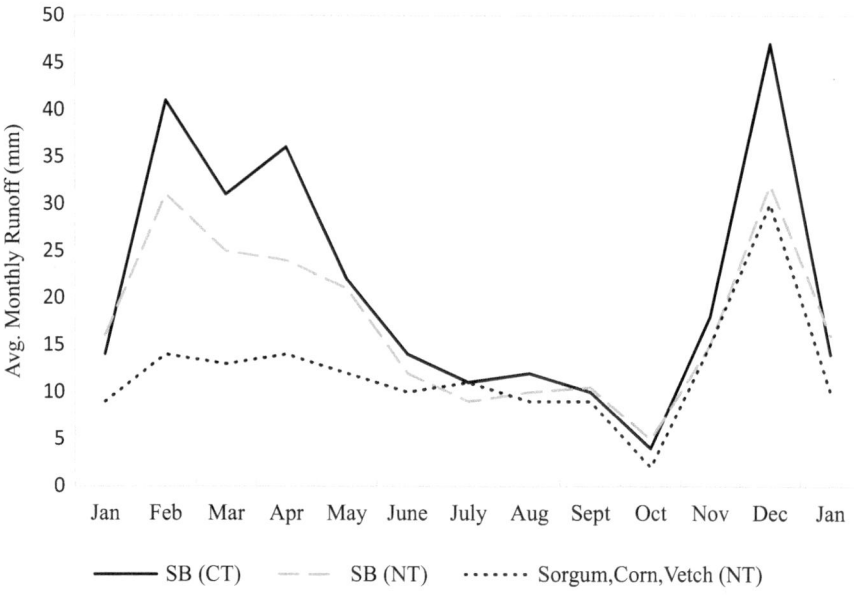

Fig. 4.6 Example of monthly runoff values with soybeans conventional tillage (SB, CT), soybeans no-till (SB-NT), and no-till sorghum/corn with winter cover crop of hairy vetch in north Mississippi (loess soil 2–6% slopes) (Meyer et al. 1997). Runoff was highest in conventionally planted soybeans, while spring runoff in no-till soybeans was about 10 mm lower in February to March

On a loess soil in northern Mississippi (2–6% slopes), after harvest of conventionally planted soybeans, runoff was highest during the winter months (February to April) compared to no-till soybeans and no-till sorghum/corn with hairy vetch as a winter cover crop (Meyer et al. 1997) (Fig. 4.6). Buyanovsky and Wagner (1986) found greater erosion losses in soybeans compared to corn due to greater soil loosening of soybean roots and less residue cover. In a 7-year study from natural rainfall plots in Iowa, Laflen and Moldenhauer (1979) found that annual soil loss was greater than corn following soybean rotation compared to soybean following corn or corn-corn rotation.

Hall et al. (1984) showed the benefit of legume cover crops to reduce water runoff and soil erosion with the added benefit of increased productivity in corn systems. They examined no-till corn planted into two legume cover crops bird's-foot trefoil (*Lotus corniculatus*) and crownvetch (*Coronilla varia*), no-till corn into stover, and conventionally planted corn. Runoff on the conventional tillage ranged from 5.5% to 22% of annual rainfall; soil loss ranged from 4.4 to 32.2 Mg ha^{-1} (2–14.4 t ac^{-1}). The no-till treatments significantly reduced runoff, erosion, and herbicide movement where runoff on the no-till plantings ranged from 0.07% to 2.5% of annual rainfall and soil loss ranged from 0 to 1.1 Mg ha^{-1} (0–0.5 t ac^{-1}).

In a cover crop study with winter rye (*Secale cereale*) in New Jersey, USA, runoff was lower for every month except February where the bare ground for no-rye

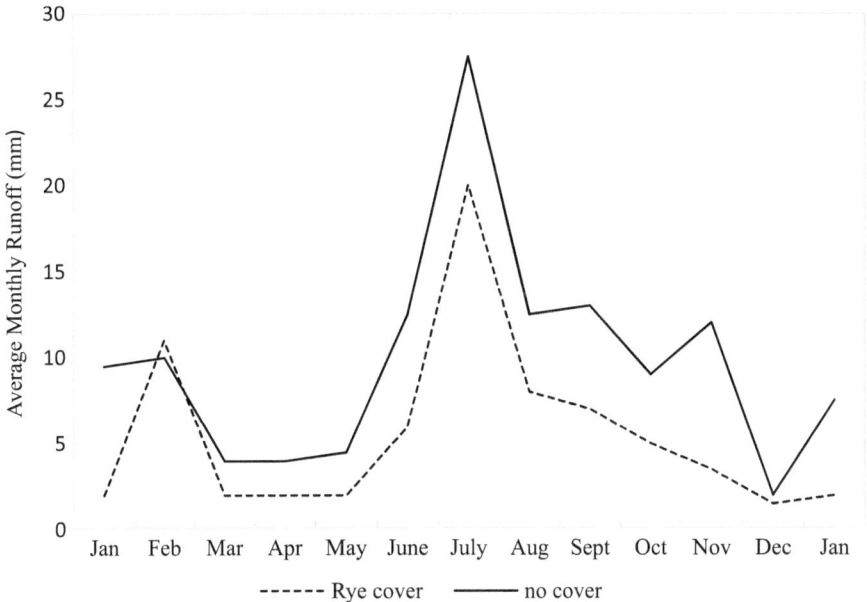

Fig. 4.7 Average monthly runoff with winter cover crop and no plant cover in New Jersey. (Data from Brill and Neal 1950)

cover treatment absorbed more water during a thaw period compared to frozen soils with the rye cover crop (Brill and Neal 1950). Erosion rates followed a similar pattern with runoff; however, during February, neither the rye nor no-rye treatment produced any sediment loss (Fig. 4.7).

In summary, cover crops can significantly reduce wind and water erosion on cropland, which can be a source of soil organic carbon loss (see Chap. 6). Over time, cover crops can enhance soil organic carbon, both at the soil surface and subsoil layer (see Chap. 6 on the role of microbes in subsoil organic carbon development) resulting in improved both hydrologic function. When hydrologic function is improved, infiltration capacity is increased as organic matter and microbial activity are the precursors to improving soil aggregate stability, and runoff and soil erosion are decreased. Cover crops can also be effective in scavenging and recycling residual soil NO_3-N from the previous crop and, upon decomposition, potentially recycle it to following crops, thus eliminating nutrient losses between crops. Legume cover crops and legume/cereal combinations also scavenge residual nutrients—the added benefit of legumes is the capture of atmospheric nitrogen in nitrogen-limited soil environments. If inorganic nitrogen is abundant, earlier planting of the subsequent crop results in more nitrogen usage than later plantings.

Cover crops are also effective in helping to control difficult and perpetual weed problems by disrupting and competing with current weed cycles. An added benefit may be less dependence and usage of herbicides and other weed control methods. Increasing subsequent crop diseases with the use of certain cover crops, especially plants closely related to the crop itself, can be offset with timing of cover crop termination. The problem of disease proliferation and transference from the cover crop to the production crop is a matter that must be carefully considered. In growing areas where there is a history of associated disease with cover crops, it is best to consult with experts in the area. Often, agronomists recommend terminating the cover crop at least 3 weeks in advance prior to planting the new crop. Mycorrhizal fungal associations with the cover crop may be a benefit in helping reduce crop disease of the subsequent crops. Disease problems, with or without the use of cover crops, can be highly variable and especially related to local conditions. It is wise to conduct prior research to establish cover crop benefits or disadvantages in specific locales. Usually choosing the right cover crop and timing planting and termination can overcome potential problems for the following crop.

Rooting Dynamics and Cover Crops

The protective capacity of surface vegetative cover is essential to reducing excess runoff and soil loss from erosion; however, belowground rooting characteristics of various common cover crops are also important to hydrologic dynamics. Many grasses and graminoid cereal crops have fibrous root systems that can enhance soil aggregation and mediate soil compaction of the soil surface layer. Legumes such as peas, and vetch have tap roots that can extend deeper into the soil profile and can help loosen compacted soils and in addition transport leaching nutrients from the subsoil to the upper soil surface where the majority of crop roots exist. Figure 4.8 shows average rooting depth for various plant life forms (Canadell et al. 1996). Crop rotations that utilize legumes for several years have the advantage of fixing

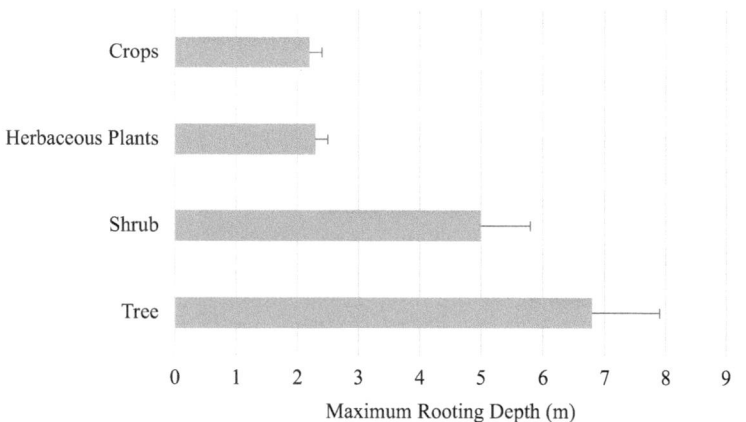

Fig. 4.8 Average maximum rooting depths (standard deviation bars) for major plant life forms (trees, shrubs, herbaceous plants, and crops). (Data from Canadell et al. 1996)

atmospheric nitrogen (N_2) and providing N to crops. Rhizobial bacteria (dominant genera—*Rhizobium*, *Mesorhizobium*, *Bradyrhizobium*, and *Ensifer*) eventually induce the formation of leguminous root nodules that are the active sites for nitrogen fixation in legumes. Plants growing in association with legumes can obtain nitrogen-rich compounds.

The growth and development of plants due to acquisition of soil nutrients and water are correlated with root morphology and physiology (Ju et al. 2015). In the early 1920s–1930s, Weaver (1919, 1926) and Weaver and Bruner (1927) provided detailed information with diagrams and photographs of root systems showing different morphological characteristics of graminoids and forbs. In develping hydrology models, on rangeland, root biomass was not highly correlated with infiltration capacity, but root morphology (differentiating fibrous and tap root classes) was characterized as a qualitative variable, significantly improving infiltration regression equations (Spaeth et al. 1996a, b). Plant roots and their distribution in the soil profile are important in ecosystem carbon balance and carbon inputs to the soil. The effective rooting zone is the depth of roots where the majority of roots are located in the soil profile. Effective rooting zone for wheat, maize, barley, and canola is estimated at ~50–100 cm, ~60–70 cm for peas, and ~50–100 cm for alfalfa (Fan et al. 2016). Rooting depths are highly variable; however, there are some amazing examples of plant rooting capabilities, e.g., alfalfa roots have been found in mineshafts at 39 m (128 ft), although on the average, the majority of alfalfa roots are found at 1.5–1.8 m (5–6 ft); mature apple tree roots can penetrate to a depth of at least 9 m (30 ft); and a 100-year-old pine was estimated to have 499 km (310 miles) of roots with 5 million root tips. Plants with deep rooting habits are commonly found in arid environments with long dry periods, e.g., Southwestern United States, where some species [e.g., mesquite *(Prosopis glandulosa)*] tap water layers at 53 m (173 ft) (Phillips 1963). In dry savannah of the Central Kalahari, shrub and tree roots have extended as deep as 140 m (459 ft.) (Jennings 1974).

Fan et al. (2016) examined root distribution with soil depth for a variety of graminoid crop and broadleaf plant species. Ranking of cumulative root distribution of graminoids (high-low) was as follows: fescue, barley, oats, maize, and wheat. Even though cumulative root distribution was lowest for wheat, rooting depth was greatest (Fig. 4.9a). The ranking of herbaceous broadleaf plants (high-low): canola, pea, soybeans, and alfalfa. Canola, pea, and soybean cumulative root distribution were very similar compared to alfalfa, which had the lowest cumulative root distribution (Fig. 4.9b).

Yamaguchi and Tanaka (1990) evaluated various root system traits of crop species (Fig. 4.10a, b). Rice had the shallowest root system (90% of the roots were in the upper 23 cm). For potato and soybeans, 90% of the root system extended 35–38 cm, 48–51 cm for wheat, and 59 cm for maize.

De Baets et al. (2011) characterized rooting morphology and rooting densities. Fodder radish and white mustard have large tap roots with lateral roots. Rye, ryegrass, and oats have an extensive fine fibrous root system, although oat roots were thicker, but less branching. Phacelia had small shallow tap roots with extensive lateral roots. About 99% of the roots comprised 10–40% of total plant biomass and predominantly located in the upper 30 cm of the soil surface (Fig. 4.11). Root

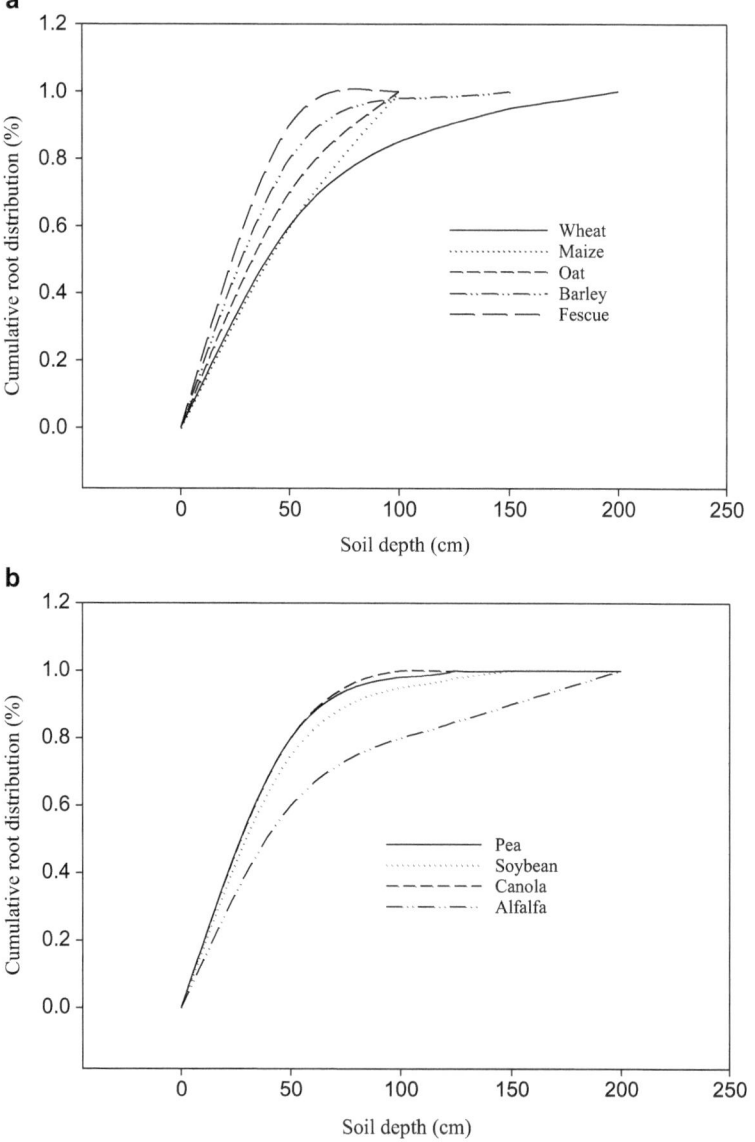

Fig. 4.9 (**a**) Cumulative root distribution with soil depth for wheat, maize, oat, barley, and fescue; (**b**) Pea, soybean, canola, and alfalfa. (Adapted from Fan et al. 2016)

densities decrease with soil depth, rye and ryegrass had the largest root densities, and phacelia and white mustard had the lowest root densities (Fig. 4.11). Tap-rooted plants (white mustard, phacelia, and radish) had less erosion-reducing potential compared to fibrous rooted plants (rye, ryegrass, oats). Radish, white mustard, and phacelia are vulnerable to winter-kill after frost and aboveground biomass and cover decrease significantly, while roots begin to decay and provided less erosion-reducing

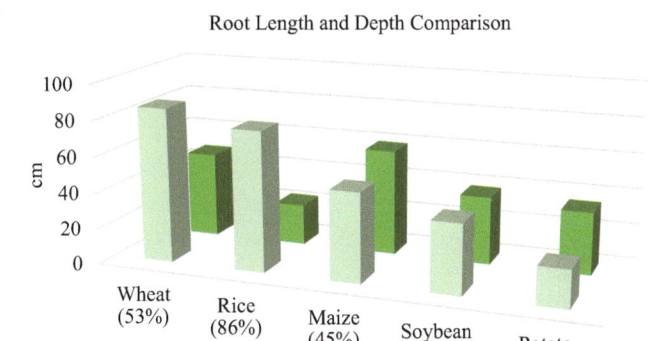

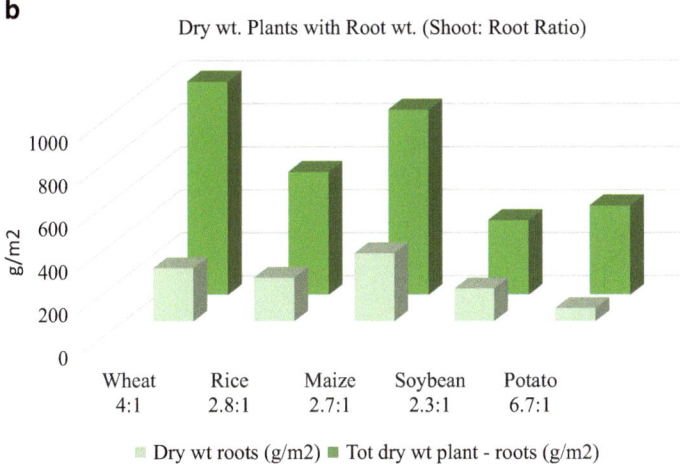

Fig. 4.10 (**a**) Root length and depth in the soil comparisons of wheat, rice, maize, soybean, and potato. Percentage of roots distributed in the top 20 cm soil is displayed. (**b**) Comparison of shoot-to-root ratios. (Adapted from Yamaguchi and Tanaka 1990)

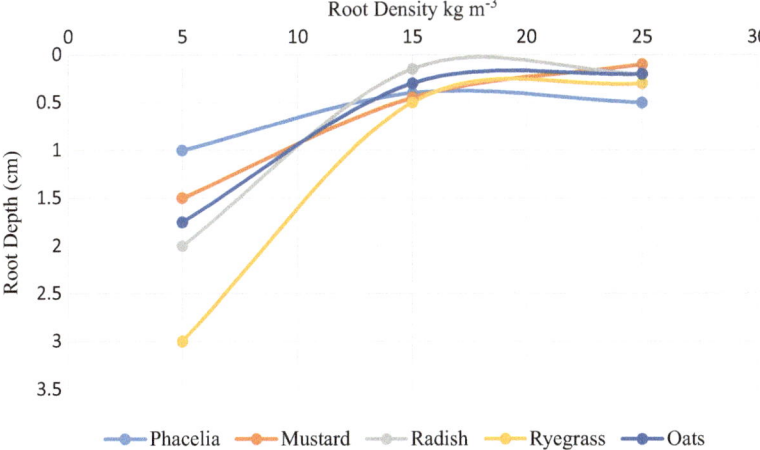

Fig. 4.11 Root density (kg m³) and distribution with soil depth for five cover crop plants (116 days after sowing). (Data adapted from De Baets et al. 2011)

capacity to the soil, compared to winter-resistant graminoids. In Fig. 4.4a, the pyramidal diagrams show that ryegrass, rye, oats, and white mustard were the most effective for reducing concentrated flow erosion.

Text Box 4.2 Water Erosion Prediction Project (WEPP) Model Runs for Iowa Corn Scenarios

Four different scenarios are modeled with USDA-ARS WEPP model: (1) fall plow corn 3 % slope, (2) fall plow corn 5% slope, (3) corn with spring disk cultivation after no-till rye cover crop 3% slope, and (4) corn no-till corn with no-till rye cover crop 3% slope. In the model output table, the minimum and maximum values are based on extremes that could occur during a 100- year period. Even though the long-term average values for soil loss and runoff may be low, over 100 years, there are storms that produce significantly more precipitation than the long-term average, which would produce significantly higher runoff and erosion rates. It is during these more intense storm events that runoff and erosion may cause long-lasting permanent damage to a field. Thus the need for constant vigilance and the advantage of using cover crops and maintaining crop stubble via reduced or no-till practices become strikingly evident.

Treatment I: Corn Grain, Fall Plow 3% Slope

Location/climate	Latitude	Longitude	Slope length (horiz) (ft)	Slope shape	Avg. slope steepness, %	Aspect
Iowa—Webster County	42.43	−94.18	200.0	Uniform	3	180

Soil Nicollet-Guckeen complex, 1–3% slopes Nicollet (5T) *Soil T value*: 5 (ton/A/yr)

Date	Operation	Vegetation	Surf. res. cov. after op, %	STIR	Fuel (gal/ac)
Apr 21, 01	Disk, tandem light finishing		9.45	19.5	0.4
Apr 28, 01	Cultivator, field 6–12 in sweeps		20.98	26	0.74
May 01, 01	Planter, double disk opnr	Corn, grain, seed	21.93	2.44	0.44
May 03, 01	Sprayer, pre-emergence		26.03	0.15	0.13
Jun 07, 01	Sprayer, postemergence, fertilizer tank mix		26.61	0.15	0.13
Oct 20, 01	Harvest, killing crop 50 pct standing stubble		88.86	0.15	1.53
Nov 01, 01	Fert applic. surface broadcast		88.39	0.06	0.16
Nov 03, 01	Plow, moldboard, 6 to 12 inch deep		11.46	65	1.87

Management	Vegetation	Yield units	# yield units, #/ac	Target yield	Yield % moisture	%Yield attained	Calibrated?
Corn grain; FP z4	Corn, grain, seed	bu/ac	110.1	112.0	16.1%	98.1	Yes

Contouring	Strips/barriers
None	None

Model output	Mean	Median	Stddev	COV	Min	Max
Precipitation (in/yr)	32.72	31.80	5.42	0.17	21.43	49.03
Sediment delivery (t/yr)	0.52	0.38	0.39	0.75	0.01	1.87
Soil loss (ton/ac/yr)	2.26	1.66	1.69	0.75	0.02	8.14
Runoff (in/yr)	5.88	5.47	2.45	0.42	1.37	13.16
Plant transpiration (in/yr)	16.82	16.64	2.49	0.15	11.73	23.09
Soil evaporation (in/yr)	9.45	9.46	1.36	0.14	6.18	14.36

SCI and STIR Output

Soil Conditioning Index (SCI)	SCI OM subfactor	SCI FO subfactor	SCI ER subfactor	Avg. annual slope STIR
0.41	1.08	−0.12	0.11	113.45

The *SCI* is the *Soil Conditioning Index* rating. If the calculated index is a negative value, soil organic matter levels are predicted to decline under that production system. If the index is a positive value, soil organic matter levels are predicted to increase under that system.

The *STIR* value is the *Soil Tillage Intensity Rating*. It utilizes the speed, depth, surface disturbance percent, and tillage type parameters to calculate a tillage intensity rating for the system used in growing a crop or a rotation. STIR ratings tend to show the differences in the degree of soil disturbance between systems. The kind, severity, and number of ground-disturbing passes are evaluated for the entire cropping rotation as shown in the management description.

Treatment II: Corn Grain, Fall Plot 5% Slope

Location/climate	Latitude	Longitude	Slope length (horiz) (ft)	Slope shape	Avg. slope steepness, %	Aspect
Iowa—Webster County	42.43	−94.18	200.0	Uniform	5	180

Soil Nicollet-Guckeen complex, 1–3% slopes Nicollet *Soil T value*: 5 (ton/A/yr)

Date	Operation	Vegetation	Surf. res. cov. after op, %	STIR	Fuel (gal/ac)
Apr 21, 01	Disk, tandem light finishing		9.53	19.5	0.4
Apr 28, 01	Cultivator, field 6–12 in sweeps		21.15	26	0.74
May 01, 01	Planter, double disk opnr	Corn, grain, seed	22.12	2.44	0.44
May 03, 01	Sprayer, pre-emergence		26.18	0.15	0.13
Jun 07, 01	Sprayer, postemergence, fertilizer tank mix		26.72	0.15	0.13
Oct 20, 01	Harvest, killing crop 50 pct standing stubble		89.08	0.15	1.53
Nov 01, 01	Fert applic. surface broadcast		88.61	0.06	0.16
Nov 03, 01	Plow, moldboard, 6–12 inch deep		11.56	65	1.87

Management	Vegetation	Yield units	# yield units, #/ac	Target yield	Yield % moisture	%Yield attained	Calibrated?
Corn grain; FP z4	Corn, grain, seed	bu/ac	110.1	112.0	16.1%	98.1	Yes

Model output	Mean	Median	Stddev	COV	Min	Max
Precipitation (in/yr)	32.72	31.80	5.42	0.17	21.43	49.03
Sediment delivery (t/yr)	1.15	0.87	0.80	0.70	0.02	3.66
Soil loss (ton/ac/yr)	5.05	3.78	3.49	0.69	0.10	15.94
Runoff (in/yr)	6.17	5.72	2.48	0.40	1.49	13.55
Plant transpiration (in/yr)	16.70	16.58	2.50	0.15	11.67	22.98
Soil evaporation (in/yr)	9.32	9.30	1.35	0.15	6.11	14.21

SCI and STIR Output

Soil Conditioning Index (SCI)	SCI OM subfactor	SCI FO subfactor	SCI ER subfactor	Avg. annual slope STIR
0.2	1.1	−0.12	−0.98	113.45

Treatment III: Corn Grain, Spring Disk Cultivation After No-Till Rye Cover Crop 3% Slope

Location/climate	Latitude	Longitude	Slope length (horiz) (ft)	Slope shape	Avg. slope steepness, %	Aspect
Iowa—Webster County	42.43	−94.18	200.0	Uniform	3	180

Soil Nicollet-Guckeen complex, 1–3% slopes Nicollet *Soil T value*: 5 (ton/A/yr)

Date	Operation	Vegetation	Surf. res. cov. after op, %	STIR	Fuel (gal/ac)
Apr 20, 01	Disk, tandem secondary		49.23	32.5	0.48
Apr 25, 01	Cultivator, field 6–12 in sweeps		51.22	26	0.74
May 01, 01	Planter, double disk opnr	Corn, grain, seed	50.86	2.44	0.44
May 11, 01	Sprayer, pre-emergence		44.4	0.15	0.13
Jun 07, 01	Sprayer, postemergence, fertilizer tank mix		40.97	0.15	0.13
Oct 20, 01	Harvest, killing crop 50 pct standing stubble		90.1	0.15	1.53
Oct 21, 01	Drill or air seeder, double disk	Small grain, winter, forage	80.66	6.34	0.36
Nov 01, 01	Fert applic. surface broadcast		80.34	0.06	0.16

Management	Vegetation	Yield units	# yield units, #/ac	Target yield	Yield % moisture	%Yield attained	Calibrated?
Corn grain; disk, rye cover crop fall N	Corn, grain, seed	bu/ac	110.1	112.0	16.1%	98.1	Yes

Contouring	Strips/barriers
None	None

Model output	Mean	Median	Stddev	COV	Min	Max
Precipitation (in/yr)	32.72	31.80	5.42	0.17	21.43	49.03
Soil loss (ton/ac/yr)	1.39	1.06	0.92	0.66	0.01	4.34
Sediment delivery (t/yr)	0.32	0.24	0.21	0.66	0.00	1.00
Runoff (in/yr)	6.24	6.03	2.46	0.39	1.36	14.13
Plant transpiration (in/yr)	16.77	16.51	2.60	0.15	11.82	23.08
Soil evaporation (in/yr)	7.82	7.79	1.20	0.15	4.98	11.23

SCI and STIR Output

Soil Conditioning Index (SCI)	SCI OM subfactor	SCI FO subfactor	SCI ER subfactor	Avg. annual slope STIR
0.69	1.16	0.33	0.45	67.79

Treatment IV: Corn Grain, No-Till Corn with No-Till Rye Cover Crop 3% Slope

Location/climate	Latitude	Longitude	Slope length (horiz) (ft)	Slope shape	Avg. slope steepness, %	Aspect
Iowa—Webster County	42.43	−94.18	200.0	Uniform	3	180

Soil Nicollet-Guckeen complex, 1–3% slopes Nicollet *Soil T value*: 5 (ton/A/yr)

Date	Operation	Vegetation	Surf. res. cov. after op, %	STIR	Fuel (gal/ac)
Apr 20, 01	Sprayer, kill crop		78.53	0.15	0.13
May 01, 01	Planter, double disk opnr, fluted coulter	Corn, grain, seed	76.23	2.44	0.54
Jun 07, 01	Sprayer, postemergence		71.25	0.15	0.13
Oct 20, 01	Harvest, killing crop 50 pct standing stubble		92.23	0.15	1.53
Oct 21, 01	Drill or air seeder, double disk	Small grain, winter, forage	83.61	6.34	0.36
Nov 01, 01	Fert. applic. anhyd knife 30 inch spacing		77	2.6	0.8
Nov 01, 01	Fert applic. surface broadcast		77	0.06	0.16

Management	Vegetation	Yield units	# yield units, #/ac	Target yield	Yield % moisture	%Yield attained	Calibrated?
corn grain; NT, anhyd after rye cover NT z4	Corn, grain, seed	bu/ac	111.1	112.0	16.1%	99.1	Yes

Contouring	Strips/barriers
None	None

Model output	Mean	Median	Stddev	COV	Min	Max
Precipitation (in/yr)	32.72	31.80	5.42	0.17	21.43	49.03
Sediment delivery (t/yr)	0.22	0.18	0.12	0.56	0.00	0.57
Soil loss (ton/ac/yr)	0.95	0.79	0.53	0.56	0.01	2.46
Runoff (in/yr)	7.15	6.83	2.54	0.35	3.10	15.50
Plant transpiration (in/yr)	9.55	7.05	8.92	0.93	0.60	24.16
Soil evaporation (in/yr)	10.36	10.29	2.45	0.24	5.81	16.10

SCI and STIR Output

Soil Conditioning index (SCI)	SCI OM subfactor	SCI FO subfactor	SCI ER subfactor	Avg. annual slope STIR
0.98	1.22	0.88	0.69	11.89

WEPP Statistical Outputs

The WEPP model produces statistical outputs based on maximum and minimum climate extreme criteria. Examine minimum and maximum precipitation, soil loss, runoff, transpiration, and soil evaporation for each of the treatments. Concerning runoff and erosion, long-term soil losses on treatments I, II, and IV are below tolerable soil loss (T) values; however, the maximum value in treatment I ($8,14 \text{ t ac}^{-1} \text{ yr}^{-1}$) significantly exceeds T with annual precipitation of 49 inches (17 inches above average).

Analysis of Fig. 4.12 data. Soil loss between Treatment II and conventionally tilled corn (fall plowed, spring disk and cultivator) on the 5% slope was double that of Treatment I at 3% slope. Runoff, plant transpiration, and soil evaporation were about equal between the conventionally tilled 3 and 5% slopes, respectively. Soil loss on Treatment III corn grain spring disk, cultivation with fall no-till rye cover crop 3% slope was 38.5% less than the conventionally tilled treatment at 3% slope. Soil loss on Treatment IV, corn grain, no-till corn with no-till rye cover crop 3% slope, was 57.9% less than Treatment I. Runoff was about equal for all four treatments, although on treatment IV, runoff was slightly higher but had the lowest soil loss of all the treatments. Since these examples are from 1-year model runs, it is difficult to explain why WEPP runoff was slightly higher on Treatment 4 with the rye winter cover crop. From an ecohydrologic perspective, rye has a dense fibrous root system; once the upper soil profile is saturated with water, water perhaps did not infiltrate as fast compared to the rough plowed treatments (I and II); therefore, runoff was slightly higher, but carried the least sediment due to consolidated soils and mulch cover. Plant transpiration was lower on treatment IV, but soil evaporation was slightly higher. Since there were a winter rye cover crop and corn stubble

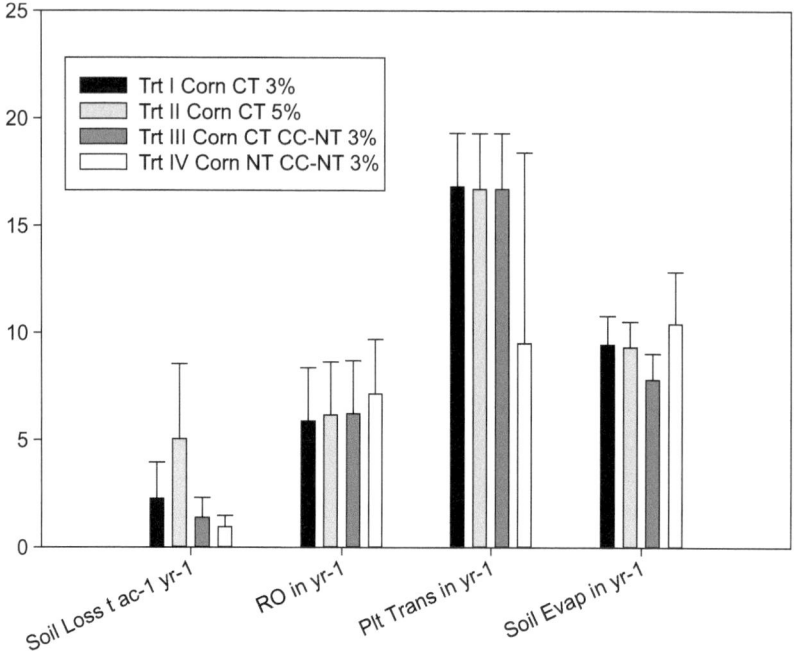

Fig. 4.12 Water erosion model estimates for four 1-year corn planting scenarios. (Data from the above tables): Trt. I corn grain, fall plow 3% slope; Trt. II corn grain fall plow 5% slope; Trt. III corn grain, spring disk, cultivation after no-till rye cover crop 3% slope; and Trt. IV corn grain, no-till corn with no-till rye cover crop 3% slope. WEPP model output is in English units; data given is soil loss (t ac^{-1} yr^{-1}), runoff (in yr^{-1}), plant transpiration (in yr^{-1}), and soil evaporation (in yr^{-1})

throughout the year, soil temperatures would be cooler, and corn plants may not transpire as much water because of more steady soil water balance and less stress. Soil evaporation was slightly increased, possibly because of higher overall available soil moisture because of ground cover throughout the year.

References

Abawi, G.S., and T.L. Widmer. 2000. Impact of soil health management practices on soilborne pathogens, nematodes and root diseases of vegetable crops. *Applied Soil Ecology* 15: 37–47.
Acharya, J., M.G. Bakker, T.B. Moorman, T.C. Kaspar, A.W. Lenssen, and A.E. Robertson. 2017. Time interval between cover crop termination and planting influences corn seedling disease, plant growth, and yield. *Plant Disease* 101: 591–600.
Bakker, M.G., J. Acharya, T.B. Moorman, A.E. Robertson, and T.C. Kaspar. 2016. The potential for cereal rye cover crops to host corn seedling pathogens. *Phytopathology* 106: 591–601.
Boswell, E.P., R.T. Koide, D.L. Shumway, and H.D. Addy. 1998. Winter wheat cover cropping, VA mycorrhizal fungi and maize growth and yield. *Agriculture Ecosystems and Environment* 67: 55–65.

Bowen, J., L. Jordan, and D. Biehle. 1991. Economics of no-till corn planted into winter cover crops. In *Proc. 1992 Southern Conservation Tillage Conference*, July 21–23 1992, Jackson, TN. Spec. Pub. 92-01, eds. M.D. Mullen, and B.N. Duck, 181–182. Tennessee Agric. Exp. Sta. Knoxville: Univ. of Tennessee.

Brill, G.D., and O.R. Neal. 1950. Seasonal occurrence of runoff and erosion from a sandy soil in vegetable production. *Agronomy Journal* 42: 192–195.

Buyanovsky, G.A., and G.H. Wagner. 1986. Post-harvest residue input to cropland. *Plant and Soil* 93: 57–65.

Canadell, J., R.B. Jackson, J.B. Ehleringer, H.A. Mooney, O.E. Sala, and E.D. Schulze. 1996. Maximum rooting depth of vegetation types at the global scale. *Oecologia* 108: 583–595.

Chen, G., and R.R. Weil. 2010. Penetration of cover crop roots through compacted soils. *Plant and Soil* 331: 31–43.

Chen, X., Z. Cui, M. Fan, P. Vitousek, M. Zhao, W. Ma, Z. Wang, W. Zhang, X. Yan, J. Yang, X. Deng, Q. Gao, Q. Zhang, S. Guo, J. Ren, S. Li, Y. Ye, Z. Wang, J. Huang, Q. Tang, Y. Sun, X. Peng, J. Zhang, M. He, Y. Zhu, J. Xue, G. Wang, L. Wu, N. An, L. Wu, L. Ma, W. Zhang, and F. Zhang. 2014. Producing more grain with lower environmental costs. *Nature* 514: 486–489.

Chew, F.S. 1988. Biological effects of glucosinolates, 155–181. In *Biologically active natural products: Potential use in agriculture*, ed. H.G. Gutler, 483. Washington, DC: American Chemical Society.

Cresswell, H.P., and J.A. Kirkegaard. 1995. Subsoil amelioration by plant-roots-the process and the evidence. *Soil Research* 33: 221–239.

Dabney, S.M. 1998. Cover crop impacts on watershed hydrology. *Journal of Soil and Water Conservation* 53: 207–213.

Dabney, S.M., J.A. Delgado, and D.W. Reeves. 2001. Using winter cover crops to improve soil and water quality. *Communications in Soil Science and Plant Analysis* 32: 1221–1250.

De Baets, S., J. Poesen, J. Meermans, and L. Serlet. 2011. Cover crops and their erosion-reducing effects during concentrated flow erosion. *Catena* 85: 237–244.

Dinnes, D., D. Karlen, D. Jaynes, T. Kaspar, J. Hatfield, T. Colvin, and C. Cambardella. 2002. Nitrogen management strategies to reduce nitrate leaching in tile-drained Midwestern soils. *Agronomy Journal* 94: 153–171.

Fan, J., B. McConkey, H. Wang, and H. Janzen. 2016. Root distribution by depth for temperate agricultural crops. *Field Crops Research* 189: 68–74.

Franklin, R.L., S.L. Ott, and W.L. Hargrove. 1989. The economics of using legume cover crops as sources of nitrogen in corn production. Faculty series 89-11, Division of Agricultural Economics, Athens, GA: The University of Georgia.

Geary, B., C. Ransom, D. Atkinson, and S. Hafez. 2008. Weed, disease, and nematode management in onions with biofumigants and metam sodium. *Hort Technology* 18: 569–574.

Grabau, Z.J., and S. Chen. 2016. Influence of long-term corn soybean crop sequences on soil ecology as indicated by the nematode community. *Applied Soil Ecology* 100: 172–185.

Grabau, Z.J., J.A. Vetsch, and S. Chen. 2017. Effects of fertilizer, nematicide, and tillage on plant–parasitic nematodes and yield in corn and soybean. *Agronomy Journal* 109: 1651–1662.

Groff, S. 2015. The past, present, and future of the cover crop industry. *Journal of Soil and Water Conservation* 70: 130–133.

Groffman, P.M., P.F. Hendrix, and D.A. Crossley. 1987. Nitrogen dynamics in conventional and no-tillage agroecosystems with inorganic fertilizer or legume nitrogen inputs. *Plant and Soil* 97: 315–332.

Hall, J.K., N.L. Hartwig, and L.D. Hoffman. 1984. Cyanazine losses in runoff from no-tillage corn in "living mulch" and dead mulches vs. unmulched conventional tillage. *Journal of Environmental Quality* 13: 105–110.

Haramoto, E.R., and E.R. Gallandt. 2004. Brassica cover cropping for weed management: A review. *Renewable Agriculture and Food Systems* 19: 187–198.

Hartwig, N.L., and H.U. Ammon. 2002. Cover crops and living mulches. *Weed Science* 50: 688–699.

Hedley, M.J., R.E. White, and P.H. Nye. 1982. Plant-induced changes in the rhizosphere of rape (*Brassica napus* var. Emerald) seedlings. *New Phytologist* 91: 45–56.

Hirsch, A.M. 2009. Brief history of the discovery of nitrogen-fixing organisms. https://www.mcdb. ucla.edu/Research/Hirsch/imagesb/HistoryDiscoveryN2fixingOrganisms.pdf

Hoffland, E., G.R. Findenegg, and J.A. Nelemans. 1989. Solubilization of rock phosphate by rape – II. Local root exudation of organic acids as a response to P-starvation. *Plant and Soil* 113: 161–165.

Hoorman, J.J., and R. Islam. 2009. Sustainable crop rotations with cover crops. Fact sheet, Ohio State University.

Isse, A.A., A.F. MacKenzie, K. Stewart, D.C. Cloutier, and D.L. Smith. 1999. Cover crop and nutrient retention for subsequent sweet corn production. *Agronomy Journal* 91: 934–939.

Jennings, C.M.H. 1974. The hydrology of Botswana. Ph. D thesis, University of Natal, South Africa.

Johnson, T., T. Kaspar, K. Kohler, S. Corak, and S. Logsdon. 1998. Oat and rye overseeded into soybean as fall cover crops in the upper Midwest. *Journal of Soil and Water Conservation* 53: 276–279.

Ju, C., R.J. Buresh, Z. Wang, H. Zhang, L. Liu, J. Yang, and J. Zhang. 2015. Root and shoot traits for rice varieties with higher grain yield and higher nitrogen use efficiency at lower nitrogen rates application. *Field Crop Research* 175: 47–55.

Kabir, Z., and R.T. Koide. 2002. Effect of autumn and winter mycorrhizal cover crops on soil properties, nutrient uptake and yield of sweet corn in Pennsylvania, USA. *Plant and Soil* 238: 205–215.

Kaspar, T.C., and M.G. Bakker. 2015. Biomass production of 12 winter cereal cover crop cultivars and their effect on subsequent no-till corn yield. *Journal of Soil and Water Conservation* 70: 353–364.

Kaspar, T.C., and J.W. Singer. 2011. The use of cover crops to manage soil. Publications from USDA-ARS/UNL Faculty. 1382. http://digitalcommons.unl.edu/usdaarsfacpub/1382

Laflen, J.M., and W.C. Moldenhauer. 1979. Soil and water losses from corn-soybean rotations 1. *Soil Science Society of America Journal* 43: 1213–1215.

Lehman, R.M., W.I. Taheri, S.L. Osborne, J.S. Buyer, and D.D. Douds Jr. 2012. Fall cover cropping can increase arbuscular mycorrhizae in soils supporting intensive agricultural production. *Applied Soil Ecology* 61: 300–304.

Lounsbury, N.P., and R.R. Weil. 2015. No-till seeded spinach after winterkilled cover crops in an organic production system. *Renewable Agriculture and Food Systems* 30: 473–485.

Lu, Y.C., K.B. Watkins, J.R. Teasdale, and A.A. Abdul-Baki. 2000. Cover crops in sustainable food production. *Food Reviews International* 16: 121–157.

Mercer, C.F., and K.J. Miller. 1997. Evaluation of 15 *Trifolium* spp. and of *Medicago sativa* as hosts of four Meloidogyne spp. found in New Zealand. *Journal of Nematology* 29: 673–676.

MESAM. 2007. (Measures against Erosion and Sensibilisation of farmers for the protection of the environment) project report on cover crops. European Interreg III project, p. 7.

Meyer, L.D., C.E. Murphree, S.M. Dabney, W.C. Harmon, and E.H. Grissinger. 1997. Effects of cropland management practices on storm runoff and erosion. In *Proceedings of the conference on management of landscapes disturbed by channel incision 20–22 May 1997*, ed. S.Y. Wang, E.J. Langendoen, and F.D. Shields Jr., 990–995. Oxford, MS: University of Mississippi.

Nearing, M.A., V. Jetten, C. Baffaut, O. Cerdan, A Couturier, M. Hernandez, and V. Souchère. 2005. Modeling response of soil erosion and runoff to changes in precipitation and cover. *Catena* 61: 131–154.

Neilsen, N.E., and H.E. Jensen. 1985. Soil mineral nitrogen as affected by undersown catch crops. In *Assessment of nitrogen fertilizer requirements*, ed. J.J. Neeteson and K. Dilz Haren, 101–110. The Netherlands: Institute of Soil Fertility.

Opperman, C., J. Rich, and R. Dunn. 1988. Reproduction of 3 root-knot nematodes on winter small grain crops. *Plant Disease* 72: 869–871.

Ott, S.L., and W.L. Hargrove. 1989. Profit and risks of using crimson clover and hairy vetch cover crops in no-till corn production. *American Journal of Alternative Agriculture* 4: 65–70.

Pantoja, J.L., K.P. Woli, J.E. Sawyer, and D.W. Barker. 2015. Corn nitrogen fertilization requirement and corn-soybean productivity with a winter rye cover crop. *Soil Science Society America Journal* 79: 1482–1495.

Phillips, W.S. 1963. Depth of roots in soil. *Ecology* 44: 424.

Pieters, A.J., and R. McKee. 1938. The use of cover crops and green-manure crops. *Soils and Men Yearbook of Agriculture 1938*, 431–444. U.S. Department of Agriculture. Washington D.C: Government Printing Office.

Reeves, D.W. 1994. Cover crops and rotations. pp. 125–172. In: J.L. Hatfield and B.A. Stewart (eds.) *Crops Residue Management. Advances in Soil Science.* Boca Raton, Florida: Lewis Publishers.

Rodale Press. 2017. https://www.rodalesorganiclife.com/garden/cover-crop-basics

Russel, D.A., and G.G. Williams. 1977. History of chemical fertilizer development 1. *Soil Science Society of America Journal* 41: 260–265.

Sainju, U.M., B.P. Singh, and W.F. Whitehead. 1998. Cover crop root distribution and its effects on soil nitrogen cycling. *Agronomy Journal* 90: 511–518.

SARE, Sustainable Agriculture Research and Education Program. 2012. In *Managing cover crops profitably*, ed. A. Clark, 3rd ed. Washington DC: Government Printing Office.

Singer, J.W., S.M. Nusser, and C.J. Alf. 2007. Are cover crops being used in the US corn belt? *Journal of Soil and Water Conservation* 62: 353–358.

Spaeth, K.E., F.B. Pierson, M.A. Weltz, and G. Hendricks, eds. 1996a. *Grazingland hydrology issues: Perspectives for the 21st century.* Denver: Society for Range Management.

Spaeth, K.E., F.B. Pierson, M.A. Weltz, and J.B. Awang. 1996b. Gradient analysis of infiltration and environmental variables as related to rangeland vegetation. *Transactions of ASAE* 39: 67–77.

Steele, M.K., F.J. Coale, and R.L. Hill. 2012. Winter annual cover crop impacts on no-till soil physical properties and organic matter. *Soil Science Society of America Journal* 76: 2164–2173.

Thorup-Kristensen, K. 2001. Are differences in root growth of nitrogen catch crops important for their ability to reduce soil nitrate-N content, and how can this be measured? *Plant and Soil* 230: 185–195.

Thorup-Kristensen, K., M.S. Cortasa, and R. Loges. 2009. Winter wheat roots grow twice as deep as spring wheat roots, is this important for N uptake and N leaching losses? *Plant and Soil* 322: 101–114.

Timper, P., R.F. Davis, and P.G. Tillman. 2006. Reproduction of Meloidogyne incognita on winter cover crops used in cotton production. *Journal of Nematology* 38: 83–89.

Unger, P.W., and M.F. Vigil. 1998. Cover crop effects on soil water relationships. *Journal of Soil and Water Conservation* 53: 200–207.

USDA-NRCS.   2018.   https://www.nrcs.usda.gov/wps/portal/nrcs/detail/soils/survey/geo/?cid=stelprdb1041925

Villamil, M.B., G.A. Bollero, R.G. Darmody, F.W. Simmons, and D.G. Bullock. 2006. No-till corn/soybean systems including winter cover crops. *Soil Science Society of America Journal* 70: 1936–1944.

Wagger, M.G., M.L. Cabrera, and N.N. Ranells. 1998. Nitrogen and carbon cycling in relation to cover crop residue quality. *Journal of Soil and Water Conservation* 53: 214–218.

Wang, K.H., R. McSorley, and R. Gallaher. 2004. Effect of winter cover crops on nematode population levels in North Florida. *Journal of Nematology* 36: 517–523.

Weaver, J.E. 1919. *The ecological relations of roots (No. 286).* Carnegie Institution of Washington: Washington, DC.

———. 1926. *Root development of field crops.* 1st ed. New York: McGraw-Hill.

Weaver, J.E., and W.E. Bruner. 1927. *Root development of vegetable crops.* 1st ed. New York: McGraw-Hill.

Weil, R., C. White, and Y. Lawley. 2009. Forage radish: A new multi-purpose cover crop for the Mid-Atlantic. In *Fact Sheet 824*. College Park: Univ. of Maryland Coop. Ext.

White, C.M., and R.R. Weil. 2010. Forage radish and cereal rye cover crop effects on mycorrhizal fungus colonization of maize roots. *Plant and soil* 328: 507–521.

White, C.M., and R.R. Weil. 2011. Forage radish cover crops increase soil test phosphorus surrounding radish taproot holes. *Soil Science Society of America Journal* 75: 121–130.

Wick, A., M. Berti, Y. Lawley, and M. Liebig. 2017. Integration of annual and perennial cover crops for improving soil health. In *Soil health and intensification of agroecosytems*, 127–150. San Diego: Academic Press.

Williams, S.M., and R.R. Weil. 2004. Crop cover root channels may alleviate soil compaction effects on soybean crop. *Soil Science Society of America Journal* 68: 1403–1409.

Yamaguchi, J., and A. Tanaka. 1990. Quantitative observation on the root system of various crops growing in the field. *Soil Science and Plant Nutrition* 36: 483–493.

Zhu, J.C., C.J. Gantzer, S.H. Anderson, E.E. Alberts, and P.R. Beuselinck. 1989. Runoff, soil, and dissolved nutrient losses from no-till soybean with winter cover crops. *Soil Science Society of America Journal* 53: 1210–1214.

Zuazo, V.H.D., and C.R.R. Pleguezuelo. 2008. Soil-erosion and runoff prevention by plant covers: A review. *Agronomic Sustainable Development* 28: 65–86.

Chapter 5
Vegetation Effects on Hydrology and Erosion: Grazinglands

Abstract Rangelands are diverse with respect to climate, soils, vegetation, and their evolutionary history of development. Most of the highly productive croplands in the United States were historically rangelands (see definitions of rangelands in Chap. 2). Rangeland soil health principles are similar in several respects to cropland principles—limit and/or reduce soil disturbance and maintain plant and litter cover; however, there are some aspects of soil health that are significantly different. Rangelands primarily occupy uncultivated soils (restored rangelands from cultivation if managed as rangeland are classified as rangeland), contain perennial vegetation of diverse species with varying above- and belowground morphology (exceptions—California annual grassland and rangelands invaded by exotic annual species), and have permanent cover and extended periods of active plant growth. Also extended dry periods and drought are common in many semiarid and arid rangeland community types. Water is the most limiting factor in rangelands, and hydrologic function is a primary factor that directly influences soil health. Microbial populations wax and wane on rangelands with available moisture, and water loss by soil evaporation and runoff diminish available water for plant growth. Plant foliar cover, biomass, growth/life forms, individual taxa, and management all influence hydrology and erosion on range- and pasturelands, but there are unique aspects that are rarely recognized by land managers. For example, other biotic components such as biological soil crusts, soil fauna, and soil microbia also play a role in hydrologic function by enhancing aggregate stability and ultimately porosity, soil surface structure, and nutrient availability for plant growth. Foliar cover with varied canopy and plant architecture reduce raindrop impact, and perennial root structure provides conduits for water infiltration. Vegetative cover during the growing season and standing dead and decumbent litter during dormant periods are essential to maintaining sustainable wind and water erosion levels. This chapter focuses on the effect of vegetation on hydrology and erosion dynamics—two important factors related to soil health.

© Springer Nature Switzerland AG 2020
K. E. Spaeth Jr., *Soil Health on the Farm, Ranch, and in the Garden*,
https://doi.org/10.1007/978-3-030-40398-0_5

Questions

- Outline the environmental links of the hydrologic cycle with soil health dynamics on grazinglands.
- What might be expected in terms of runoff and erosion if the plant community shifts from perennial to annual grasses?
- Why do various plant life forms have different effects on hydrology and erosion on grazinglands?
- What is the effect of prescribed and/or wildfire on soil surface physical characteristics?
- Discuss the implications of light, moderate, and heavy grazing on range- and pasturelands?
- What are the critical environmental factors to monitor when using specialized grazing systems?
- What is the effect of heavy vs. moderate stocking during a season where soils are typically wet and saturated?
- What components of the hydrologic cycle can be managed by ranchers to maximize available water for plant growth?
- Are there significant benefits for infiltration capacity between a rotational grazing system and a continuous stocking system?
- What are the hydrologic effects of mechanical brush control in a particular rangeland plant community?
- Do different shrubs and grasses affect infiltration and runoff differently?

Soil Health Concepts on Rangeland

Rangeland comprises over two-thirds of the nation's watershed area (FAO 1990) and provides a significant part of its water. The increasing importance of water has added a new dimension in range management strategies. In the Southwestern and Western United States, rangeland watersheds are the source of most surface water flow and aquifer recharge. Management on these lands can have a positive or negative effect with respect to plant cover and compositional change, which ultimately influences water quality and quantity. Since the need for clean water is critical, and rangelands comprise a vast watershed area in the United States, it is of prime importance that policies and activities are formulated and implemented to arrest resource degradation. In order to accomplish the goal of maintaining and improving the nation's rangelands, management must be directed at sustaining plant and all biotic components, hydrologic, and soil health—the three attributes of rangeland health (Pellant et al. 2019).

Soil health focuses on three main points: "capacity to function, sustainability, and meeting human needs" (Brown and Herrick 2016). Some unique principles associated with soil health on rangelands are (1) maintain or increase native plant diversity, and preferably reduce invasive plants that can dominate composition and/or shade understory plants; (2) limit and/or reduce soil disturbance; (3) maintain vegetative cover; and (4) encourage and extend active plant growth. In comparison to the principles of cropland soil health, grazinglands focus on maintaining and/or improving the composition of perennial plants. The California annual grassland presents a different set of circumstances in that the vegetation is dominated by annual species; therefore, annual grass management should focus on adequate standing dead and litter cover during the dormant period. Derner et al. (2018) point out that these four principles are dependent on the dynamics of ecological site processes, which pose challenges for land managers to "assess the applicability of these principles to their natural resource concerns."

On rangelands, the application of soil health principles is limited by scientific evidence (Brown and Herrick 2016), and the development for process-based soil health indicators lags behind cropland (Derner et al. 2018). The approach to evaluating soil health on rangelands is dynamically different from cropland in many respects and "should not focus on the improvement of soil health on lands where potential is limited but rather to forward science-based management for improving grazing lands' resilience to environmental change and sustainability" (Derner et al. 2018). This can be accomplished by concentrating on (1) ecological processes (hydrologic and nutrient cycling) associated with grazing management instead of maximum short-term profits from livestock production; (2) management goals based on specific objectives integrated with maintenance and monitoring of applicable soil health attributes; (3) integrate soil health attributes in a holistic way with appropriate variables associated with rangeland; (4) enhance communication and technical knowledge of rangeland soil health among land producers/managers, resource professionals, and scientists; and (5) create a "living laboratory network" of rangeland soil health case studies (Derner et al. 2018).

Text Box 5.1 Ten Differences Between Rangeland and Cropland Factors that Influence Soil Health
Note existing rangelands are typically those that were not suitable for cultivation or have been re-established because of high erosivity and low crop potential.

1. In contrast to croplands, which on the average either receive adequate rainfed precipitation or are irrigated, many arid and semiarid rangelands are characteristic of high inter- and intra-annual variability in timing and amount of precipitation, which is the key factor that drives ecological dynamics including hydrologic function, soil and surface stability, and biotic integrity, and response to management. Microbial populations that are active with carbon cycle "wax and wane" with available soil moisture content; therefore, many cropland soil health assessments are not applicable to rangeland. Rangeland and cropland are unique in their capacity, and evaluations of soil health should be based on their particular dynamics.

(continued)

2. Existing rangeland soils are highly variable ranging from shallow to very deep (the best rangeland is usually in some form of cultivation) with lower tolerable soil loss rates than cropland, especially on fragile arid and semi-arid rangelands. Much of the remaining uncultivated Great Plains prairies occur on soils with soil properties (shallow, steep, droughty, gravely and/or rocky, wet, saline, and high pH) that are unsuitable for crop production.

3. Many rangeland soils in the southwest and western rangelands have lower soil organic matter content; however, the true prairie soils of the Great Plains with higher organic matter are now some of the most productive soils for cropland purposes.

4. Soil organic carbon sequestration in arid and semiarid rangeland environments can be more significant than on cropland because of perennial root systems and large biomass turnover of aboveground plant litter. Rangeland ecosites with perennial shrubs and trees have substantial carbon storage above- and belowground. Forests (1.2–1.4 Pg C yr^{-1}) and cropland (0.4–1.2 Pg C yr^{-1}) have the largest potentials for sequestering carbon; although grazinglands (range and pasturelands) can contribute up to 10% of the sink capacity. On a global perspective, rangelands occupy about half of the world's land area, 10% of the terrestrial biomass, and 10–30% of the soil organic carbon (Schlesinger 1977). An average estimate of globally sequestered soil carbon on rangelands is 0.5 Pg C yr^{-1} (Schlesinger 1977; Scurlock and Hall 1998).

5. Rangelands include many different types of vegetation life forms (trees, shrubs, forbs, grasses, and biotic crusts) and growth forms (bunch, sod, vine, stoloniferous, rhizomatous, tap and fibrous root systems) compared to monocultures in cropland settings. Carbon cycles, turnover, and sequestration between perennial rangeland vegetation and cropland are significantly different.

6. Perennial vegetation on rangelands is typically associated with less soil disturbance than cultivated croplands and is typically associated with more stable soil aggregates with associated soil organic matter and microbial populations. Rangeland soils are typically not compacted to the extent that may exist in cropland agriculture. Also, cropland soils often have a plow or plow pan layer that persists after cultivation and may inhibit infiltration capacity and root development of plants.

7. Soil heterogeneity can be more prevalent on existing rangeland because of geomorphic topographic influences that inhibit cultivation. This heterogeneity promotes plant species diversity and associated microbial mycorrhizal associations.

8. Erosion control on rangeland is highly dependent on maintaining plant foliar cover (>60% cover; according to ecological site potentials; some rangeland plant communities such as greasewood (*Sarcobatus vermiculatus*) have low interspace plant cover).

(continued)

9. Biological soil crusts can be prevalent in rangeland plant communities, especially in semiarid and arid shrub ecosites with plant interspaces. They provide soil surface stability against wind and water erosion and are active in nutrient cycling.

10. Arid and semiarid rangelands have slower decomposition and carbon turnover rates because of more erratic climate and precipitation compared to temperate grasslands, savannahs, and shrublands. Cropland environments (rainfed and irrigated) supply more consistent moisture that is necessary to maintain microbial populations associated with decomposition and organic matter turnover. Management associated with soil health and soil organic matter restoration may not be possible in many arid and semiarid rangeland plant communities due to inherent soil carbon dynamics, wind and water erosion, and changes from perennial to annual plant dominance. If major depreciating vegetative state changes have crossed critical thresholds, they may be permanent and irreversible. On croplands that still have good production potential (adequate rainfall and/or irrigation), improvement of soil organic matter and microbial mycorrhizae interactions are significantly more probable with the use of reduced and no-till practices and cover crops.

In natural plant communities, the hydrologic condition of a site is the result of complex interactions of soil and vegetation factors, which influence interception, infiltration, evaporation, transpiration, percolation, surface runoff, soil water storage, soil erosion, and deposition of sediment. Every plant-soil complex exhibits a characteristic infiltration pattern (Gifford 1985). Natural plant communities are not homogeneous, even within seemingly continuous unbroken expanses of grass. Mosaic patterns and patchiness prevalent in most natural rangeland plant communities are spatially heterogeneous and temporally dynamic. Spatial and temporal variability of soil and vegetation characteristics also strongly influence hydrology and erosion on rangelands (Blackburn et al. 1990; Wilcox and Breshears 1994; Reid et al. 1999; Pierson et al. 2002a, b; Madsen et al. 2008).

Research has demonstrated a significant correlation between vegetative cover and kinds of vegetation with soil erosion, infiltration, and runoff (Pearse and Wooley 1936; Osborn 1950; Mazurak and Conrad 1959; Dee et al. 1966; Rauzi et al. 1968; Blackburn and Skau 1974; Blackburn 1975, 1984; Hanson et al. 1978; Gifford 1984; Swanson and Buckhouse 1984; Blackburn et al. 1986, 1990, 1992; Synman and Van Rensburg 1986; Johnson and Gordon 1988; Thurow et al. 1988; Wilcox and Wood 1988; Spaeth 1990; Spaeth et al. 1996a, b; Pierson et al. 2002b). Plant cover on cropland and grazinglands can be viewed in several ways: canopy cover, foliar cover, and basal cover. Canopy cover is an abstract view of the plant canopy where it is an estimate of the area occupied by the plant (the whole area) but ignores gaps or holes viewed from a vertical projection (Fig. 5.1). Canopy cover can also be

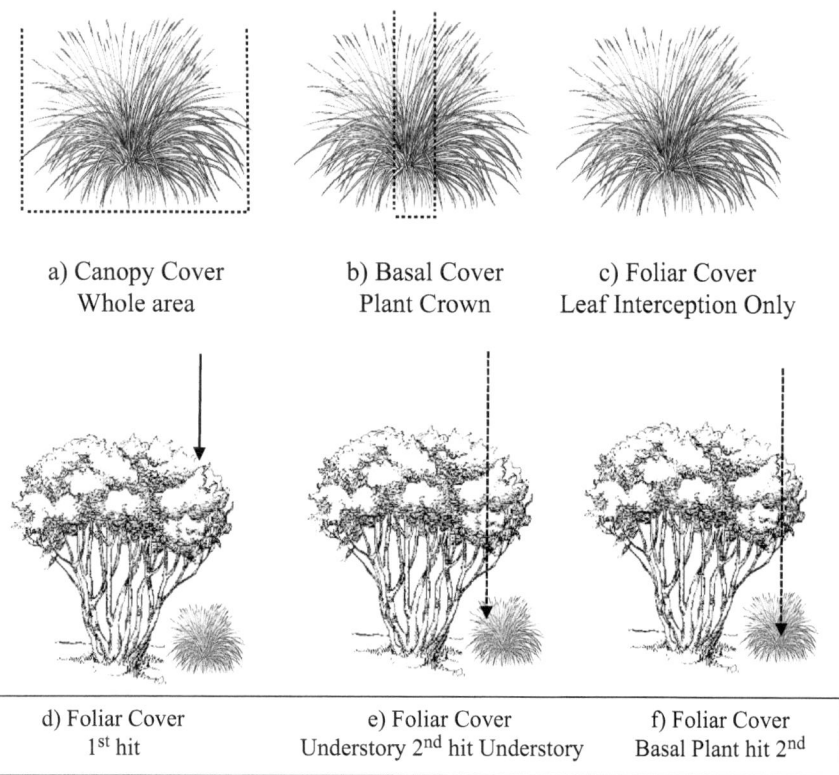

a) Canopy Cover   b) Basal Cover   c) Foliar Cover
   Whole area   Plant Crown   Leaf Interception Only

d) Foliar Cover   e) Foliar Cover   f) Foliar Cover
   1st hit   Understory 2nd hit Understory   Basal Plant hit 2nd

Fig. 5.1 Examples of (**a**) canopy cover, (**b**) basal cover-plant crown, (**c**) foliar cover (leaf interception only, no gaps), (**d**) line point intercept (LPI) foliar cover first hit, (**e**) LPI second hit understory foliar cover bunchgrass, and (**f**) LPI foliar cover second hit basal plant hit

viewed as the vertical projection of the outer perimeter or edges of the plant. Foliar cover is more specific as it is the vertical projection of exposed leaf area covering soil. If a pin was lowered through the plant canopy, foliar cover is recorded when the pin intercepts a plant part. Foliar cover does not include gaps or openings in the plant canopy. Basal plant cover is related to the crown of the plant, the proportion of the plant at ground level or which extends into the soil. Plant canopy cover occupies an outline of the projected area, foliar cover is specific to a plant part that would intercept a raindrop, and basal cover occupies the least area as a plant crown at the ground surface.

Although plant cover is an important variable in protecting the soil against runoff and erosion, individual plant species and the amount of standing biomass and ground cover (basal plants, mulch litter, rock cover, and biotic soil crusts) are also influential to hydrology at the site level. Biomass is a three-dimensional variable including length, width, and height; whereas, plant foliar cover is usually recorded in one dimension (a plant part hit with a sampling pin). In rangeland rainfall simulation studies, standing biomass often produces better correlations for prediction of infiltration, runoff, and sediment loss (Spaeth 1990; Spaeth et al. 1996a; Pierson et al. 2002) (see California annual grass case study, this chapter).

Hydrologic assessments across rangeland plant communities are unique, and no single factor ever varies alone. Variables such as above- and belowground plant morphology, total production, production of individual plant species, total canopy cover, canopy cover of individual plant species, plant architecture, sod-forming growth form vs. bunchgrass growth form, interspace, shrub coppice, soil physical properties, and soil chemical properties are not consistently correlated in natural rangeland plant communities. Plant life forms such as tallgrasses, mid grasses, shortgrasses, sedges, forbs, shrubs, half-shrubs, and trees exhibit varying influences on infiltration and runoff dynamics. Infiltration is commonly highest under trees and shrubs and decreases progressively in the following order: bunchgrass, sod grass, and bare ground (Thurow et al. 1986; Blackburn et al. 1986) (Fig. 5.2; Table 5.1). Infiltration rates and capacity can also vary because of seasonality and antecedent moisture conditions. Annual grasses can provide sufficient cover to protect the soil

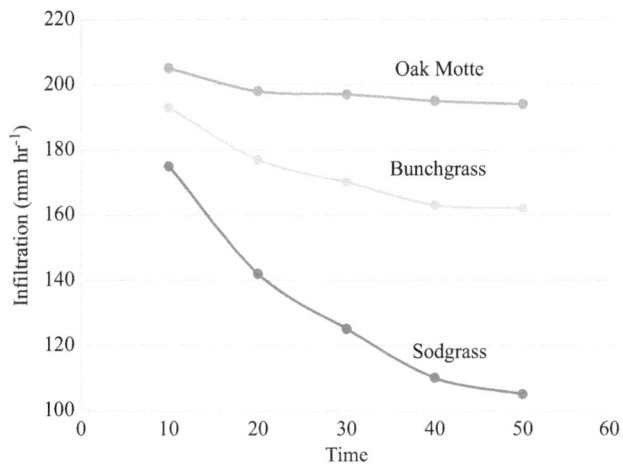

Fig. 5.2 Average infiltration rates for three vegetation types, Edwards Plateau, Texas. (Data Thurow et al. 1986)

Table 5.1 Summary of canopy interception, interrill erosion, runoff, and erosion from oak, bunchgrass, sod grass, and bare ground dominated areas, Edwards Plateau, Texas

	Oak motte	Bunchgrass	Sod grass	Bare ground
% Canopy interception	7	–	–	–
% Grass and litter interception	–	0.5	0.4	0.0
% Litter interception	12	–	–	–
Interrill erosion (kg ha^{-1}; lbs ac^{-1})	0.0	200 (179)	1402 (1250)	6005 (5358)
% Surface runoff	0.0	24	45	75
% Infiltration	81	75.5	54.6	25

Based on 10 cm (4 in) rainfall rate in 30 minutes (data Blackburn et al. 1986)

surface but are vulnerable to wildfire, which usually denudes the site causing high risk to accelerated runoff and high erosion rates. In annual grass communities, foliar cover of grass species may be high; however, rainfall simulation experiments show that the amount of biomass on the site (with high cover) is more closely correlated with higher infiltration and lower runoff (see Fig. 5.10) (Spaeth 2018). Foliar cover associated with soil surface protection from accelerated soil erosion on rangelands varies from 20% in Kenya (Moore et al. 1979) to 100% for some Australian conditions (Costin 1959); however, the majority of studies indicate that cover of 50–75% is probably sufficient (Wood and Blackburn 1981; Gifford 1985; Weltz et al. 1998; Pierson et al. 2011, Pierson and Williams 2016; Williams et al. 2014, 2016; Cadaret et al. 2016a, b) to prevent degradation from accelerated soil erosion processes.

Semiarid shrub-dominated rangelands throughout the Western United States have significant spatial and temporal variations with regard to hydrologic and erosion processes (Thurow et al. 1986, 1988; Blackburn et al. 1992; Pierson et al. 2001, 2002a, b). Sagebrush steppe physiognomy is characterized by shrub coppice and interspace microsites with uniform distribution (Spaeth et al. 1994). Many shrub and half-shrub species are associated with coppice dunes or mounds composed of litter and wind-transported soil. Shrub coppice dunes have significantly different infiltration and erosion rates compared to interspace areas (Blackburn 1975; Blackburn et al. 1992; Wood and Blackburn 1981; Schlesinger et al. 1999; Pierson et al. 2002b; Madsen et al. 2008; Glenn and Finley 2010). Shrub coppices, compared to interspaces, are typically associated with higher organic matter, soil water content, biotic soil crusts, overall biomass, aggregate stability, silt content in the soil surface, and infiltration capacity (Blackburn et al. 1992). Increased runoff is anticipated after wildfire (Moody and Martin 2001; Moffet et al. 2007, 2008) due to a lack of vegetative cover and elevated water-repellent properties of soils after fire, which results in reductions in infiltration capacity (Salih et al. 1973; Debano and Krammes 1966; DeBano et al. 1998; DeBano 2000; Doerr et al. 1996, 2000, 2009; Glenn and Finley 2010) that tend to diminish in succeeding years (MacDonald and Huffman 2004).

Blackburn (1975) compared infiltration in shrub coppice and interspace zones in big sagebrush (*Artemisia tridentata*) plant community (Fig. 5.3). Coppices under dry and field capacity conditions had higher infiltration rates (often 3–4 times greater) than interspace areas. Pierson et al. (2002b) investigated fire effects on sagebrush shrub coppice and interspace hydrologic dynamics and erosion potential (Fig. 5.4). For example, during the fire year (1999), fire-induced water repellency was pronounced on burned shrub coppices, and initial infiltration was reduced by 28% compared to non-burned coppices. Cumulative erosion was also four times higher on burned vs. non-burned coppices. One year following the wildfire, the effect of water repellency was beginning to recede. The dynamics of infiltration in juniper coppices is also distinct: Madsen et al. (2008) measured an eightfold increase in unsaturated hydraulic conductivity across a distance gradient extending outward from the canopy edge.

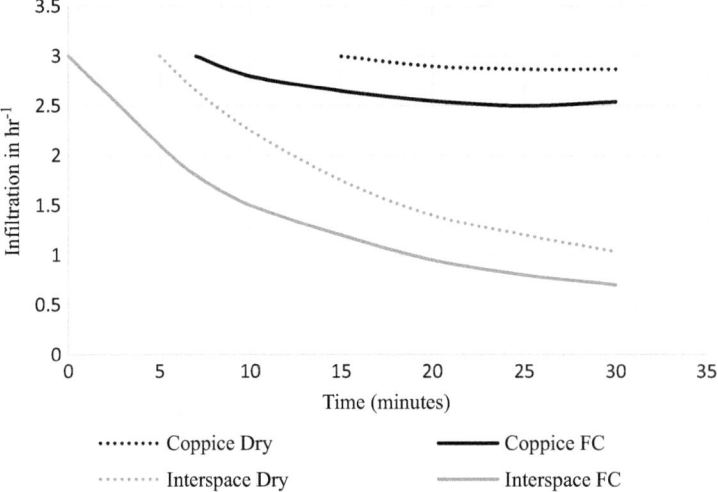

Fig. 5.3 Comparison of infiltration curves for coppice and interspace (dry conditions) and coppice and interspace at field capacity. Big sagebrush (*Artemisia tridentata*) community, Duckwater Watershed, Nevada. (Data Blackburn 1975)

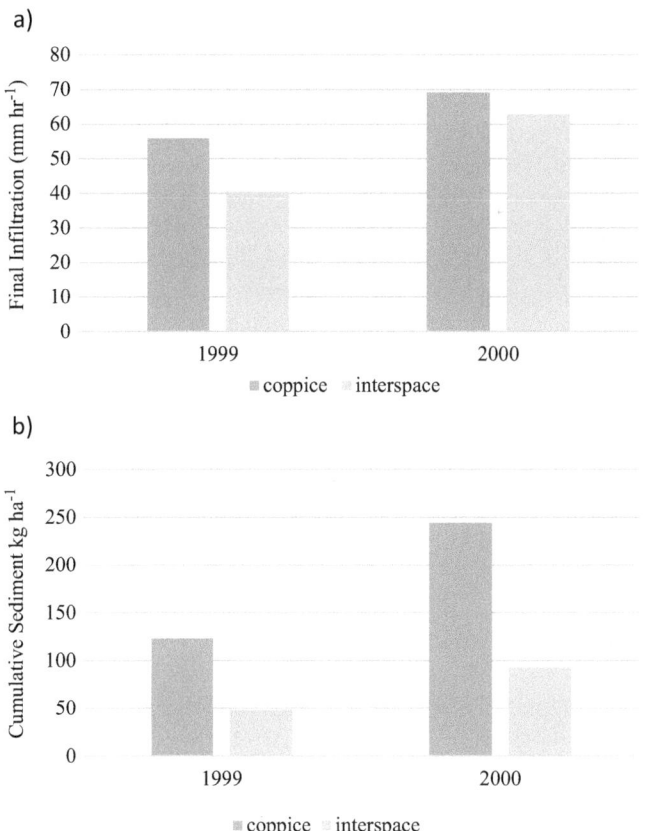

Fig. 5.4 (a) Average final infiltration rates on sagebrush coppice and interspace (Year of fire 1999 and 1 year following wildfire, 2000). (b) Cumulative sediment losses on sagebrush coppice and interspace. Mountain big sagebrush site, Denio, Nevada. (Data Pierson et al. 2001)

Text Box 5.2 Quick Quotes: Documented Vegetation Effects on Infiltration, Runoff, and Erosion

- For every watershed and site within the watershed, there exists a critical point of deterioration due to surface erosion. Beyond this critical point, erosion continues at an accelerated rate, which cannot be overcome by the natural vegetation and soil stabilizing forces. Areas that have deteriorated beyond this critical point continue to erode even when man-caused disturbance is removed (Satterlund 1972).
- "Each plant-soil complex will exhibit a characteristic infiltration pattern" (Gifford 1985). Fifty to 75% plant cover is most likely sufficient to minimize sheet or interrill erosion and maximize infiltration. The impact of vegetative cover to infiltration is not constant from one range-soil complex to another. In semiarid climates, vegetal cover has a minimal influence on infiltration: the erosion process is more complex and is a function of plants, soils, and storm dynamics.
- "The kinds and amounts of vegetation also influence hydrologic variables" (Wood and Wood 1988). The interactions of many soil variables influence the hydrology of a site.
- Vegetational cover influences many hydrological processes including interception, infiltration, evaporation, transpiration, percolation, surface runoff, soil water storage, soil erosion, and deposition of sediment (Dadkah and Gifford 1980).
- "Vegetation cover and standing biomass are generally positively correlated to infiltration and negatively related to interrill erosion" (Rauzi et al. 1968; Blackburn et al. 1986; Thurow et al. 1988; Blackburn et al. 1990).
- The amount of plant material on the soil surface is important in absorbing the impact of raindrops. Erosion and soil surface sealing are also reduced (Rauzi 1960). "Regardless of soil type, water-intake rates depend on the type of plant cover, the amount of standing vegetation, and the amount of mulch material on the ground" (Rauzi 1960).
- Infiltration is usually highest under trees and shrubs and decreases progressively in the following order: bunchgrass, sod grass, and bare ground (Blackburn 1975; Wood and Blackburn 1981; Blackburn et al. 1986; Knight et al. 1984; Thurow et al. 1986).
- In Kansas, Linnell (1961) found that infiltration was higher when little bluestem (*Schizachyrium scoparium*) was the dominant species compared to sideoats grama-dominated sites.
- Plant community types were an important consideration in determining infiltration rate on a loamy blue grama (*Bouteloua gracilis*) range site on the High Plains of Texas (Spaeth 1990). Infiltration rates were highest in the historic plant community, blue grama (*Bouteloua gracilis*) stand, and invasive stands of broom snakeweed (*Gutierrezia sarothrae*). Infiltration rate and capacity were lower in the blue grama/buffalograss (*Bulbilis dac-*

(continued)

tyloides) stands. The lowest infiltration rates and capacities occurred in the buffalograss (periphery of prairie dog towns) and three-awn (*Aristida oligantha*)/Texas tumblegrass (*Schedonnardus paniculatus*) (in active prairie dog towns). There were positive and negative correlations of similarity indices to infiltration capacity: historic reference stands of blue grama (high SI, positive correlation), whereas, three-awn/tumblegrass stands, representative of low SI showed a negative correlation. During the first 15 minutes of the rainfall simulation, variables related to entire and individual species foliar cover, biomass, and litter were most influential in infiltration regression equations. After 15 minutes, soil physical properties (bulk density, aggregate stability) were more influential in predicting infiltration.

- On shortgrass steppe in west Texas, infiltration rates were highest in coppice dunes under broom snakeweed, followed by clumps of blue grama and sideoats grama (*Bouteloua curtipendula*), transition areas of blue grama and sideoats grama, transition areas of broom snakeweed-bare ground, and bare ground. Sediment concentrations were variable between and within treatments; however, they were highest during the first 10 minutes of the rainfall simulation. Interrill erosion rates were lower under broom snakeweed coppice dunes, followed by clumps of blue and sideoats grama, transition areas of blue and sideoats grama-bare ground, transition areas of broom snakeweed-bare ground, and bare ground of interspaces between plants. Regression analysis identified bare ground, grass basal cover, biotic crust cover, organic carbon, and sand and clay content as the most influential factors affecting infiltration rates, sediment concentration, runoff, and interrill erosion (Gutierrez-Castillo 1994).
- Kentucky bluegrass sod cover from 0% to 43%, with a 10% slope was evaluated with sediment rate yields. Rainfall was applied on a sandy loam at 115 mm h^{-1}. Sediment yield was neither linear nor exponential but increased as vegetative cover decreased from 43% to 15%. Sediment yield was greatest with zero percent plant cover; however, a threshold was observed at 15% vegetative cover, where the effectiveness of cover below 15% was significantly diminished—sediment yield reached a degree of equilibrium. The authors concluded that "costly efforts to slightly increase or maintain cover of vegetation below about 15% in semiarid and arid regions appears not to be a valid approach to erosion control" (Rogers and Schumm 1991).
- On a silty clay loam, Dee et al. (1966) found that water infiltrated three times faster under blue grama and silver bluestem (*Bothriochloa saccharoides*) than areas dominated by annual weeds such as summer cypress (*Kochia scoparia*) and windmill grass (*Chloris verticillata*). Buffalograss stands were associated with the lowest infiltration rates (21 mm hr^{-1}; 0.83 in hr^{-1}), while blue grama stands had the highest infiltration rates (63.5 mm hr^{-1}; 2.5 in hr^{-1}).

(continued)

- Mazurak and Conrad (1959) studied the effects of different grasses and fertilization on various warm and cool season grasses at three locations in Nebraska. There was variation among the three sites for infiltration rates of various grasses; however, western wheatgrass (*Pascopyrum smithii*) and big bluestem (*Andropogon gerardii*) were consistently associated with high infiltration rates for each site. Buffalograss was associated with slow infiltration rates on the fine sandy loam site. In general, infiltration rates from lowest to highest for the various grass species were as follows: buffalograss, smooth bromegrass (*Bromus inermis*), blue grama, sideoats grama, crested wheatgrass (*Agropyron cristatum*), western wheatgrass, and big bluestem.
- Plant growth form can have a dramatic effect on infiltration. Pearse and Wooley (1936) compared bare plots with fibrous and tap-rooted species. Compared to the bare plots, the fibrous-rooted plant was associated with a 127% increase in infiltration, whereas tap-rooted species was associated with a 51% increase.
- On Arizona and Nevada rangelands, a 39% decrease in average infiltration rates was observed by removing the vegetative canopy cover. Also, when rock and gravel cover was removed from the clipped sites, infiltration decreased an additional 34%. A 60% decrease in average infiltration rates occurred between the vegetated and bare soil areas. Vegetative canopy cover and surface cover of rock and gravel exhibited similar influences on final infiltration rates. Final infiltration rates decreased significantly as vegetative canopy cover and soil surface rock and gravel cover decreased (Lane et al. 1987).
- Infiltration of water was nearly three times greater in paloverde (*Cercidium microphyllum*) and creosote bush (*Larrea tridentata*) understory (shrub coppice zones) compared to interspace zones. Bulk density was lower and organic matter content was higher in understory coppice zones compared to interspaces (Lyford and Qashu 1969).
- On semiarid rangeland (slopes 16–70%) in the Guadalupe Mountains of southeastern New Mexico, Wilcox and Wood (1988) studied the effects of slope, vegetation, rock, and soil characteristics of curlyleaf muhly (*Muhlenbergia setifolia*)/sideoats grama; cliff muhly (*M. palycaulis*)/wavyleaf oak (*Quercus undulata*); blue grama/one-seed juniper (*Juniperus monosperma*); and needle-and-thread grass (*Stipa comata*)/blue grama communities. Plant cover was generally better correlated with infiltrability than plant biomass.

Text Box 5.3   Case Study on Ecological Site: State and Transition State Changes and Effects on Hydrology in Shortgrass Prairie

Climate

ESD Climate Description: Climate is semiarid dry steppe; summers are hot with winters being generally mild with numerous cold fronts that drop temperatures into the single digits for 24–48 hours. Temperature extremes are common rather than the exception. Humidity is generally low and evaporation high. Wind speeds are highest in the spring and are generally southwesterly. Canadian and Pacific cold fronts come through the region in fall, winter, and spring with predictability, and temperature changes can be rapid. Mean average precipitation is 53 cm (21 in); the majority comes in the form of rain and during the period from May through October. Snowfall averages around 38 cm (15 in), but ranges from 20 cm (8 in) to as much as 91 cm (36 in). Rainfall in the growing season often comes as intense showers of relatively short duration. Long-term droughts occur on the average of once every 20 years and may last as long as 5–6 years (during these drought years, precipitation is reduced by 50–60%). Based on long-term records, approximately 60% of years are below the mean rainfall, and approximately 40% are above the mean. May, June, and July are the main growth months for perennial warm-season grasses, whereas forbs make their growth somewhat earlier. Average frost-free days are 205 and freeze-free days 210.

The ecological site, Deep Hardland Loamy 40–53 cm (16–21 in) precipitation (R077CY022TX), major land resource area (MLRA): 077C-Southern High Plains, Southern Part is used to demonstrate plant composition effects on infiltration. On this site, plant foliar cover is important in reducing runoff and erosion; however, individual plant species also have a profound impact on hydrology. The reference plant community (Fig. 5.5b) is shortgrass prairie grassland dominated by blue grama (*Bouteloua gracilis*) (60–70% composition by weight) and buffalograss (*Bulbilis dactyloides*) (15–25% composition by weight). Other shortgrass species and a variety of forbs comprise the remaining plant composition. Typically, forbs contribute around 5–8% of the total production. A few woody species, cholla cactus (*Cylindropuntia imbricata*), prickly pear (*Opuntia* spp.), broom snakeweed (*Gutierrezia sarothrae*), or occasional yucca (*Yucca* spp.) will be present, usually only 1–2% of the total plant community. Although honey mesquite (*Prosopis glandulosa*) is not a native component species on this ecological site, it can be invasive. The Deep Hardland ecological site can exhibit high plant species richness and diversity (Spaeth 1990).

With continued heavy grazing pressure, the plant community shifts to an equal distribution of blue grama and buffalograss (Fig. 5.5a, phase 1.2). In community phase 1.2, the soil can become more compacted, and subsequently, rainfall infiltration capacity is reduced and runoff increases. Further long-term grazing pressure can result in a transition to State 2.1, where

(continued)

a)

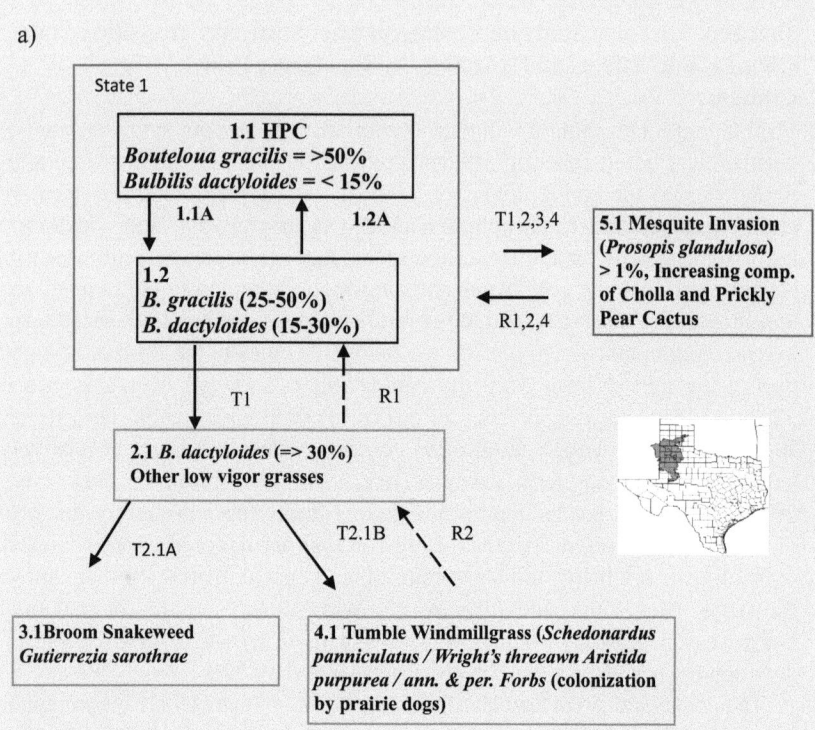

State 1

1.1 HPC
Bouteloua gracilis = >50%
Bulbilis dactyloides = < 15%

1.1A ↓   ↑ 1.2A

1.2
B. gracilis (25-50%)
B. dactyloides (15-30%)

T1,2,3,4 →

R1,2,4 ←

5.1 Mesquite Invasion
(*Prosopis glandulosa*)
> 1%, Increasing comp.
of Cholla and Prickly
Pear Cactus

T1 | R1

2.1 B. dactyloides (=> 30%)
Other low vigor grasses

T2.1A   T2.1B   R2

3.1 Broom Snakeweed
Gutierrezia sarothrae

4.1 Tumble Windmillgrass (*Schedonardus
panniculatus / Wright's threeawn Aristida
***purpurea / ann. & per.* Forbs (colonization**
by prairie dogs)

Legend

1.1A   Lack of Prescribed Grazing, heavy grazing use

1.2 A   Prescribed grazing, above average spring, summer precipitation

T1   Traversed threshold, Transition from State 1 to 2; Lack of Prescribed Grazing, drought, prairie dog use

R1   Restoration from State 2.1 to 1.2; Tenuous; time factor could be decades; Climate pulse—above average summer
         precipitation; Prescribed grazing; Periodic deferment from grazing

T2.1A   Traversed threshold to State 3 due to stand deterioration, significant broom snakeweed increases > 2% cover

T2.1B   Traversed threshold to State 4 due to prairie dog colonization, significant bare ground increase

R2   Restoration from State 4.1 to 2.1 is tenuous, time factor could be decades, Climate pulse—above average summer
         precipitation; prescribed grazing; Periodic deferment from grazing

T1234   From any state, 1,2,3,4: Mesquite Invasion > 1%, increasing comp. of Cholla and Prickly Pear Cactus

R124   Restoration from State 5.1 to 1.2 will require one or more of the following conservation practices (brush management,
         prescribed burning, and prescribed grazing).

Fig. 5.5 (**a**) State and Transition Diagram for Ecological Site Deep Hardland Loamy 6.3–8.3 cm (16–21 in). PZ (R077CY022TX), major land resource area (MLRA): 077C-Southern High Plains, Southern Part. Site Muleshoe National Wildlife Reserve, Muleshoe, Texas. State 1 (reference plant community) and possible alternative states with recovery pathways and types of conservation practices needed to maintain or restore to respective states. (**b**) Photos of states 1.1, 1.2, 2.1, and 4.1. (Photos by author)

(continued)

Fig. 5.5 (continued)

buffalograss dominates the stand. Once buffalograss dominates the stand, the transition to State 1 can be long-term (decades) because of the ecohydrologic dynamics of buffalograss (see Rangeland Hydrology and Erosion Model (RHEM) modeling results and discussion). The dominant buffalograss State 2.1 also occurs as a transition from State 4.1, which results from prairie dog colonization and abandonment. This transition may take decades and depends on climate and management of the site.

Combinations of long-term heavy grazing pressure and drought can facilitate the increase of the native half-shrub broom snakeweed. Sandier soil pockets and components within the ecological site are also more conducive to broom snakeweed invasion (Spaeth 1990). This low-growing (less than 0.5 m tall) suffrutescent plant is poisonous and is considered undesirable by many landowners because it suppresses growth of other native grasses and forbs. Allelopathy may be a factor as it is correlated with reduced grass and forb production, which enhances its life cycle (Lowell 1980). Plant diversity is low in stands with dominance of broom snakeweed (Spaeth 1990).

Mesquite and cholla cactus can be invasive on this ecological site (State 5.1). Once this state becomes established, gains momentum, and woody

(continued)

densities increase, more stringent applications of conservation practices will be necessary (brush management, prescribed burning, and grazing). The economic inputs to convert State 5.1 to 2.1 can be high.

Black-tailed prairie dogs (*Cynomys ludovicianus*) often referred to as "ecosystem engineers" and "keystone species" (Lawton and Jones 1995; Power et al. 1996) in shortgrass prairie can have a profound effect on grassland structure, composition, and ecosystem dynamics (Winter et al. 2002; Fahnestock et al. 2002). Where prairie dogs are abundant, grassland vegetation can be altered dramatically with extensive and persistent burrow systems. Prairie dogs have intrinsic biological value in grasslands; colonies can provide refugia for subdominant grasses, forbs, and shrubs (Coppock et al. 1983); soil structure and chemistry can be modified; nutrients can be altered (Whicker and Detling 1988); and modifications in habitat can benefit other grassland animals (Clark et al. 1982; Lomolino and Smith 2003). Although the disturbance regime can be extreme in active prairie dog colonies, floristic richness can be high, even greater than State 1.1 (Bonham and Lerwick 1976; Klatt and Hein 1978; Coppock et al. 1983; Martinson et al. 1990; Spaeth 1990; Fahnestock and Detling 2002). Soil surface physical and chemical condition changes created by prairie dog colonization also have a significant effect on decreasing infiltration capacity, soil water storage, and increased runoff and erosion (see RHEM modeling results and discussion) (Table 5.2).

Figure 5.8a–d provides an overview of plant communities and summary of precipitation, runoff, sediment yield, and soil loss rates for the annual average and 2-, 5-, 25-, 50-, and 100-year runoff recurrence intervals. For the Deep Hardland Loamy ecological site, hydrology and soil loss are highly variable across the respective states. As management and climate affect cover, production, and species composition, significant ecological (species composition) and hydrologic changes occur—sometimes abruptly or subtly. The decline of foliar plant cover and production affect the hydrologic regime; however, plant life/growth forms, such as tallgrasses, mid grasses, shortgrasses, forbs, shrubs, half-shrubs, and trees, and their compositional differences on a site greatly influence infiltration and runoff dynamics. Infiltration is usually highest under trees and shrubs and decreases progressively in the following order: bunchgrass, sod grass, and bare ground (Carlson et al. 1990; Thurow 1991; Weltz and Blackburn 1995).

Individual plant species also have a profound effect on hydrology and erosion dynamics, i.e., different species of grasses, forbs, and shrubs (USDA-NRCS 2003; Spaeth 1996a, b). Field studies have documented infiltration capacity with individual species composition. Dee et al. (1966) found that water infiltrated three times faster in blue grama and silver bluestem (*Bothriochloa saccharoides*) stands than areas dominated by annual weeds such as summer cypress (*Kochia scoparia*) and windmill grass (*Chloris verticillata*). Blue grama terminal infiltration capacity was about four times higher

(continued)

Table 5.2 RHEM model inputs for evaluation of hydrologic impact of transitions from one ecological state to another ecological state for Deep Hardland Loamy 40–53 cm (16–21 in) PZ (R077CY022TX) site, Berda loam in the surface horizon (RHEM parameterization can be English or metric)

Input parameter	Reference State 1.1	State Phase 1.2 Bogr/Buda	State 2.1 Buda	State 3.1 Gusa	State 4.1 Scpa/ Arol
Soil texture	Clay	Clay	Clay	Clay	Clay
Soil water saturation (%)	25	25	25	25	25
Slope length (ft)	100	100	100	100	100
Slope shape	Linear	Linear	Linear	Linear	Linear
Slope steepness (%)	2	2	2	2	2
Foliar canopy cover (%)					
Bunch grass foliar cover (%)	90	45	0	25	5
Forbs and/or annual grass foliar cover (%)	5	5	5	10	5
Sod grass foliar cover (%)	5	50	90	10	5
Woody foliar cover (%)	0	0	0	0	90
Ground surface cover %					
Basal cover (%)	10	6	5	1	1
Rock cover (%)	0	0	0	0	0
Litter cover (%)	30	20	5	0	10
Biological Cross Cover (%)	0	0	0	0	0

than buffalograss stands, holding soil type constant. Figure 5.6 shows comparative infiltration rates derived from rainfall simulation experiments for various ecological states and phases (Spaeth 1990). Initial infiltration capacity from the onset of rainfall to 25 minutes was slightly different for the reference state—blue grama (Bogr) and perennial broom snakeweed (Gusa) stands; however, long-term infiltrability (near-saturated hydraulic conductivity) was the same. The Gusa stands, indicative of low similarity index values, higher percentage of bare ground, low graminoid and forb cover, and high sub-shrub cover had infiltration rates similar to the reference Bogr stands (representative of high similarity index). This demonstrates that the Gusa stands, representative of low biotic integrity and similarity index [with significant changes in plant functional groups (graminoid-to-woody), high composition of an invasive plant, and loss of native grass cover] still maintain adequate hydrologic function. However, soil loss was higher in Gusa stands compared to the reference stands (1.1 Bogr and 1.2 Bogr/Buda) due to higher bare ground between snakeweed canopies (Fig. 5.6). What factors may be responsible for the near-identical infiltration curves for the reference Bogr sands and the Gusa stands? The answer most likely is due to the morphology of the plants and coppice dune formation if present. Field studies show that infiltration capacity in

(continued)

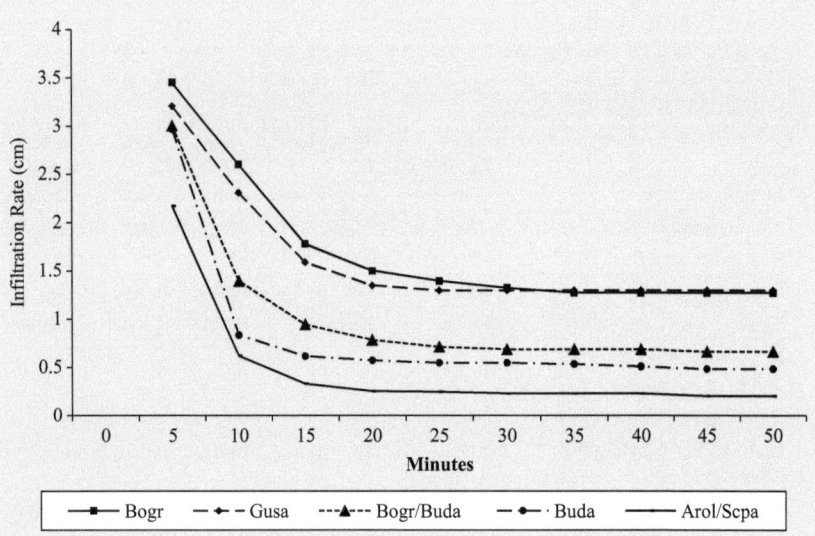

Fig. 5.6 Comparative infiltration on five ecological states associated with a Deep Hardland Loamy ecological site, Berda loam soil in west Texas. Reference State 1.1 Bogr = blue grama; State phase 1.2 Bogr/Buda (blue grama and buffalograss); State 2.1 Buda = buffalograss; State 3.1 Gusa = perennial broom snakeweed; and State 4.1 Arol = perennial three-awn, Scpa = Texas tumblegrass

bunchgrass stands have inherently higher rates compared to sod grass stands (Mazurak and Conrad 1959; Dee et al. 1966; Spaeth 1990, 1996a, b; Pierson et al. 2002). Root morphology in grasses has a significant impact on infiltration. Some shrubs and half-shrubs are associated with coppice dunes or mounds composed of litter and wind transported soil. Coppice dunes form under broom snakeweed plants, which create a zone of high infiltrability and low runoff. Field experiments show that surface soil organic carbon, bulk density, percentage silt, and infiltration and interrill erosion rates are significantly higher for shrub-coppice compared to shrub-interspace areas (Blackburn 1975; Johnson and Gordon 1988; Blackburn et al. 1990, 1992; Spaeth 1990; Blackburn et al. 1992a, b).

Infiltration capacity of State phase 1.2 is different from the reference community 1.1, where blue grama is the dominant species (Fig. 5.6). State phase 1.2 is representative of increasing buffalograss, where the ratio of blue grama and buffalograss is close to 1:1. As buffalograss increases in the stand, infiltration capacity decreases. This is also evident in State 2.1, where buffalograss occurs almost in a monoculture (Fig. 5.6). Dominant stands of buffalograss (state 2.1; Fig. 5.6) are common around the periphery of active prairie dog colonies and in pastures where grazing has been consistently heavy. Buffalograss is a short shoot plant (grazing tolerant plant with protected

(continued)

meristematic tissue, growing points) that is more tolerant to drought and hot temperatures than blue grama (Weaver 1954) and reproduces sexually (seed) and vegetatively (surface runners-stolons). Research shows that buffalograss also exhibits a dense shallow fibrous root system (root pan) that is correlated with significantly reduced infiltration capacity (Spaeth 1990, 1996a, b). In some grass stands, where roots are found in the inter-aggregate pores, water repellent compounds form on soil aggregates and soil structural peds as a result of decaying organic matter and the production of humic and fulvic acids (Bisdom et al. 1993; Dekker and Ritsema 1996). Ritsema et al. (1998a, b) state that water repellency is considered a plant-induced soil property. Sources of water repellent compounds include accumulated plant-derived organic matter from mulch, decomposing roots, and plant material and root exudates (Doerr et al. 1996; Czarnes et al. 2000). Particulate organic matter contains plant and microbial produced compounds such as waxes (Franco et al. 2000; Schlossberg et al. 2005); humic acids (Spaccini et al. 2002); a presence of a protective water-repellent lattice of long-chain polymethylene compounds around soil aggregates (Shepherd et al. 2001); aliphatic C present in organic matter (Ellerbrock et al. 2005); mycorrhizal and saprobic soil fungi (Bond and Harris 1964; Paul and Clark 1996; Hallett and Young 1999; White et al. 2000; Rillig 2004); basidiomycete fungi (Bond and Harris 1964; Fidanza 2003); fungal proteins such as hydrophobins (Rillig 2005; Rillig and Mummey 2006); and fatty acids, fulvic acids, extracellular enzymes, and polysaccharides (Bisdom et al. 1993; Kostka 2000; Eynard et al. 2006).

State 4.1 was produced by prairie dog colonization. Although plant cover was minimal in the active colonies, plant species diversity was greater than all the contrasting states associated with this ecological site because of disturbance factors (Fig. 5.5) (Spaeth 1990). Several authors have shown that floristic richness increases with prairie dog and other rodent disturbance (Bonham and Lerwick 1976; Klatt and Hein 1978; Coppock et al. 1983; Martinson et al. 1990; Fahnestock and Detling 2002). In State 4.1, infiltration capacity is significantly reduced, and erosion potential is higher than any of the other states represented in this ecological site (Fig. 5.6).

In summary, the extent of vegetation cover and individual plant species (within a life/growth form or contrasting growth habit) can be primary factors that influence spatial and temporal variability of surface soil processes controlling infiltration and interrill erosion rates on rangeland (Fig. 5.7).

RHEM Risk Assessment Analysis
RHEM outputs the risk assessment data, as shown in Fig. 5.8e and Table 5.3, same data, different format. Interpretations are as follows: there is a 50% chance that soil loss will be less than 0.49 t ac^{-1} in the Bogr state, a 3% chance in Bogr/Buda, and zero % in Buda, Gusa, and Arol/Scpa (Fig. 5.8, Table 5.3). There is a 30% chance that soil loss will be within 0.49 and 0.72 t ac^{-1} in the

(continued)

Fig. 5.7 (**a**) Rangeland hydrology and erosion model estimated average annual precipitation and runoff for representative states (see state and transition model Fig. 5.6). (**b**) Average sediment yield and soil loss. Deep Hardland ecological site near Muleshoe, Texas. Model output in English units

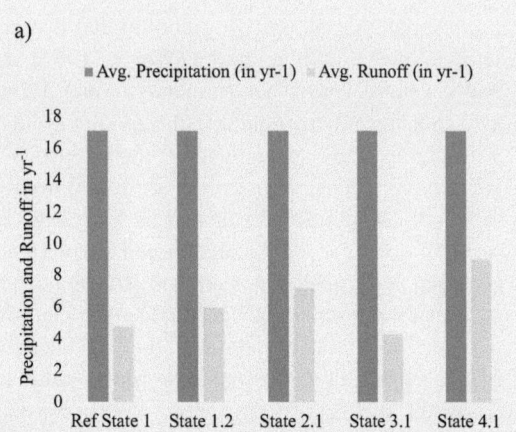

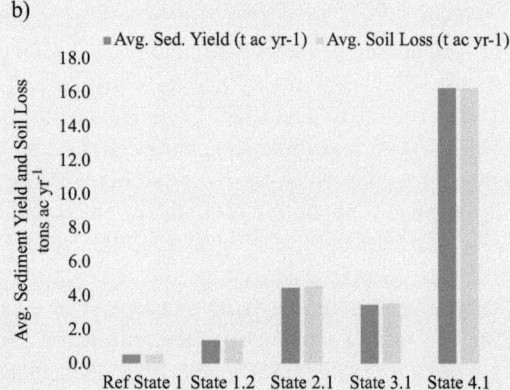

Bogr state, 6% in Bogr/Buda. In the Bogr/Buda state, there is a 5% chance that soil loss will exceed 1.03 t ac^{-1}, whereas the probability of soil loss exceeding 1.03 t ac^{-1} is high in Bogr/Buda (70%), Buda (97%), Gusa (100%), and Arol/Scpa (100%). Table 5.4 shows a frequency analysis of soil loss on this site for return period storms 2 thru 100 years.

Summary

Analysis of the RHEM simulation runs on the Deep Hardland Loamy 406–533 mm (16–21 in precipitation) ecological site provides a basis for interpreting the impacts of vegetative canopy cover, surface ground cover, and topography on runoff as well as sediment detachment, transport, and deposition in overland flow for each ecological state. RHEM can predict runoff and erosion as a function of vegetation structure and behavior of different plant community phases and the amount of cover for the different states.

(continued)

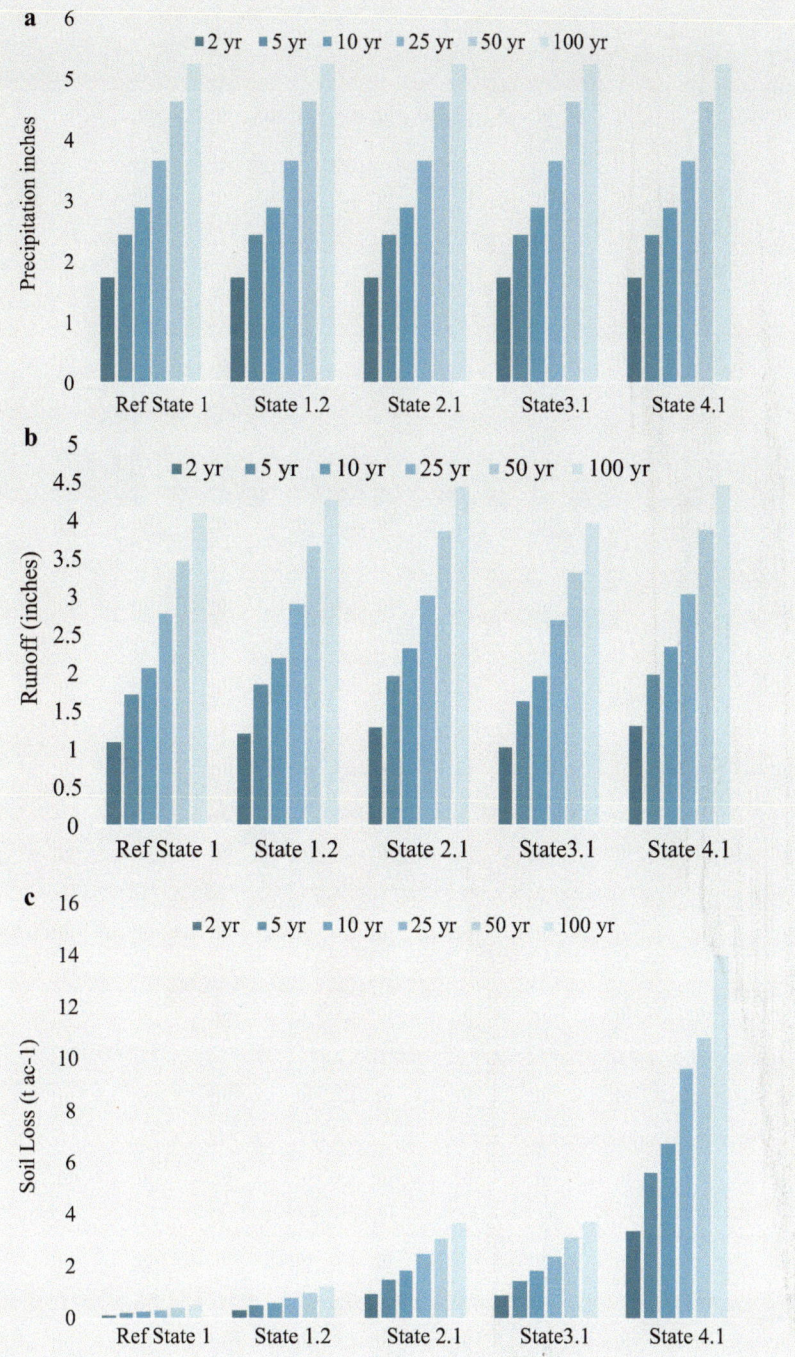

Fig. 5.8 Rangeland hydrology and erosion model estimated return period estimates (2–100 year). (**a**) precipitation; (**b**) estimated runoff; (**c**) estimated soil loss; (**d**) estimated sediment yield; and (**e**) probability of occurrence for yearly soil loss for all scenarios using erosion classes of low (50%), medium (80%), high (95%), and very high (>95%)

(continued)

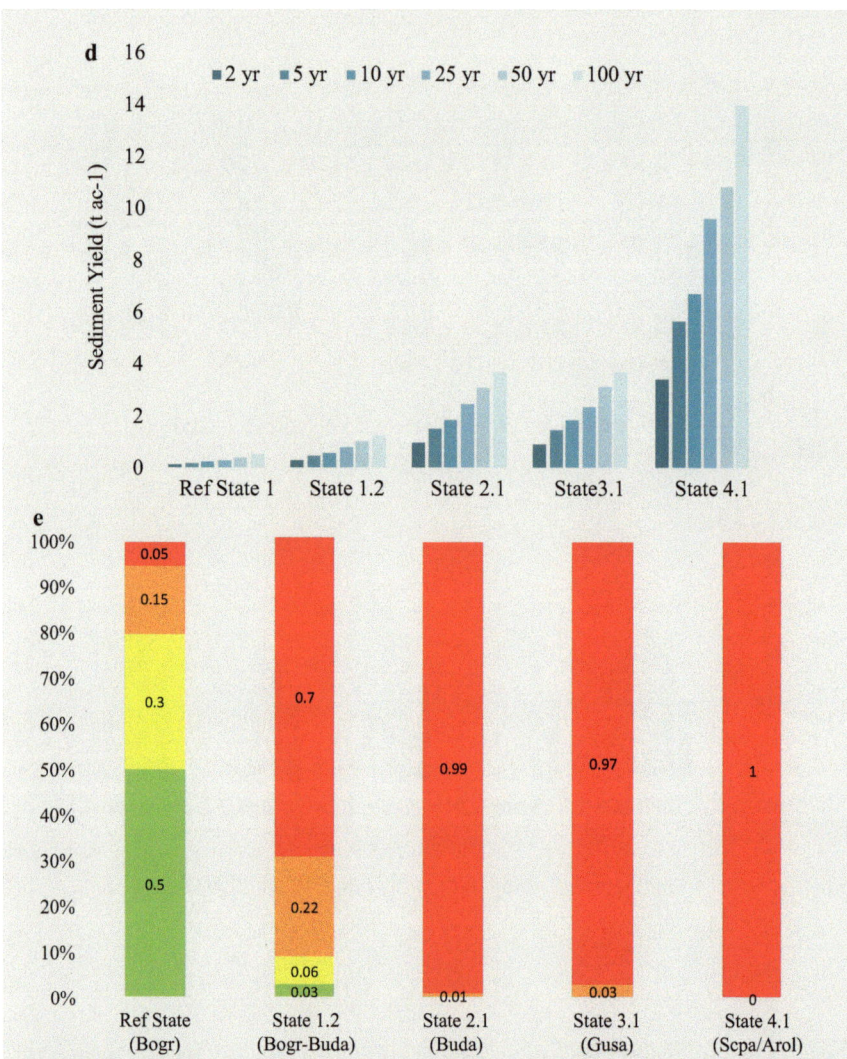

Fig. 5.8 (continued)

Table 5.3 RHEM risk assessment of accelerated soil erosion. (RHEM output in English units)

Range of annual soil loss (ton ac⁻¹)	Bogr	1.2 Bogr/Buda	2.1 Buda	3.1 Gusa	4.1 Scpa/Arol
Low $x < 0.49$	0.5	0.03	0	0	0
Medium). $49 \le X < 0.72$	0.3	0.06	0	0	0
High $72 \le X < 1.03$	0.15	0.22	0.01	0	0
Very high $X > 1.03$	0.05	0.7	0.97	1	1

(continued)

Table 5.4 Frequency analysis of annual soil loss (ton ac^{-1} year) by return period for Deep Hardland ecological site

Return period	State 1.1 Bogr	State 1.2 Bogr-Buda	State 2.1 Buda	State 3.1 Gusa	State 4.1 Arol-Scpa
2	0.118	0.306	0.962	0.899	3.388
5	0.19	0.48	1.504	1.445	5.592
10	0.239	0.599	1.853	1.834	6.68
20	0.301	0.756	2.328	2.309	8.274
30	0.328	0.878	2.613	2.557	9.685
40	0.36	0.91	2.733	2.677	10.33
50	0.37	0.918	2.758	2.781	10.782
60	0.409	1.024	3.074	3.135	10.869
70	0.425	1.057	3.164	3.193	11.881
80	0.434	1.076	3.22	3.212	12.976
90	0.461	1.146	3.428	3.434	13.604
100	0.492	1.224	3.663	3.699	13.915

There are significant differences in estimated annual soil erosion rate between the ecological states on this ecological site. The drivers are plant composition, largely the interaction between the two dominant C4 grass species, blue grama and buffalograss. As buffalograss increases in the stand, infiltration capacity will decrease. The causative factors are associated with root morphological differences between blue grama and buffalograss and the degree of water repellency found in buffalograss stands. Water repellent compounds appear to be associated with stands of buffalograss, although more research is needed to confirm the dynamics. Prairie dog activity has a profound effect on biotic integrity, hydrologic function, soil and surface stability, and similarity index calculations. A high degree of bare ground and significant changes in plant composition are associated with prairie dog colonization. Runoff and soil loss can be extreme in Arol/Scpa. Broom snakeweed stands and the reference state, blue grama, exhibit the highest infiltration capacity on this site; however, broom snakeweed stands have significantly higher soil loss because of a depauperate understory and high bare ground in shrub interspaces.

High intensive convective storms can have a significant impact on this site. During 5-, 10-, 25-, 50-, and 100-year storms, where there is a high short burst of rainfall, a significant amount of runoff and soil loss will occur. Management of this site should strive to maintain a higher ratio of blue grama to buffalograss. The threshold where increasing buffalograss begins to affect infiltration capacity is around 30%. Infiltration experiments on this site show that plant-related variables such as cover, biomass, and species composition largely influence infiltration dynamics during the early phases of rainfall (0–15 minutes), whereas soil-related variables such as bulk density, aggregate stability, and porosity influence infiltration as storms progress >15 minutes.

Text Box 5.4 Water Budget Comparisons Between Grass Stands Dominated by Warm (C4) and Cool (C3) Grasses with Bunch and Sod-Forming Growth Habits

Table 5.5 is an example of a modeled water budget grass stands of varying composition for a loamy soil in Major Land Resource Area (MLRA 106; Nebraska and Kansas Loess-Drift Hills). Stand I represents tallgrass prairie reference state (historic plant community) dominated by native C4 bunchgrasses. Stand II is dominated by Kentucky bluegrass (*Poa pratensis*) a sod-forming C3 species, and Stand III is dominated by smooth bromegrass (*Bromus inermis*) a C3 sod form with subdominant Kentucky bluegrass. Note the hydrologic differences between native bunchgrasses and invasive sod-forming grasses. Twenty percent of the total precipitation was runoff in Stand I (tallgrasses) and 40% and 30% on sites dominated by Kentucky bluegrass and smooth bromegrass, respectively.

Table 5.5 Water budget examples in MLRA 106, Nebraska and Kansas Loess-Drift Hills. Loamy site, 635 mm (25 in) average annual precipitation

Plant composition (%)			
	Stand I	Stand II	Stand III
Little bluestem (C4, bunch)	30–50		
Big bluestem (C4 bunch)	15–30		
Prairie dropseed (C4 bunch)	10		
Porcupine grass (C3 bunch)	40		
Sideoats grama (C4 bunch)	5		
Grasses (subdominants) Blue grama (C4 bunch, sod) Sedges (C3 bunch) Prairie June grass (C3 bunch) Buffalograss (C4 sod)	5		
Kentucky bluegrass (C3 sod)	0	75	25
Smooth bromegrass (C3 sod)	0	25	75
Hydrologic data			
	Stand I Inches (%)	Stand II Inches (%)	Stand III inches (%)
Precipitation (in)	25	25	25
Infiltration	19.3 (77)	13 (52)	17 (68)
Runoff	5 (20)	11.2 (45)	7.5 (30)
Grass and litter interception	0.13 (0.05)	0.10 (0.04)	0.15 (0.06)
Evapotranspiration (ET)	18.1 (94)	18.3 (95)	18.3 (95)
Soil evaporation	11.5 (60)	11.5 (60)	11.5 (60)
Plant transpiration	6.5 (34)	6.7 (35)	6.7 (34)
Deep percolation	0.6 (2.5)	1 (4)	2 (0.05)
Change in soil water (affected by antecedent soil moisture)	0	−1.4	−0.6

Stands I, II, and III vary with respect to species composition. Field data from rainfall simulations collected by NRCS and modeled with Simulation of Production and Utilization of Rangelands (SPUR) and Rangeland Hydrology and Erosion Model (RHEM). (Modeled data in English units)

Text Box 5.5 Infiltration rates for various grass species (cool and warm season and bunch and sod forming)
(data from Mazurak and Conrad 1959). Bunchgrass and sod-forming grasses can have a significant impact on infiltration rate on range and pasturelands (Fig. 5.9).

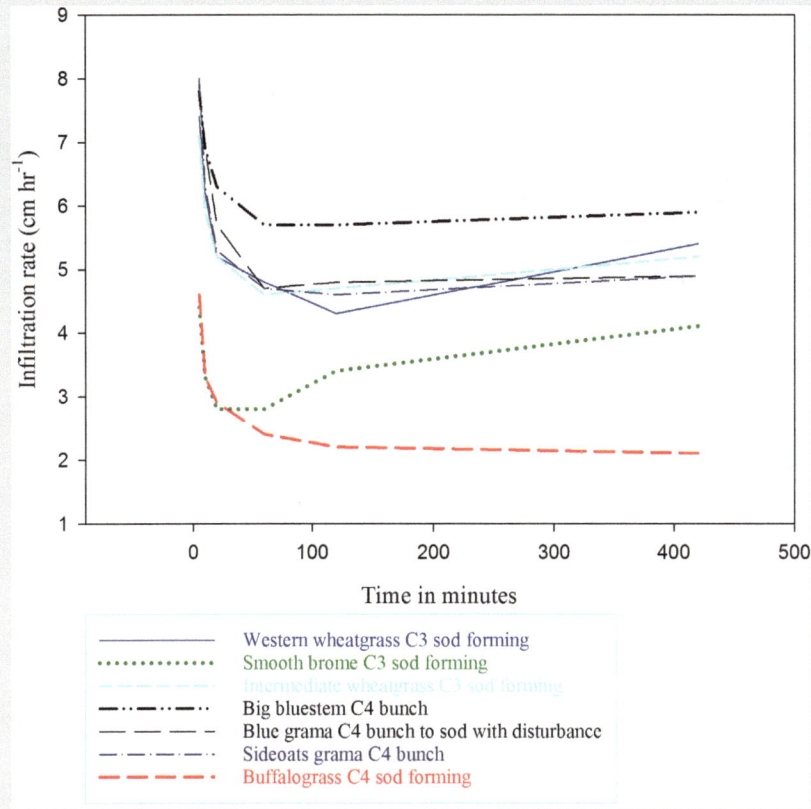

Fig. 5.9 Infiltration rates with different grass species plantings, including sod-forming and bunchgrass species (data Mazurak and Conrad 1959)

Text Box 5.6 Case Study: Plant Cover and Biomass Effects on Infiltration Rate in California Annual Grassland
At Stemple Creek, California, hydrologic comparisons with a rainfall simulator were made between two adjacent grazed pastures with remaining standing biomass: Pasture I [1120–1681 kg ha^{-1} (1000–1500 lbs ac^{-1} residue)] and pasture II [2242–2802 kg ha^{-1} (2000–2500 lbs ac^{-1} residue)]. The soil component was identical, and there were no differences in bulk density. These experimental sites were adjacent (20 m apart) separated by a fence. Both pastures had similar annual grass composition [California oatgrass (*Danthonia californica*), rye brome (*Bromus secalinus*), rattail fescue (*Vulpia myuros*), ripgut brome (*Bromus diandrus*), and filaree (*Erodium cicutarium*)] and >95% foliar cover. The pasture with the greater amount of biomass had a higher overall infiltration rate, and final steady-state infiltration was almost 70% higher than the pasture with lower biomass. Rainfall interception was not a major factor during the 60-minute simulation. Additional root biomass may have been a major factor in the infiltration capacity differences between the two pastures. Remaining filaree biomass was higher in the pasture with greater biomass, and this tap-rooted species may have provided more micropores for infiltration (Fig. 5.10).

Fig. 5.10 Rainfall simulation experiment, infiltration rate vs. time at Stemple Creek site, California. Adjacent plots, soil type identical, with two biomass residue levels

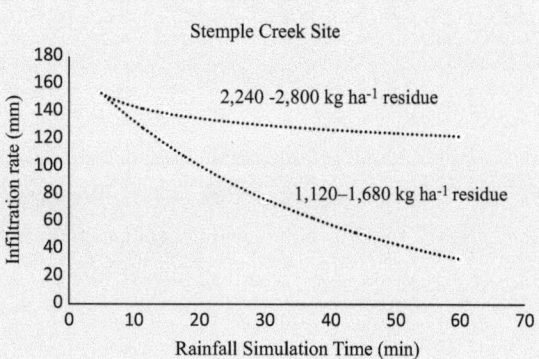

Text Box 5.7 Case Study: Comparative Infiltration and Soil Loss between Annual and Perennial Grasses in the California Annual Grassland (Tables 5.6 and 5.7)
Rainfall simulations (60 minutes) were conducted on two adjacent ungrazed sites (holding soil constant): site I was an annual grass stand, and site II was a stand of native purple needlegrass (*Nassella pulchra*). Runoff occurred after 3 minutes on the annual stand and was minimal after 60 minutes on the needlegrass site. Cumulative infiltration was 90 mm greater on the perennial needlegrass site. Cumulative sediment yield was negligible on the needlegrass site and 233 kg ha^{-1} (207 lbs ac^{-1}) on the annual grass site.

(continued)

Table 5.6 Rainfall simulation experiment at Upper Stoney Watershed, California

Location	Treatment	Soil component	Foliar cover (%)	Rainfall simulation rate (mm hr⁻¹)	Steady-state infiltration (mm hr⁻¹)	Cumulative infiltration (mm hr⁻¹)	Equiv. total runoff volume (m³ ha hr⁻¹)	Cumulative sediment yield (kg ha⁻¹)
Upper Stoney watershed	Annual grass stand	Zamora loam 6% slope	75	101.6	0.2	6.1	955.0	233.8
Upper Stoney watershed	Native perennial grass stand, purple needlegrass	Zamora loam 6% slope	100	101.6	96.5	96.5	50.8	0.1

Treatments were annual grass stand and native perennial purple needlegrass (*Nassella pulchra*). Adjacent plots, soil type identical

Text Box 5.8 Water Intake Comparisons for Grain Crop vs. Seeded Perennial Grass

Mazurak et al. (1960) compared grain crop infiltration with various age seeded perennial grass stands (intermediate wheatgrass (*Agropyron intermedium*) and smooth bromegrass (*Bromus inermis*)). Infiltration rates of grass seedlings starting at 4 years of age were significantly different from grain-fallow system. Infiltration increased with age of grass stands; the authors discuss possible effects due to more developed root systems (Fig. 5.11).

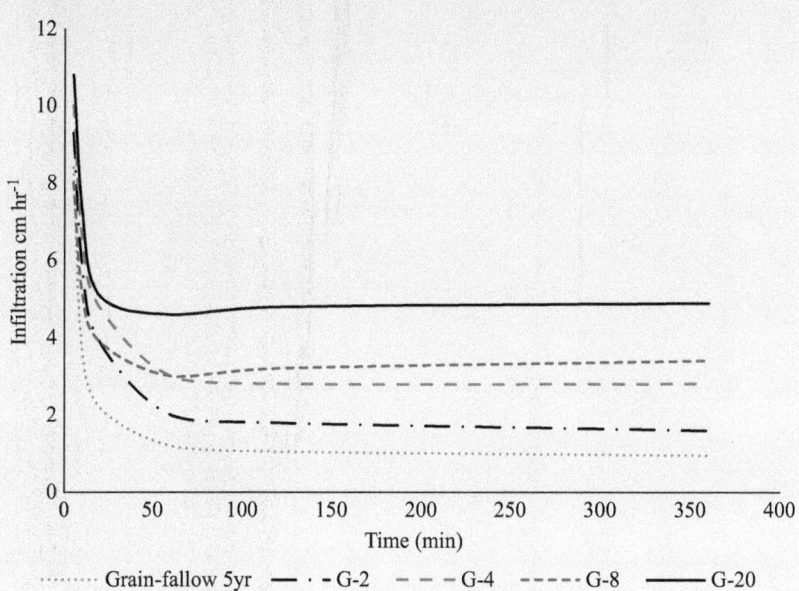

Fig. 5.11 Comparison of infiltration of grain crop with different age intermediate wheatgrass and smooth bromegrass pastures. Notes: plots hayed with 44 kg ha⁻¹ (40 lbs ac⁻¹) ammonium nitrate in spring, not grazed. (age of stand-in years: G-2 = 2 yr. perennial grass . . . G-20 = 20 yr. perennial grass)

Text Box 5.9 Hydrologic and other Environmental Gradients Associated on Mixed-grass prairie with Clubmoss (*Lycopodium dendroideum*) Invasion

This study shows how different plant composition between native and invasive plant species are associated with different runoff rates (Fig. 5.12).

All three sites, same section, fenced in 1925, mostly hayed until 1958 and then divided into 4 (65 ha; 160 ac units). No record of fertilization, prescribed burning, or brush control. All sites have occasional use by antelope. Summer and fall season-long grazing in a one of 3-year rest cycle (unknown when this grazing system started).

Site information:

- Three mixed-grass prairie sites (430 mm, 17 in avg. ppt)
- 9.6 km (6 mi) NW of Killdeer, ND (June 1992)
- Sandy Ecological Site
- Parshall fine sandy loam (study on 11% slope)
- Rainfall simulation 6.3 cm hr^{-1}, 2.5 in hr^{-1} for 1 hr
- 3 × 9 m (10 × 30 ft) plots, 6 plots per site, 3 sites on the same soil and ecological site
- Complete soil descriptions and samples by plot
- Complete vegetation cover, production by plot

Fig. 5.12 Clubmoss root system on rainfall simulation site (photo by author)

(continued)

Dominant plant species ranking by simulation groups. Similarity index is the percent of similarity of existing stand compared with historic reference plant community (calculations based on weight). Species data is kg ha^{-1} (lbs ac^{-1}).

Table 5.7 Dominant plants on three different plant communities holding soil and ecological site constant. Each group consisted of six rainfall simulation plots

Group 1		Group 2		Group 3	
Similarity index	60%	Similarity index	52%	Similarity index	43%
Needle and thread	235.4 (210)	Clubmoss	391 (349)	Sedge	409 (365)
Prairie sandreed	176 (157)	Sedge	372 (332)	Blue grama	287 (256)
Prairie June grass	22.2 (49)	Blue grama	156 (139)	Clubmoss	236.5 (211)
Blue grama	55 (45)	Prairie June grass	116.5 (104)	Prairie sandreed	93 (83)
Club moss	38 (34)			Needle and thread	90.8 (81)

Data values are kg ha^{-1} (lbs ac^{-1})

Prairie sandreed (*Calamovilfa longifolia*), Prairie June grass (*Koeleria macrantha*), blue grama (*Bouteloua gracilis*), club moss (*Lycopodium dendroideum*), sedge spp. (*Carex* spp.), and needle-and-thread grass (*Hesperostipa comata*)

Gradient analysis is an analytical approach to summarize the structure and variation of vegetation across a landscape in terms of gradients. An indirect approach, indirect gradient analysis (IGA) (ter Braak 1987) was used, which requires no prior knowledge about how environmental factors affect plant composition or vice versa. The two-step method of IGA was used in this study: (1) ordination was performed using non-metric multidimensional scaling (NMS) (PC-ORD; McCune and Mefford 2011) with Sorensen distance measure to extract the dominant pattern of plant variation in community composition (main matrix) and (2) analyze by correlation, the ordination pattern to measured environmental variables (secondary matrix). In Fig. 5.13 the joint plot arrows identify higher values for the respective gradients, while the opposite extension of the arrow is associated with the lowest values.

The ordination revealed a number of gradients along axis 1 (Fig. 5.13). For example, the second group of simulation plots (H2) dominated by clubmoss, sedge, blue grama, and prairie June grass were significantly correlated with

(continued)

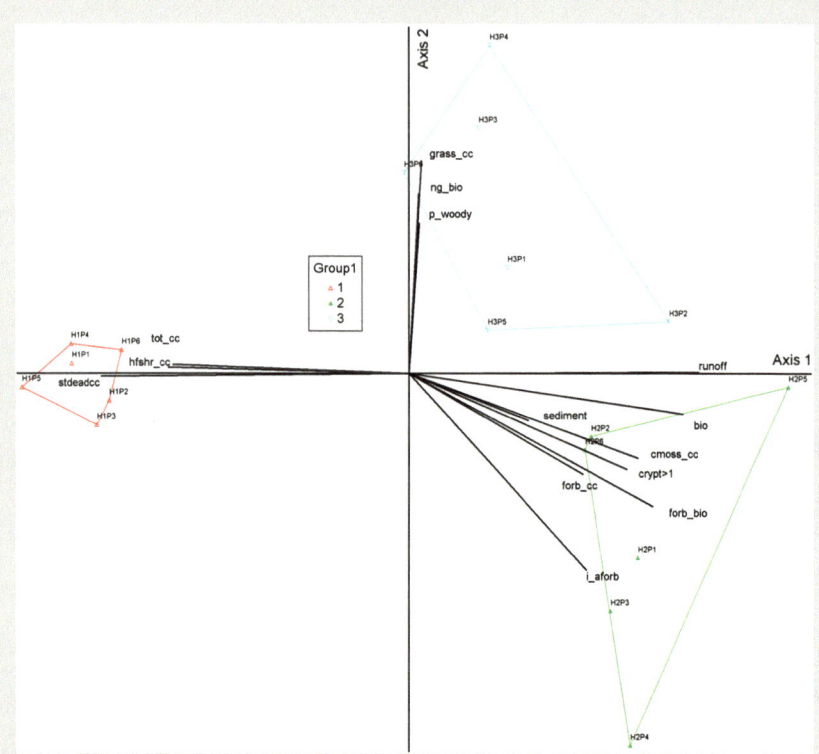

Fig. 5.13 Hfshr_cc = half-shrub canopy cover, Tot_cc = total canopy cover, Stdeadcc = standing dead canopy cover, N_grass = native grass, Ro_rain = runoff/rainfall ratio, Basal_gc = plant basal cover, Grass_cc = grass canopy cover, Bio = biomass; Forb_ cc = forb canopy cover, Forb_bio = forb biomass, cmoss_cc = club moss canopy cover, Runoff = runoff (dry run condition), crypt > 1 = cryptogam >2.54 cm

higher runoff. Other variables associated with group 2 were forb canopy cover and biomass, cryptogams greater than 25.4 mm, and basal plant cover. Group 1, associated with the lowest runoff and the vegetation, was most similar to the historic plant reference community with a similarity index of 60%. The dominant plants in group 1 were needle and thread, prairie sandreed, prairie June grass, and blue grama—all native perennial grasses. Club moss composition was lowest in group 1 (38 kg ha⁻¹; 34 lbs ac⁻¹) (Table 5.7). Group 3 is intermediate between groups 1 and 2 (391 kg ha⁻¹; 349 lbs ac⁻¹) and was dominated by sedges (gradient arrow showing high grass canopy cover), blue grama, and club moss (236.5 kg ha⁻¹; 211 lbs ac⁻¹). In summary, club moss-dominated plots in group 2 were correlated with higher runoff, whereas the simulation plots with native cool season grasses exhibited the lowest runoff.

Fire Dynamics on Hydrology and Erosion

On many range and forest plant communities, fire is a natural disturbance and has been instrumental in the formation of these ecosystems (Keeley and Rundel 2005); however, interruption of natural fire frequencies and increased fire frequency associated with wildfires since European settlement has altered many plant communities which have transitioned to "no-analog, novel, or emerging ecosystems" (Fuhlendorf et al. 2011). Table 5.8 shows some fire frequencies for selected range and forest plant ecosystems. Note that fire natural fire frequency can be short as in the California annual grassland, tallgrass, and southern short or mixed-grass prairies. In comparison, natural fire frequency in the California coastal redwoods is greater than 1000 years. Rangelands in transition because of combinations of natural and anthropogenic disturbances have caused significant changes on native plant community structure, especially with the presence of exotic introduced annual grasses and forbs, and soil physical properties related to hydrology and erosion.

Table 5.8 Fire frequency comparisons in range and forest plant communities

Vegetation community	Fire severity[a]	Fire regime characteristics			
		% of fires	Mean interval (years)	Minimum interval (years)	Maximum interval (years)
Pacific NW grassland					
Bluebunch wheatgrass	Replacement	47%	18	5	20
	Mixed	53%	16	5	20
Pacific NW shrubland					
Mountain big sagebrush (high elev)	Replacement	100%	20	10	40
Wyoming big sagebrush steppe	Replacement	89%	92	30	120
California shrubland					
California grassland	Replacement	100%	2	1	3
Chaparral	Replacement	100%	50	30	125
California forested					
Coast redwood	Replacement	2%	≥1000		
	Surface or low	98%	20		
Southwest grassland					
Desert grassland with shrubs and trees	Replacement	85%	12		
	Mixed	15%	70		
Plains mesa grassland	Replacement	81%	20	3	30
	Mixed	19%	85	3	150

(continued)

Table 5.8 (continued)

Vegetation community	Fire severity[a]	Fire regime characteristics			
		% of fires	Mean interval (years)	Minimum interval (years)	Maximum interval (years)
Northern and central rockies forested					
Douglas-fir (cold)	Replacement	31%	145	75	250
	Mixed	69%	65	35	150
Grand fir-Douglas-fir-western larch mix	Replacement	29%	150	100	200
	Mixed	71%	60	3	75
Ponderosa pine (Black Hills, low elevation)	Replacement	7%	300	200	400
	Mixed	21%	100	50	400
	Surface or low	71%	30	5	50
Plains grassland					
Central tallgrass prairie	Replacement	75%	5	3	5
	Mixed	11%	34	1	100
	Surface or low	13%	28	1	50
Northern mixed-grass prairie	Replacement	67%	15	8	25
	Mixed	33%	30	15	35
Southern shortgrass or mixed-grass prairie	Replacement	100%	8	1	10
Northeast woodland					
Pine barrens	Replacement	10%	78		
	Mixed	25%	32		
	Surface or low	65%	12		
Eastern woodland mosaic	Replacement	2%	200	100	300
	Mixed	9%	40	20	60
	Surface or low	89%	4	1	7

LANDFIRE Rapid Assessment Vegetation Models. Courtesy of US Forest Service https://www.fs.fed.us/database/feis/fire_regime_table/fire_regime_table.html
[a]Fire severities
Replacement: Any fire that causes greater than 75% top removal of a vegetation-fuel type, resulting in general replacement of existing vegetation; may or may not cause a lethal effect on the plants
Mixed: Any fire burning more than 5% of an area that does not qualify as a replacement, surface, or low-severity fire; includes mosaic and other fires that are intermediate in effects
Surface or low: Any fire that causes less than 25% upper layer replacement and/or removal in a vegetation-fuel class but burns 5% or more of the area

   High fire temperatures from prescribed burning or wildfires on forested, chaparral, and rangeland communities can cause significant increases in soil water repellency resulting in increased runoff and erosion (Pierson et al. 2008). On rangelands, hydrologically important soil physical factors can be altered by various aspects of fire (duration and temperatures) (DeBano et al. 1998; Shakesby and Doerr 2006).

The degree of water repellency or hydrophobicity formed after fire is highly dependent upon soil temperatures and the amount of organic matter present on-site (Pierson and Williams 2016). During fire, the combustion of organic matter at the soil surface is vaporized as heat radiates downward in the soil profile causing organic substances to be translocated downward along a temperature gradient until they cool and condense, which causes a transitional hydrophobic layer parallel to the soil surface (DeBano et al. 1998; DeBano 2000; Shakesby 2011). Hydrophobicity also increases with reductions in soil moisture (Doerr et al. 2000; Pierson et al. 2008).

Fire temperatures have varying effects on humic acids in organic matter. Humic acids and organic compounds (long-chain aliphatic hydrocarbons) are lost at temperatures below 100 °C (212 ° F). At temperatures between 100 and 198 °C (388 °F), nondestructive distillation of volatile organic substances occurs, and at temperatures between 198 and 299 °C (570 °F), about 85% of the organic substances are destroyed by destructive distillation. Soil organic matter is consumed at 500 °C (932 °F) (DeBano et al. 1998)—temperatures which are commonly reached in grassland and shrubland fires (Wright and Bailey 1982; Miller et al. 2013). The duration and temperature of the fire can distill and translocate organic material and other substances downward along a temperature gradient until they cool and condense, which causes a transitional hydrophobic layer parallel to the soil surface, which can form a non-wettable hydrophobic layer (DeBano et al. 1996; Shakesby 2011). Fuels that burn quickly (e.g., grass) or very hot (brush piles) generally do not form a hydrophobic layer in the soil. Water repellent layers in the soil are most common in shrub communities where fires burn from 5 to 25 minutes. This situation is also inherent in chaparral communities where 90% of the decomposed organic matter is usually lost as smoke and ash, and the remaining material is distilled downward and condensed in the soil. Ash on the soil surface can clog surface soil pores and reduce infiltration and increase runoff and erosion (Neary et al. 1999).

Fire temperatures of 175–270 °C (347–518 °F) enhance water repellency, whereas soil temperatures above 270–400 °C (518–752 °F) break down water repellent factors (DeBano et al. 1976; Doerr et al. 2004). The extent of soil water repellency on rangelands is dependent on the amount of plant biomass and residues, soil surface organic matter, and soil temperatures reached during burning (DeBano et al. 1970; Doerr et al. 2004; Pierson and Williams 2016). On rangelands, the combustion of organic matter on the soil surface and subsurface (depending on depth) can dramatically alter soil structure, increase bulk density, decrease porosity and infiltration, and soil aggregation and stability (Pierson et al. 2001; Hubbert et al. 2006; Shakesby and Doerr 2006; Andreu et al. 2001; Mataix-Solera et al. 2011).

The thickness and depth of a hydrophobic layer depend on the intensity and duration of the fire, soil water content, and soil physical properties. Hydrophobic layers form in dry soils more than in wet soils; coarse-textured soils are more likely to become water repellent than fine-textured soils (Pierson et al. 2002, 2008). In sagebrush ecosystems, water repellency was generally greater on unburned hillslopes and had a greater impact on infiltration capacity than fire effects (Pierson et al. 2008). Fire-induced reduction in infiltration was the result of the combined effect of canopy and ground cover removal and the presence of naturally occurring water repellent soils. Pierson et al. (2008) summarized: "removal of ground cover likely increased the spatial connectivity of runoff areas from strongly water repellent soils. The results indicate that for coarse-textured sagebrush landscapes with high pre-fire soil water repellency, post-fire increases in runoff are more influenced by fire removal of ground and canopy cover than fire effects on soil water repellency and that the degree of these impacts may be significantly influenced by short-term fluctuations in water repellent soil conditions."

Water repellency is also a common phenomenon on dry soils in many other unburned range and forest vegetation types [Sierra Nevada vegetation (Hussain et al. 1969; Meeuwig 1971); California chaparral (Teramura 1980); alpine-subalpine zone (Barret and Slaymaker 1989); Mediteranean ecosystems (Lavee et al. 1995) *Pinus* and *Eucalyptus* forest (Doerr et al. 1998); pine forest Colorado Front Range (Huffman et al. 2001); sagebrush steppe (Pierson et al. 2002); and Amazonian pastures, Brazil (Johnson et al. 2005)]. Research shows that numerous factors influence the development of water repellency: the presence of shrub canopy (Madsen et al. 2008; Glenn and Finley 2010); dry conditions (Bond and Harris 1964; Jungerius and de Jong 1989; Dekker et al. 1998); wetting and drying cycles (Dexter 1988); soil moisture (King 1981; Chan 1992; Doerr et al. 2000); and transformation of organic matter (Giovannini et al. 1983; De Jonge et al. 1999).

In summary, high-intensity rainfall events occurring after burns can result in high risk of accelerated runoff and flooding, debris-flow events, high erosion and sedimentation in water courses, damage to property, and loss of life. The dynamics of runoff and erosion in response to fire is highly dependent on the intensity of the burn, fuel type, soil, climate, time of burn, topography, and vegetation.

Fire effects on hydrology implicate several dynamics:

- Reduced infiltration, increased runoff, and soil surface protection
- Alteration of physical, chemical, and biological factors
- Exacerbation, alteration, and formation of soil water repellency (hydrophobicity)

Fire effects and erosion affects:

- Increased rain splash and soil erosion
- Altered concentrated flow processes

Grazing Effects on Rangeland Hydrology and Erosion

There are a considerable number of studies and reviews concerning grazing impacts on hydrologic processes on rangeland (Gifford and Hawkins 1978; Branson et al. 1981; Blackburn et al. 1992; Warren et al. 1986a, b, c; Warren 1987; Savory 1988; Thurow 1991; Spaeth et al. 1996a, b; Asner et al. 2004; Holechek et al. 2010). Grazing affects the soil surface, plant species composition, and ultimately hydrology dynamics of the site, field, and watershed (Fig. 5.14). The amount of disturbance on a site from hoof action by livestock depends on soil type, soil water

Fig. 5.14 Model depicting the effect of grazing practices on the soil surface and subsequent results on plant communities, hydrology, energy, and nutrient cycles and erosion and sedimentation dynamics. All the properties are dynamic and are interactive

Short-and Long-Term Effect of Grazing Practices and Potential for Modification of

⇓ ⇑

Soil Surface Properties

⇓ ⇑

Plant Species Structure and Composition

⇓ ⇑

Hydrology (Infiltration and Runoff)

⇓ ⇑

Energy and Nutrient Cycles

⇓ ⇑

Carbon, Erosion and Sedimentation Dynamics

content, seasonal climatic conditions, and vegetation type. Repetitive and continuous high-intensity trampling results in a cascading effect: bulk density increases (compaction) and soil aggregates degrade, especially on lighter textured soils. This results in lower infiltration, higher runoff, and greater potential for erosion. If intensive grazing occurs on wet soil, soil aggregate stability is further damaged, resulting in an impermeable surface layer and possible development of a physical crust. On some soils, a physical vesicular crust can occur, which impedes infiltration.

Grazing at any level triggers an immediate vegetation response and imposes either subtle or drastic changes in soil physical, chemical, and biological factors. Depending on the level of grazing (light to heavy), there are numerous implications that managers should be aware of (Fig. 5.15). These changes may be entirely natural as many rangeland plant communities evolved with ungulate grazing (see Mack and Thompson 1982) or they may be disruptive depending on the severity of the perturbation.

Influence of Livestock Grazing and Trampling on Hydrology

Many grasslands and rangeland plant communities evolved with ungulates, some more intensively than others (Mack and Thompson 1982). Mack and Thompson (1982) characterized two grassland provinces dominated by perennial grasses, the *Agropyron* province (dominated by bunch or caespitose grasses) west of the Rocky Mountains and the *Bouteloua* province (dominated by rhizomatous/stoloniferous species) throughout the northern and southern Great Plains. Before and after the

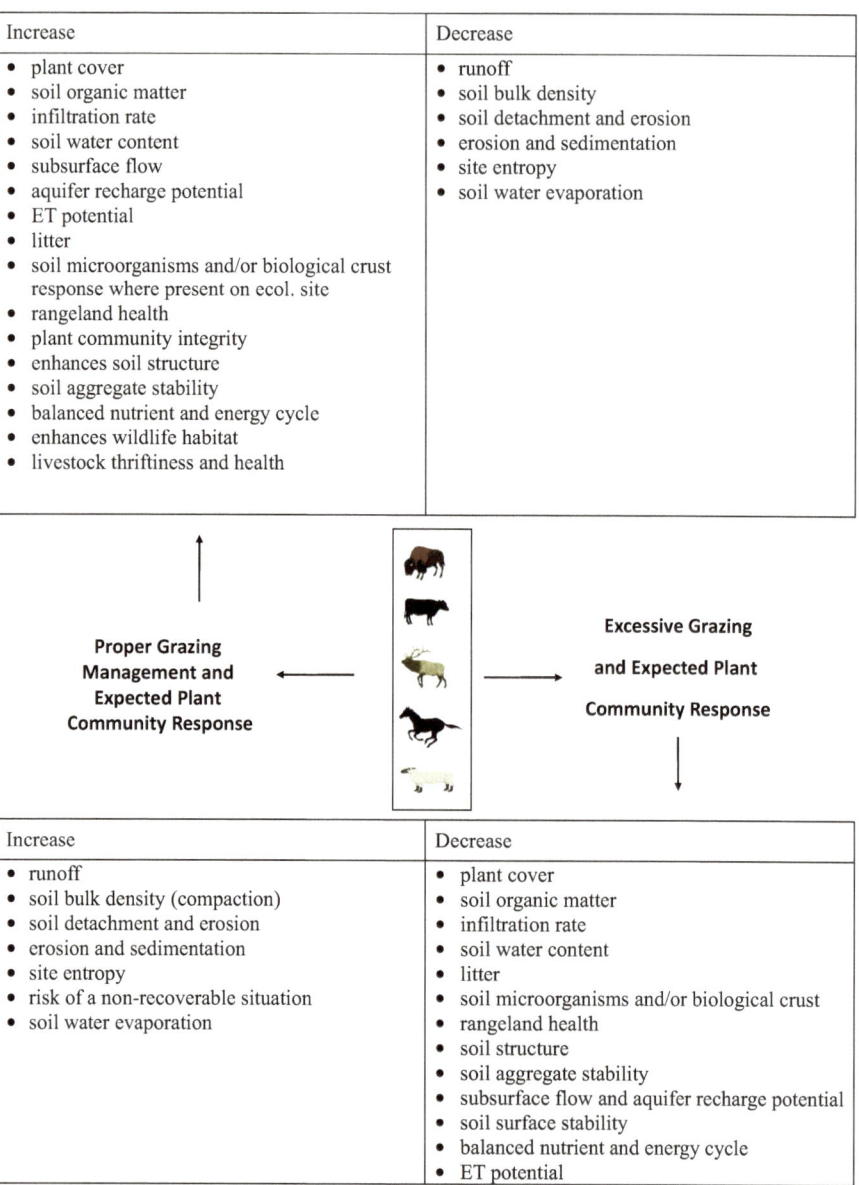

Increase	Decrease
• plant cover • soil organic matter • infiltration rate • soil water content • subsurface flow • aquifer recharge potential • ET potential • litter • soil microorganisms and/or biological crust response where present on ecol. site • rangeland health • plant community integrity • enhances soil structure • soil aggregate stability • balanced nutrient and energy cycle • enhances wildlife habitat • livestock thriftiness and health	• runoff • soil bulk density • soil detachment and erosion • erosion and sedimentation • site entropy • soil water evaporation

Proper Grazing Management and Expected Plant Community Response

Excessive Grazing and Expected Plant Community Response

Increase	Decrease
• runoff • soil bulk density (compaction) • soil detachment and erosion • erosion and sedimentation • site entropy • risk of a non-recoverable situation • soil water evaporation	• plant cover • soil organic matter • infiltration rate • soil water content • litter • soil microorganisms and/or biological crust • rangeland health • soil structure • soil aggregate stability • subsurface flow and aquifer recharge potential • soil surface stability • balanced nutrient and energy cycle • ET potential

Fig. 5.15 Diagrammatic representation of grazing and the relationship with the soil surface. Heavy grazing use by livestock may (1) compact the soil, (2) have a negative impact on soil structure, (3) mechanically disrupt soil aggregates, (4) reduce soil aggregate stability, and (5) destroy biological soil crusts, which may be essential to erosional stability. Infiltration capacity is generally reduced with increased grazing intensity mainly through vegetation removal and composition change, changes to soil structure, bulk density, and aggregate stability

Wisconsin glaciation, *Bison (Bison bison)* were prolific and numerous throughout the *Bouteloua* province (may have exceeded 40 million animals at the time of European contact). In contrast, there is not much evidence of large herds of bison west of the Rockies (Durrant 1970); few prehistoric records of bison existed on the Columbia Plateau of Washington and Oregon with a regional decline to extinction by 2500 BP (Schroedl 1973).

With the introduction of livestock on US grasslands and ranges came the irrevocable impact on plant composition change in more arid rangelands, especially the *Agropyron* province where these endemic native grass species did not evolve with heavy grazing and therefore are sensitive to grazing pressure (Mack and Thompson 1982) (Table 5.9). In the *Bouteloua* province, one bovid essentially replaced the other (bison to cattle). The main change that occurred with cattle herds is frequency of disturbance rather than the type of disturbance. In more recent times, fire frequency has also changed dramatically; in some locations it has been essentially eliminated, excepting the inescapable wildfires that occur. In the *Agropyron* province, dramatic changes occurred compared to the *Bouteloua* province where the

Table 5.9 Physiological characteristics and grazing tolerance of key dominant species in the *Agropyron* and *Bouteloua* provinces in the United States

Species	Sensitivity to grazing	Growth habit	Photosynthetic pathway	Flowering/ veg. culm ratio	Meristem position
Bluebunch wheatgrass (*Pseudoroegneria spicata*)	Very sensitive in late spring	Caespitose, rhizomatous at higher elev.	C3	High	Elevates early
Idaho fescue (*Festuca idahoensis*)	Sensitive to light grazing	Caespitose	C3	High	Elevates early
Sandberg bluegrass (*Poa sandbergii*)	Sensitive to light grazing	Caespitose	C3	High	
Kentucky bluegrass (*Poa pratensis*)	Delayed response if any	Rhizomatous	C3	Low	At ground level throughout the season
Blue grama (*Bouteloua gracilis*)	Delayed response if any	Tufted, but forming a sod	C4	Low	At ground level throughout the season
Western wheatgrass (*Pascopyrum smithii*)	Less sensitive than PSSP	Rhizomatous	C3	Low	Elevated early
Buffalograss (*Bulbilis dactyloides*)	Delayed response in any	Stoloniferous stolon forming a dense sod	C4	Low	At ground level throughout the season

introduction of domestic cattle herds and changes in frequency of grazing rapidly changed the composition of many grazed native ranges with the introduction of Eurasian exotic annual grasses and forbs. The more fragile arid rangelands were more susceptible to exotic plants as many of them are winter annuals. Cheatgrass (*Bromus tectorum*) was first collected in Pennsylvania, 1861; Washington, 1893; Utah, 1894; Colorado, 1895; and Wyoming, 1900 (Yensen 1981). Field bromegrass (*Bromus arvensis*), another common exotic annual C3 brome species is also native to Eurasia. It is found in all states within the United States with the exception of Hawaii and Alaska and in the Canadian provinces of British Columbia, Alberta, Saskatchewan, Manitoba, Ontario, Quebec, and Prince Edward Island (USDA-NRCS Plants Database). Field bromegrass is also characterized as a winter annual that germinates in fall and survives throughout the winter as "rosettes" (Baskin and Baskin 1981). Medusahead wild rye (*Taeniatherum caput-medusae*), also a Eurasian C3 introduction (Hungary through Ukraine to Tadzhikistan), made its début in the United States in the 1880s near Roseburg, Oregon. The spread of medusahead has been slower than cheatgrass (Pyke 1999) but is now becoming increasingly prominent on western US rangelands (Turner et al. 1963; Hilken and Miller 1994; USDA-NRCS 2018). Unlike cheatgrass and field brome, medusahead has a high ash and silica content (72–89%), which makes it less palatable as the plant matures (Swnenson et al. 1964).

By the turn of the century, irreversible changes to the ecological and biological structure of the *Agropyron* province were well underway. Native grass species could no longer recolonize disturbed areas once the exotic annual grasses were present. With the advent of heavy grazing, the transformation from perennial bunchgrass was rapid, and when annual grass dominance occurs, ecosystem function can be compromised (Vitousek et al. 1997). On US rangelands, nonnative exotic plants can negatively impact biotic integrity, ecosystem resilience and stability, composition and structure, natural fire cycles, diversity, soil biota, vegetation production, forage quality, wildlife habitat, soil physical properties, organic matter dynamics, carbon balance, nutrient and energy cycles, and hydrology and erosion dynamics in many unique ways (Evans et al. 2001; Pierson et al. 2002; Ehrenfeld 2003; Ogle et al. 2003, 2004; Brooks et al. 2004; Norton et al. 2004; Belnap et al. 2005; Hooper et al. 2005; Sommer et al. 2007; Boxell and Drohan 2008; Herrick et al. 2010; Davies 2011).

Modification of plant composition, cover, and biomass on a site can singularly or collectively affect infiltration dynamics and may accelerate erosion. Stocking rates that continuously exceed moderate stocking (continuous, season long, and intensive systems) will ultimately decrease infiltration and increase runoff, erosion, and sediment loss (Rhodes et al. 1964; Rauzi and Hanson 1966; Rauzi and Smith 1973; Blackburn 1984; Wood and Blackburn 1984; Gifford and Hawkins 1978; Owens et al. 1982; Warren et al. 1986a, b, c; Warren 1987; Thurow et al. 1988; Naeth et al. 1990; Thurow 1991).

In general, the effects of livestock grazing on hydrologic resilience are associated with the degree to which grazing pressure affects surface soil physical properties and alteration of spatial and temporal variations in canopy and ground cover (Gifford and Hawkins 1978; Thurow 1991, 1986, 1988; Trimble and Mendel 1995; Wood and Blackburn 1981). Grazing pressure that substantially reduces vegetation and ground cover and/or compacts and continually disturbs the soil surface will likely result in increased loss of water storage, organic matter, and soil through water and wind erosion processes (Field et al. 2011; Greene et al. 1994; Castellano and Valone 2007; Trimble and Mendel 1995). It is difficult to differentiate between the influences of moderate grazing and light grazing on hydrology and erosion in the literature. Light and moderate grazing have small impacts on rangeland hydrology (Gifford and Hawkins 1978). Continuous grazing at light and moderate levels is hydrologically superior to other more intensive grazing systems (Warren et al. 1986a; Thurow et al. 1988). Deferred rotation systems generally maintain hydrologic parameters similar to ungrazed areas (Blackburn et al. 1980). High-intensity grazing, particularly over multiple sustained years, can alter plant composition such that the biotic structure triggers long-term site degradation through abiotic-driven losses of water and soil resources (Greene et al. 1994; Ludwig et al. 2007; Schlesinger et al. 1990; Turnbull et al. 2008, 2012; van de Koppel et al. 1997; Warren et al. 1986b, c). Short duration grazing had no beneficial effect on the hydrology at the Ft. Stanton Experimental Ranch in New Mexico, and moderate continuous grazing was superior to heavy continuous and short-duration grazing (Weltz and Wood 1986).

Grazing systems such as intensive rotation grazing (IRG), short-duration grazing methods, high intensity-short duration, or other forms of heavy stocking rotation systems ("mob grazing, ultrahigh stocking densities") have been highlighted in the popular literature to be advantageous to hydrology, animal performance, litter decomposition, and organic matter dynamics, lessening carbon footprints, and cattle production profits. Some intensive grazing systems such as those mentioned above (Goodloe 1969; Savory 1978, 1979, 1988; Savory and Parsons 1980) and mob grazing have been subject to controversy: managers claim purported benefits supported by anecdotal evidence; however, under scrutiny of scientific experimental evidence, the systems may not be sustainable in the long term in more humid pastoral settings, and not sustainable in semiarid and arid rangelands. Rangeland vegetation types vary greatly in terms of resistance and resilience to grazing (see discussion of *Agropyron* and *Bouteloua* provinces above). Under most intensive short-duration grazing systems, livestock are concentrated in small paddocks for short periods, creating a concentrated herd effect with intensive trampling of the soil and plants. Proponents claim this grazing system mimics the grazing dynamics of historic bison herds where hoof action disturbs the soil surface and helps incorporate litter for microbial breakdown and in the process enhances soil surface organic matter and infiltration of rainfall. In truth, hoof action may quicken contact of standing litter with the soil, but subsequent decomposition and soil organic sequestered would be equal to standing dead litter collapsing and eventually falling to the soil surface. Organic materials begin to decompose when residues come in contact with the soil and when mineralization by soil microbes, coupled with adequate soil moisture and complementary temperatures begins; different gases such as CH_4, NH_4^+, NO_3^-, SO_4^{-2}, H_2S, and HPO_4^{-2} and especially CO_2 are emitted (Al-Kaisi and Lal 2017). The rate of mineralization is influenced by biotic and abiotic factors, and nutrients released vary significantly among the different fractions of soil organic matter. On average, after 1 year, about 17–35% of the carbon added as plant residue is incorporated into SOM (Stewart et al. 2017).

Trampling intensity is correlated with compaction and changes in bulk density. Warren et al. (1986a) demonstrated experimentally that repeated high-intensity trampling decreased soil aggregate stability and increased bulk density, which reduced infiltration rates, and increased surface runoff and interrill erosion (Table 5.10). When moist silty clay soils were trampled, soil aggregates were largely destroyed by compaction into the impermeable surface layer. In summary, Warren et al. 1986a found that infiltration rate decreased and sediment production increased significantly on the silty clay soil that was devoid of vegetation following trampling that typifies intensive rotation grazing systems. Warren et al. (1986a) also found that 30 days deferment from grazing was insufficient to allow for hydrologic recovery.

The level of disturbance to a site from hoof action by livestock depends on soil texture, soil structure, soil water content, seasonal climatic conditions, and vegetation type. In summary, repetitive and continuous high-intensity trampling increases bulk density (compaction) and breaks down soil aggregates resulting in lower infiltration, higher runoff, and potential for erosion (Gifford and Hawkins 1978; Warren et al. 1986a, b, c: Greenwood and McKenzie 2001; Teague et al. 2011). If this action occurs on wet soils, soil aggregate stability can be severely reduced, resulting in an impermeable surface layer. Freeze-thaw cycles and wet-dry cycles can result in ameliorating surface soil compaction from livestock depending on the time since disturbance and soil texture (Weltz et al. 1989).

Table 5.10 Mean infiltration rate and interrill erosion in relation to trampling intensity and water content on the Edwards Plateau, Texas

	Infiltration rate (mm hr^{-1})			
Stocking intensity	Before trampling	Trampled dry	Before trampling	Trampled moist
0	166	166	160	160
1X	147	132	133	130
2X	137	106	115	83
3X	134	101	109	82
	Interrill erosion (kg ha^{-1})			
0	976	976	2007	2007
1X	1892	3824	2998	2752
2X	2272	4605	3542	5048
3X	2211	7078	4057	7465

Simulated rainfall 15 cm (6 in) in 45 minutes (Warren et al. 1986a). Symbols IX indicates moderate stocking intensity, 2X double the moderate-intensity, 3X triple the moderate-intensity, soil texture was silty clay

Text Box 5.10 Case Studies Relating to Hydrology and Grazing
Infiltration and Erosion Response to Trampling Intensity (Warren
et al. 1986a)
Warren et al. (1986a) state: "the idea that benefit can be derived from short-term, high-intensity physical impact of livestock is the principal foundation upon which many proponents of intensive, rotational grazing systems base their support. The physical impact is believed to chip or churn the soil surface and break up surface crusting without compacting the soil, thus improving water infiltration and reducing erosion (Goodloe 1969; Savory and Parsons 1980) ... this study showed that repeated high-intensity trampling, at levels typical of intensive rotational grazing systems, was detrimental to soil properties which are normally well correlated to infiltration rate and sediment production. The detrimental impact generally increased as stocking rate increased. Trampling a dry soil did, indeed, chip and churn the soil surface. However, the hoof action reduced the size of naturally occurring soil aggregates and increased the density of the surface soil layer. Trampling on moist paddocks destroyed existing soil aggregates and led to the creation of a flat, comparatively impermeable surface layer composed of dense, unstable clods. Both of these conditions were detrimental to infiltration rate and sediment production."

In summary, trampling activity by grazing animal hooves reduces infiltration by altering soil surface physical factors: bulk density or compaction, breakdown of soil aggregates, and reduction of porosity. Intense trampling as a result of doubled or tripled stock intensities in smaller paddocks for short periods (creating a herd effect) has been hypothesized as enhancing infiltration and reducing erosion. Research to date by rangeland hydrologists has not supported the idea that increased intensity of trampling enhances infiltration capacity.

Effects of Intense, Short-Duration Grazing on Microtopography in a
Chihuahuan Desert Grassland (Nash et al. 2004)
On mesquite (*Prosopis glandulosa*) rangeland in Chihuahuan Desert (Jornada Experimental Range), intense high stocking rate was associated with "virtual elimination of the fine-scale microtopography." Intense grazing resulted in erosion of coppice or micromounds with interfiling of intermound depressions (interspace). Losses of coppice dunes and vegetation cover and height from short-duration intense grazing exposed soils and increased risk of erosion.

Livestock, Soil Compaction, and Water Infiltration Rate (Castellano and
Valone 2007)
Hoof action in short-term high-intensity grazing systems has been proposed as a management mechanism to stimulate infiltration capacity to increase and maintain productivity on grasslands. This grazing system suggests that intense

(continued)

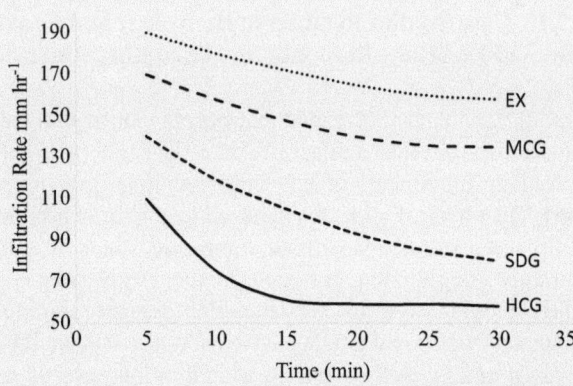

Fig. 5.16 Mean infiltration rates from four grazing treatments on the Edwards Plateau, Texas (6-year study). HCG = continuous grazed at 1.75 × the moderate-intensity; SDG = short-duration rotation (14 pasture, 1 herd; 4 days on, 50 days rest); MCG = continuously grazed at moderate intensity; EX = livestock enclosure, no grazing (Thurow et al. 1988)

short-term hoof action increases infiltration, litter decomposition, and seed incorporation. On San Simon Valley, Arizona grasslands, Castellano and Valone (2007) found that short-term, intense hoof action reduced water infiltration.

Comparison of Grazing Treatments on Infiltration and Interrill Erosion (Thurow et al. 1988)

The key components affecting infiltration rate, i.e., aggregate stability, bulk density, standing crop, litter biomass, and litter cover, are sensitive to grazing management systems (Thurow et al. 1988). It is imperative for the land manager and conservationists to assist in planning grazing systems to recognize temporal responses of infiltration and interrill erosion to the planned grazing system as unwanted shifts to alternate states (consult state and transition model in ESD).

Thurow et al. (1988) reported that the continuously grazed at moderate intensity (MCG) and the high-intensity low-frequency (HILF) grazing system resuled in infiltration rates and erosion that were acceptable; however, infiltration rates decreased and interrill erosion increased on heavy continuous (HCG) and heavily stocked short-duration grazing (SDG) pastures (Fig. 5.16). Infiltration and erosion on HCG and SDG were not gradual but occurred in a more dramatic stair-step pattern. Also, infiltration and interrill erosion rates did not recover after drought conditions when normal precipitation patterns returned. Heavy stocking rate and climatic factors were the primary factors in reducing infiltration and increasing erosion.

(continued)

Lack of Vegetation Change After 13 Years of Livestock Grazing Exclusion in a Sagebrush Semidesert in West Central Utah (West et al. 1984)

West et al. (1984) reported, after 13 years of livestock deferment, that desirable native perennial vegetation did not re-establish despite a trend of increased precipitation over the length of the study. They concluded that the site had transitioned to a stable shrub-dominated site. The concept that removing livestock would return the plant community to the original sagebrush-native shrub-grass assemblage is unlikely. Therefore, direct manipulation of the site is mandatory if a rapid return to the desired plant community is desired. Belnap et al. (2009) reported that grazed watersheds in southeast Utah had significantly more soil loss from wind than ungrazed watersheds. When comparing soil losses among the sites, they determined that biological soil crusts were the most important in predicting site stability followed by perennial plant cover.

Infiltration Rates and Sediment Production as Influenced by Grazing Systems in the Texas Rolling Plains (Pluhar et al. 1987)

Two production scale grazing treatments were sampled at Throckmorton Texas to evaluate their impact on hydrologic processes. Treatments were year-long continuous grazing stocked at a moderate rate (MC), 16-paddock rotational grazing treatment stocked at a heavy rate (RG-16), moderately stocked pasture, 3-herd deferred rotation treatment (DR-3), and ungrazed exclosure (EX). Sediment production was lowest in the exclosure. Compared to the RG-16 treatment (sediment = 63 kg ha^{-1} (56 lbs ac^{-1}) and infiltration 82 mm hr^{-1}; 3.2 in hr^{-1}), sediment production was least (33 kg ha^{-1}; 29 lbs ac^{-1}) and infiltration rate greatest (89 mm hr^{-1}; 3.5 in hr^{-1}) in the MC treatment. As aboveground biomass and cover increased, there was a corresponding increase in infiltration rates and a decrease in sediment production. The RG-16 treatment had higher sediment and lower infiltration rates than MC treatment.

Before grazing treatments, infiltration rates and sediment production in the RG-16 and DR-3 treatments before grazing were not significantly different from those in the MC treatment. However, subsequent grazing caused a significant decline in infiltration rates and an increase in sediment production in both treatments (as a function of removal of aboveground biomass, cover, and proportion of the area occupied by midgrasses). Midgrass stands had higher infiltration rates and lower sediment production than shortgrass.

Grazing Effects on Plant Cover, Soil, and Microclimate in Fragmented Woodlands in Southwestern Australia (Yates et al. 2000)

This study evaluated the impacts of livestock grazing on native plant species cover, litter cover, soil surface condition, surface soil physical and chemical properties, surface soil hydrology, and near-ground and soil microclimate in remnant *Eucalyptus salmonophloia* woodlands, Western Australia (loamy sand over clay).

(continued)

"Heavy grazing by sheep was associated with a decline in native perennial cover and an increase in exotic annual cover, reduced litter cover, reduced soil cryptogam cover, loss of surface soil microtopography, increased erosion, changes in the concentrations of soil nutrients, degradation of surface soil structure, reduced soil water infiltration rates, and changes in near-ground and soil microclimate. The results suggest that livestock grazing changes woodland conditions and disrupts the resource regulatory processes that maintain the natural biological array in *E. salmonophloia* woodlands."

Intensive livestock trampling typical of multi-pasture rotational grazing systems had a negative impact on soil physical properties. The deleterious effects tended to increase as stocking rate increased. Trampling on dry soil disrupted naturally occurring aggregates and compaction of the surface soil layer. Trampling on moist soil disturbed existing aggregates and led to the creation of a flat, comparatively impermeable surface layer composed of dense, unstable clods. Bulk density was higher in heavily grazed woodlands, which increased soil penetration resistance and a loss of perennial vegetation cover, and biological soil crusts. Perennial vegetation provides a feedback mechanism where organic litter provides protection against raindrop splash and surface sealing effects.

Runoff on Pasture

Alderfer and Robinson (1947) reported high surface runoff from heavily grazed pastures in Pennsylvania, whereas ungrazed pastures produced little or no runoff. The cause was attributed to a lack of soil cover interacting with surface compaction of the 0–25.4 mm (0–1 in) soil surface layer.

In a moderately grazed sloping pasture in the upper Mississippi Valley, mean peak flows were 1.75 mm hr^{-1} (0.07 in hr^{-1}) compared to 0.5 mm hr^{-1} (0.02 in hr^{-1}) on ungrazed watersheds. Removal of grazing after a 3-year period reduced total runoff to within 10% of ungrazed pasture (Sartz and Tolstead 1974).

Erosivity of Fescue Pasture

Barnett et al. (1972) found that erosion on closely grazed fertilized tall fescue (*Festuca arundinacea*) on gentle slopes of the Southern Piedmont Region was the same as when the grass was allowed to reach a height of >75 mm (3.0 in).

Grazing and Infiltration Capacity on Rough Fescue Pasture in Calgary, Canada

On a foothills rough fescue (*Festuca campestris*) site near Calgary, Canada, steady-state infiltration rates were 1.5–2 times greater in light grazed and control treatments than in moderate, heavy, and very heavy grazed treatments (Naeth et al. 1990). On fescue (*Festuca hallii*) pasture near Edmonton, Canada, initial infiltration rates were lowest in June grazed treatments and highest in autumn light grazed and control treatments (Naeth et al. 1990).

(continued)

Yearlong Grazing on Pastures (Owens et al. 1996)
On pastureland in Ohio, the highest annual soil loss values (2.5 Mg ha^{-1}; 1.12 t ac^{-1}) occurred on unimproved pastures grazed yearlong where cattle had direct access to riparian areas. Rotational summer grazing with >90% grass cover had trace amounts of soil loss.

Evaluation of Grazing Effects on Runoff and Soil Loss on Ohio Pastureland (Owens et al. 1982)
In Eastern Ohio, Owens et al. (1982) evaluated a 5-year cattle grazing program on fertilized pasture. Cattle were rotated through three pastures during the summer grazing period. One pasture was grazed and used for winter feeding. Runoff from summer grazed pastures increased slightly compared with the long-term average. Reduction of vegetative cover in the summer grazing/ winter feeding area resulted in a large increase in soil loss and runoff 132 mm yr^{-1} (5.2 in yr^{-1}) compared to the three summer grazed pastures 15 mm yr^{-1} (0.6 in yr^{-1}). Eighty-one percent of the soil loss occurred when the vegetation was dormant. Large runoff events produced most of the runoff volume and erosion; however, the number of large events that occurred was a small percentage of the total number of events.

Impact of Cattle Treading on Hill Land and Soil Damage Patterns (Sheath and Carlson 1998)
"Animal treading in grassland ecosystems is known to affect the condition of both soil and vegetation. Much of this knowledge centres on dry and/or rangeland conditions where soils and vegetation are considered to be very fragile. Treading action can result in reduced soil water infiltration and increased water runoff." Soil surface damage was greatest on animal tracks and easy contoured areas. For rapid recovery of damaged paddocks, continued grazing of cattle during spring should be avoided.

MOB Grazing and Effects on Compaction, Forage Quality, and Hydrology
Unfortunately, there is a lack of scientific studies related to MOB grazing. Most of the positive or negative claims about MOB grazing are from popular literature. Mob grazing (MOB) is a more intensive type of rotational grazing and is characterized by ultrahigh stocking densities, short durations (one day or less between rotations), and long rest periods (usually for at least 45 days of growth). Highly intensive grazing treatments have more flexibility of management options in humid climates where there is adequate precipitation to regenerate forage species. Semiarid and arid rangelands are not conducive to long-term success from MOB grazing due to a multiplicity of factors (erratic climate, plant species growing points and grazing resilience, and change to soil physical factors affecting hydrology and erosion). The scientific literature on rangeland intensive stocking systems does not support the claims of increased infiltration, organic matter, and plant response. Many semiarid and arid native perennial rangeland grasses are not resilient to continual heavy grazing, compared to more temperate prairies (tallgrass).

Summary

Water and hydrologic processes are the main drivers in plant ecosystems. If hydrologic function is compromised, i.e., infiltration reduced and runoff increased, all other systems including nutrient, energy, and organic carbon cycles will be altered. Ignorance of these effects in high-intensity grazing treatments is a "recipe for disaster" with sometimes unrecoverable consequences, including vegetation state change and impacts on livestock health.

Grazing affects vegetation stature and composition and soil surface factors, which can subsequently affect the hydrologic cycle (Fig. 5.16). On a watershed scale, livestock grazing at intensified levels can decrease plant cover, and soil aggregate stability and increase compaction and soil crusting. Improper grazing intensities, over longer periods, can and often does alter plant composition which may seriously affect the hydrology of a watershed. With continuous heavy stocking, plant composition will ultimately change over time, perhaps subtlety at first where the manager does not notice immediate changes. Soil physical factors consistently change in many high-intensity grazing treatment studies. Increases in bulk density and compaction and decreased soil aggregate stability are common with high-intensity grazing treatments. If erosion is occurring at a higher than sustainable rate, the dynamics of the organic matter cycle may change resulting in altering the stability of soil aggregates, soil microbial populations, and activity associated with organic matter and soil health.

Grazing systems and hydrologic impacts vary according to vegetation type: management specialists should consult references for particular grazing trials (see Blackburn 1984; Blackburn et al. 1980, 1981; Wood and Blackburn 1981; Warren et al. 1986a; Weltz and Wood 1986; Warren 1987; and Holechek et al. 2010). The literature is clear concerning high-intensity grazing: infiltration rates and capacity are reduced, water runoff is increased, and interrill erosion increases. The main factors associated with hydrologic decline on rangeland are due to changes in:

- Declining plant cover.
- Plant species composition.
- Accelerated erosion leading to a decline in surface soil horizon dynamics.
- Increased incidence of invasive species.
- Decline of grazing intolerant perennial grasses.
- Reduced biomass production, deterioration of soil aggregates, and soil biological crusts.
- Increased bulk density.
- Increased soil surface water evaporation.
- Decreasing water holding capacity and available moisture.
- Soil microorganism populations may change between soil fungi and bacteria may change. Bacterial populations in the soil can be depressed by dry conditions, soil acidity, soil compaction, and lack of organic matter. Soil microorganism populations are important indicators of soil health as they respond rapidly to environmental changes and site conditions because of their direct relationship

with conditions of the site. Changes in microbial populations can occur and precede detectable changes of soil physical and chemical properties—an early warning system to soil degradation and decreasing soil health (Nielsen et al. 2002). Soil organisms are inherent to many soil processes such as organic matter and nutrient cycles, which are prominent aspects of soil health. Plant vigor, productivity, and reproductive capability are "first responders" to soil health, which are immediate indicators to imbalances on rangeland, an aspect that cannot be determined soil physical and chemical measures alone.

Grazing effects on hydrology and erosion studies on range and pastureland summarize the following:

- Species composition changes can have a positive or negative effect on hydrology and depend on the individual species involved.
- Hydrology studies consistently show that ungrazed areas and study exclosures have the lowest runoff rates compared to the grazing systems in the respective study areas.
- The reaction to the impact of trampling varies with stocking rate, soil type, soil water content, time of grazing and seasonal climatic conditions, and vegetation type.
- On heavier textured soils, trampling impact on wet soils can break down soil aggregates, and impermeable surface layers can develop.
- "Deferred rotation systems" with adequate rest periods generally maintain hydrologic parameters similar to ungrazed areas. Adequate rest periods vary with soil type and vegetation types. Monitoring soil surface conditions should be done on a site-specific basis.
- Watershed research data suggests that soil and plant health can be maintained and improved with light and moderate continuous grazing. On lighter textured soils, there is little hydrologic response between light and moderate continuous grazing on rangeland hydrology.
- Short-duration high-intensity grazing is associated with higher sediment production compared to moderate continuous grazing. The reduced standing vegetation and plant cover associated with this system appear to be the cause of the increased sediment production. No definite hydrologic advantage of increased stocking density via manipulation of pasture size and numbers has been documented in the scientific literature.

  - Caution needs to be exercised concerning short-duration high-intensity systems. Soil surface physical properties, mineral, microbiotic crusts, and plant species composition must be monitored carefully. Rangeland plant communities are unique, and plant-soil interactions are complex and are not consistent from one vegetation—soil type to the next. This makes it difficult for the land manager since consistency in hydrologic response is not well documented for many plant-soil complexes. Frequent on-site monitoring is essential.

- On continuous heavily grazed pastures, research indicates that removal of grazing after a 3-year period reduced total runoff to within 10% of ungrazed pastures.

- In Midwestern pasturelands, the majority of soil loss occurs when the vegetation is dormant. Large runoff events (usually a small percentage of the total number of rainfall events) produce most of the runoff volume and erosion; however, these events are the ones we must be concerned with in regard to soil erosion.

Soil Erosion and Sediment Production on Watersheds

A serious recurring problem in the United States and throughout the world is the loss of soil resources on productive and functioning watersheds as well as dysfunctional watersheds. Soil loss impacts both on-site and off-site watershed functions. On-site effects include the changes in soil structure, a decline in organic matter and nutrients in the soil, and a reduction of available soil moisture (Gregersen et al. 2007) with overall net impacts on productivity and decline in value of natural resources within watershed landscapes (Ffolliott et al. 2013). Losses of soil resources from otherwise productive and well-functioning watersheds are often a recurring problem confronting hydrologists and watershed managers. Sediment is the product of soil erosion, and its source can be from upland sheet and rill erosion, gully erosion, soil mass movement, or channel erosion. Sediment that is deposited in a stream channel is dependent on:

- The proximity of the source of the erosion to the channel
- Shear forces acting on soil and rock
- Size and distribution of sediment particles
- Transport of sediment from one part of the watershed to another and eventually into a major stream channel

Not all of the eroded soil that accumulates as sediment in watercourses is transported through and out of the given watershed in response to storm events. Eroded soil can be deposited at the base of hillslopes, on stream terraces, or buffered in riparian zones before deposition to a stream channel (Neary et al. 2012). The consequences of sediment and nutrient loads from upland watersheds are reductions in river flow capacity, thus increasing risk of flooding in river basins and pollution that impacts water quality. The consequential increases of sediment in the streamflow from upland watersheds often reduce the capacity of rivers to deliver high-quality water to downstream users, increase the risk of flooding in river basins, reduce or block the flow of water through irrigation systems, and shorten the expected operational life of downstream reservoirs. Increased soil erosion and sedimentation rates can also impact a variety of ecosystem services from watersheds such as quality water supply, groundwater and aquifer recharge, effective nutrient cycling, and biodiversity of plants and animals.

On agricultural watersheds (cropland, pastureland), ranges of 1–30% of the estimated erosion is delivered to rivers (Robinson 1979). It is estimated that about 8% of all erosion from cropland is deposited in estuaries and the ocean; however, cropland soil erosion is highly variable from site to site (Office of Technology and

Assessment 1982). Smaller watersheds generally have higher sediment delivery ratios than larger watersheds. In the United States, estimates suggest that between 5% and 10% of water eroded soil ends up in the Gulf of Mexico or oceans (Robinson 1979). Estimates of sediment delivery should be tempered by judgment and consideration of other influencing factors such as soil texture, relief, type of erosion, sediment transport system, and deposition areas.

Average sediment delivery ratios (SDR) for various sized watersheds are:

- 25-acre watershed—30–90% (SDR)
- 2400-acre watershed—10–50% (SDR)
- 10,000 mi^2—5% (SDR)

Hydrologic Effects of Range Improvement Practices

Management practices that are conducive to increased production usually increase transpiration, while surface runoff and water yields are reduced (Boughton 1970). Many researchers have reported increases in infiltration following mechanical range improvement practices, i.e., root plowing, vibratilling, pitting, etc. by creating a macroporous surface which are able to store more water (Branson et al. 1966; Wight 1976; Tromble 1976; Neff and Wight 1977; Gonzales and Dodd 1979; Bedunah 1982; Bedunah and Sosebee 1985).

Bedunah and Sosebee (1985) studied the results of site manipulation on infiltration on a mesquite/buffalograss community in West Texas. Seven treatments were applied: (1) foliar spray, (2) shred, (3) control, (4) grub between trees, (5) grub trees, (6) kleingrass (*Panicum coloratum*) planting, and (7) vibratill. Vibratilling resulted in the highest infiltration rates. Shredding trees was ranked second for increasing infiltration rates. The shredding treatment was associated with increases in litter and standing crop. Removal of mesquite trees by foliar spraying, mechanical grubbing, or planting to Kleingrass did not increase water infiltration rates compared to the control treatment.

In the last 150 years, the proliferation of trees and shrubs on rangelands worldwide has had a significant impact on land cover at the expense of perennial grasses (Archer et al. 2011). With shrub encroachment, estimates suggest that for every millimeter of precipitation above 300 mm, aboveground net primary production increases by 0.6 g C m^{-2} yr^{-1} (Barger et al. 2011). Research from the Walnut Gulch Experimental Watershed and the Jornada Experimental Range have demonstrated significant differences in hydrology and water erosion between grasslands and shrublands (Wainwright et al. 2000). In general, higher splash detachment and inter-rill erosion rates are higher in shrublands compared to grassland sites (Abrahams et al. 1988; Parsons et al. 1996). However, there are interesting dynamics in shrub coppice zones with significant litter accumulations under the canopy. Higher infiltration rates, greater organic matter and nutrient accumulation, greater aggregate stability, lower bulk density, and greater biological activity are associated with coppice mounds (Pierson and Williams 2016).

When grass production is lost due to woody plant encroachment, there is often a rapid decline in grass cover and production. Above- and belowground productivity, litter inputs, rooting depth, distribution, biomass changes, hydrology, microclimate, and energy balance are altered as woody plant encroachment is active (Archer et al. 2011).

Brush control on rangeland can be accomplished by one or more means such as prescribed burning, herbicides, and selecting the proper class of grazing animal. Generally, brush control on watersheds is done for two reasons: (1) to increase available water to other usually more desirable forage plants, which can include seeding as part of the management action, and (2) to increase runoff water for off-site use by replacing deep-rooted shrubs with more shallow-rooted grasses and/or forbs which consume less water. Overall broad sweeping conclusions about the hydrologic impacts of brush control are difficult because of the interactions of climate, weather, vegetation composition before and after treatment, soil type, shrub control methods, density and type of shrubs, understory vegetation, timing of shrub control, and management after treatment. Brush control impacts will vary over time and from one rangeland plant community type to another because of these natural variations. Improvements in hydrologic response following brush control are not automatic and depend upon the factors listed above.

Riparian Vegetation and Grazing

Riparian zones occur along the interface between aquatic and terrestrial ecosystems. Riparian ecosystems generally make up a minor portion of the landscape in terms of land area but are extremely important components in the planning and management of the rangeland or pastureland unit. It is important to recognize that management and condition of the transitional zone (inactive floodplains, terraces, meadows, etc.) and upland sites are critical to the health of riparian ecosystems because these are areas of runoff and recharge. Excessive runoff and gully erosion on uplands ultimately have a profound impact on the riparian zone and stream corridor.

A well-planned grazing system that provides periodic rest can alleviate many of the problems associated with livestock in riparian areas. Continuous season-long grazing is the most damaging grazing regime to riparian sites because livestock congregate and spend most of their time in riparian zones. Riparian zones, compared to more rugged, steep upland sites in the western United States, provide available and easily assessable water, forage, and shade. Excessive livestock impacts, i.e., heavy grazing and trampling, affect riparian-stream habitats by reducing or eliminating riparian vegetation, changing streambank and channel morphology, increasing stream sediment transport, and lowering the surrounding water tables.

Livestock are perceived as a major cause of riparian degradation in the west, which has resulted in accelerated concerns from resource users. In addition to forage for livestock, riparian areas often amount to cover 1–2% of the summer rangeland area but produce about 20% of the summer forage. Riparian areas have high value for fisheries habitat, wildlife habitat, recreation, transportation routes, precious metals, water quality, and timing of water flows.

Rehabilitation of riparian zones can include rotation grazing schemes, complete exclusion of livestock, changes in type or class of animal, and techniques to improve livestock distribution (salt placement, development of watering areas away from the riparian zone, fencing, herding, alternate turnout dates). Rest rotation is one of the most practical means of restoring and maintaining riparian zones. Under moderate stocking, rest rotation can improve riparian vegetation and physical stability. Where livestock grazing is compatible in a particular riparian area, grazing management practices must allow for regrowth of riparian plants and should leave sufficient vegetation cover for maintenance of plant vigor and streambank protection.

Streamside use of herbaceous forage in riparian areas in summer grazed pastures should be used judiciously (not more than 50% by weight), and in the intermountain region, riparian plant communities have limited regrowth potential after midsummer. "Rule of thumb" stubble heights proposed by some grazing guides 10 cm (4.0 in) may or may not be adequate for certain species. Technical guides published by USDA-Natural Resources Conservation Service and University extension bulletins should be consulted for the dominant species on the site. Fall grazing should be monitored carefully because little or no regrowth potential remains. Utilization should be monitored on a per weight basis for native species or by stubble height (as per state technical guides) for pasture or domestic species.

Summary and Conclusions

On range and pasturelands, hydrologic dynamics are the result of complex interactions between climate, soil, and vegetation factors, which influence interception, infiltration, evaporation, transpiration, percolation, surface runoff, soil water storage, soil erosion, and deposition of sediment. Rangeland vegetation is not homogeneous, even within seemingly continuous unbroken expanses of grass. Mosaic patterns and patchiness prevalent in most natural rangeland plant communities are spatially heterogeneous and temporally dynamic. Spatial and temporal variability of soil and vegetation also strongly influence hydrology and erosion on rangelands. Every grazingland plant-soil complexes exhibit characteristic hydrologic patterns.

Plant species can have a profound influence on the hydrology of grazinglands (both range and pasture). Plant canopy, foliar, and ground cover are essential factors in protecting the soil surface from erosion in grazingland plant communities. However, plant productivity in grasslands is also an important factor to consider as it is three dimensional and often can be an additional factor to consider in managing runoff and erosion. In many shrub communities such as juniper, sagebrush, and mesquite, coppice dunes directly under the shrubs are safe zones with high infiltration, low runoff, and low erosion. In comparison, the shrub interspaces, which may be inhabited by other subdominant shrubs and grasses and forbs, transport water (overland flow paths) and exhibit lower infiltration, higher runoff, and higher rates of soil detachment and transport. As shrub and tree overstories increase on rangelands, understory vegetation tends to decrease with higher potential increases in runoff and erosion.

Vegetative cover is a significant factor in protecting soils, which ultimately influences watershed health and water quality and quantity. However, land managers should also contemplate the fact that different plant life and growth forms and individual species also have a profound impact on hydrology. Rangeland hydrology models such as the Rangeland Hydrology and Erosion Model (RHEM) require data inputs of cover by plant life and growth form. Plant composition, soil characteristics (organic matter), and the soil flora and fauna are indisputably crucial to soil health on rangelands; however, hydrology is the driver that determines plant growth, survivability of plant species, and, in a more indirect way, soil health. There are many dynamics that differentiate and affect soil physical properties on cropland and grazinglands (see Text Box 5.1). Microbial populations in arid and semiarid rangeland plant community's wane and wax with available moisture, which affects decomposition rates. Plant community composition and production, soil development, and intrinsic soil health are inextricably linked and dependent on the climate and hydrologic dynamics at the ecological site level.

References

Abrahams, A.D., A.J. Parsons, and S.H. Luk. 1988. Hydrologic and sediment responses to simulated rainfall on desert hillslopes in southern Arizona. *Catena* 15: 103–117.

Alderfer, R.B., and R.R. Robinson. 1947. Runoff from pastures in relation to grazing intensity and soil compaction. *Journal American Society Agronomy* 39: 948–958.

Al-Kaisi, M.M., and R. Lal. 2017. Conservation agriculture systems to mitigate climate variability effects on soil health. In *Soil health and intensification of agroecosytems*, ed. M.M. Kaisi and B. Lowery, 79–107. London: Academic Press.

Andreu, V., A.C. Imeson, and J.L. Rubio. 2001. Temporal changes in soil aggregates and water erosion after wildfire in a Mediterranean pine forest. *Catena* 44: 69–84.

Archer, S.R., K.W. Davies, T.E. Fulbright, K.C. McDaniel, B.P. Wilcox, and K.I. Predick. 2011. Brush management as a rangeland conservation strategy: a critical evaluation. In *Conservation benefits of rangeland practices: assessment, recommendations, and knowledge gaps*, ed. D.D. Briske, 105–170. Washington, D.C: USDA-NRCS.

Asner, G.P., J. Andrew, L.P. Elmore, L.P. Olander, R.E. Martin, and A.T. Harris. 2004. Grazing systems, ecosystem responses, and global change. *Annual Review Environmental Resources* 29: 261–299.

Barger, N.N., S.R. Archer, J.L. Campbell, C.Y. Huang, J.A. Morton, and A.K. Knapp. 2011. Woody plant proliferation in North American drylands: A synthesis of impacts on ecosystem carbon balance. *Journal of Geophysical Research: Biogeosciences* 116: 1–17.

Barnett, A.P., E.R. Beaty, and A.E. Dooley. 1972. Runoff and soil losses from closely grazed fescue, a new concept in grass management for the Southern Piedmont. *Journal of Soil and Water Conservation* 27: 168–170.

Barret, G., and O. Slaymaker. 1989. Identification, characterization, and hydrological implications of water repellency in mountain soils, Southern British Columbia. *Catena* 16: 477–489.

Baskin, J.M., and C.C. Baskin. 1981. Ecology of germination and flowering in the weedy winter annual grass *Bromus japonicus*. *Journal of Range Management* 34: 369–372.

Bedunah, D.J. 1982. Influence of some vegetation manipulation practices on the biohydrological state of a depleted Deep Hardland range site. Ph.D. Dissertation, Texas Tech University, Lubbock Texas.

Bedunah, D.J., and R.E. Sosebee. 1985. Influence of site manipulation on infiltration rates of a depleted west Texas range site. *Journal of Range Management* 38: 200–205.

Belnap, J., J.R. Welter, N.B. Grimm, N. Barger, and J.A. Ludwig. 2005. Linkages between microbial and hydrologic processes in arid and semiarid watersheds. *Ecology* 86: 298–307.

Belnap, J., R.L. Reynolds, M.C. Reheis, S.L. Phillips, F.E. Urban, and H.L. Goldstein. 2009. Sediment losses and gains across a gradient of livestock grazing and plant invasion in a cool, semi-arid grassland, Colorado Plateau, USA. *Aeolian Research* 1: 27–43.

Bisdom, E.B.A., L.W. Dekker, and J.F.T. Schoute. 1993. Water repellency of sieve fractions from sandy soils and relationships with organic material and soil structure. *Geoderma* 56: 105–118.

Blackburn, W.H. 1975. Factors influencing infiltration and sediment production of semi-arid rangelands in Nevada. *Water Resources Research* 11: 929–937.

———. 1984. Impacts of grazing intensity and specialized grazing systems on watershed characteristics and responses. In *Developing strategies for rangeland management*. National Research Council/National Academy of Sciences., 927–983. Boulder: Westview Press.

Blackburn, W.H., and C.M. Skau. 1974. Infiltration rates and sediment production of selected plant communities and soils of Nevada. *Journal of Range Management* 27: 476–480.

Blackburn, W.H., R.W. Knight, M.K. Wood, and L.B. Merrill. 1980. Watershed parameters as influenced by grazing. In *Proc. of the Symposium on Watershed Management*, 552–572. Boise: American Society Civil Engineers.

Blackburn, W.H., T.L. Thurow, and C.A. Taylor. 1986. Soil erosion on rangeland. Symposium on the use of cover, soils, and weather data in rangeland monitoring, Orlando, February 12, 1986. Texas Agr. Experiment Station TA-2119.

Blackburn, W.H., F.B. Pierson, and M.S. Seyfried. 1990. Spatial and temporal influence of soil frost on infiltration and erosion of sagebrush rangelands. *Water Resources Bulletin* 26: 991–997.

Blackburn, W.H., F.B. Pierson, C.L. Hanson, T.L. Thurow, and A.L. Hanson. 1992. The spatial and temporal influence of vegetation on surface soil factors in semiarid rangelands. *Transactions of the ASAE* 35: 479–486.

Blackburn, W.H., F.B. Pierson, C.L. Hanson, T.L. Thurow and A.L. Hanson. 1992a. Variations in rangeland hydrologic/erosion processes. *Transactions of the ASAE* 35: 479–486.

Blackburn, W.H., F.B. Pierson, C.L. Hanson, T.L Thurow, and A.L. Hanson. 1992b. The spatial and temporal influence of vegetation on surface soil factors in semiarid rangelands. *Transactions of the ASAE* 35: 479–486.

Bond, R.E., and J.R. Harris. 1964. The influence of the microflora on physical properties of soils I. Effect associated with filamentous algae and fungi. *Australian Journal of Soil Research* 2: 111–122.

Bonham, C.D., and A. Lerwick. 1976. Vegetation changes induced by prairie dogs on shortgrass range. *Journal of Range Management* 29: 221–225.

Boughton, W.C. 1970. Effects of land management on quantity and quality of available water. Aust. Water Resources Council Research Project 68/2. University of New South Wales, Water Research Laboratory, Manly Vale, N.S.W., Rep. No. 120.

Boxell, J., and P.J. Drohan. 2008. Surface soil physical properties and hydrological characteristics in *Bromus tectorum* L. (cheatgrass) versus *Artemisia tridentata* Nutt. (big sagebrush) habitat. *Geoderma* 149: 305–311.

Branson, F.A., R.F. Miller, and I.S. McQueen. 1966. Contour furrowing, pitting and ripping on rangelands of the western United States. *Journal of Range Management*. 19: 182–190.

Branson, F.A., G.F. Gifford, K.G. Renard, and R.F. Hadley. 1981. *Rangeland hydrology*. Dubuque: Kendall Hunt Publishing Co.

Brooks, M.L., C.M. D'antonio, D.M. Richardson, J.B. Grace, J.E. Keeley, J.M. DiTomaso, and D. Pyke. 2004. Effects of invasive alien plants on fire regimes. *Bioscience* 54: 677–688.

Brown, J.R., and J.E. Herrick. 2016. Making soil health a part of rangeland management. *Journal of Soil and Water Conservation* 71: 55A–60A.

Cadaret, E.M., K.C. McGwire, and S.K. Nouwakpo. 2016a. Vegetation canopy cover effects on sediment erosion processes in the Upper Colorado River Basin Mancos Shale Formation, Price, Utah. *Catena* 147: 334–344.

Cadaret, E.M., S.K. Nouwakpo, and K.C. McGwire. 2016b. Vegetation effects on soil, sediment erosion, and salinity transport processes in the Upper Colorado River Basin Mancos Shale formation. *Catena* 147: 650–662.

Carlson, D.H., T.L. Thurow, R.W. Knight, and R.K. Heitschmidt. 1990. Effect of honey mesquite on the water balance of Texas Rolling Plains rangeland. *Journal of Range Management* 43: 491–496.

Castellano, M.J., and T.J. Valone. 2007. Livestock, soil compaction and water infiltration rate: Evaluating a potential desertification recovery mechanism. *Journal of Arid Environments* 71: 97–108.

Chan, K.Y. 1992. Development of seasonal water repellence under direct drilling. *Soil Science Society America Journal* 56: 326–329.

Clark, T.W., T.M. Campbell III, D.G. Socha, and D.E. Casey. 1982. Prairie dog colony attributes and associate vertebrate species. *Great Basin Naturalist* 42: 572–582.

Coppock, D.L., J.K. Detling, J.E. Ellis, and M.I. Dyer. 1983. Plant-herbivore interactions in a North American mixed-grass prairie. I. Effects of black-tailed prairie dogs on intraseasonal aboveground plant biomass and nutrient dynamics and plant species diversity. *Oecologia* 56: 1–9.

Costin, A.B. 1959. Studies in catchment hydrology in the Australian Alps. 1. Trends in soils and vegetation. Studies in catchment hydrology in the Australian Alps. 1. Trends in soils and vegetation.

Czarnes, S., P.D. Hallett, A.G. Bengough, and I.M. Young. 2000. Root and microbial-derived mucilages affect soil structure and water transport. *European Journal of Soil Science* 51: 435–443.

Dadkah, M., and G.F. Gifford. 1980. Influence of vegetation, rock cover, and trampling on infiltration rates and sediment production. *Water Resource Bulletin* 16: 979–986.

Davies, K.W. 2011. Plant community diversity and native plant abundance decline with increasing abundance of an exotic annual grass. *Oecologia* 167: 481–491.

De Jonge, L.M., O.H. Jacobsen, and P. Moldrup. 1999. Soil water repellency: Effects of water content, temperature, and particle size. *Soil Science Society of America Journal* 63: 437–442.

DeBano, L.F. 2000. The role of fire and soil heating on water repellency in wildland environments: A review. *Journal of Hydrology* 231–232: 195–206.

DeBano, L.F., and J.S. Krammes. 1966. Water repellent soils and their relation to wildfire temperatures. *International Association of Hydrological Sciences* 11 (2): 14–19.

DeBano, L.F., L.D. Mann, and D.A. Hamilton. 1970. Translocation of hydrophobic substances into soil by burning organic litter. *Soil Science Society America Proceedings* 34: 130–133.

DeBano, L.F., S.M. Savage, and A.D. Hamilton. 1976. The transfer of heat and hydrophobic substances during burning. *Proceedings of the Soil Science Society of America* 40: 779–782.

DeBano, L.F., P.F. Ffolliott, and M.B. Baker Jr. 1996. Fire severity effects on water resources. In *Effects of Fire on Madrean Province Ecosystems—A Symposium Proceedings, General Technical Report RM vol. 289*, ed. P.F. Ffolliott, L.F. DeBano, M.B. Baker Jr., G.J. Gottfried, G. Solis-Gorza, C.B. Edminster, D.G. Neary, L.S. Allen, and R.H. Hamre, 77–84. United States Department of Agriculture. Ft Collins: Forest Service Rocky Mountain Forest and Range Experimental Station.

DeBano, L.F., D.G. Neary, and P.F. Ffolliott. 1998. *Fire's effects on ecosystems*. New York: Wiley, 352 p.

Dee, R.F., T.W. Box, and E. Robertson. 1966. Influence of grass vegetation on water intake of Pullman silty clay loam. *Journal of Range Management* 1: 77–79.

Dekker, L.W., and C.J. Ritsema. 1996. Preferential flow paths in a water repellent clay soil with grass cover. *Water Resources Research* 32: 1239–1294.

Dekker, L.W., C.J. Ritsema, K. Oostindie, and O.H. Boersma. 1998. Effect of drying temperature on the severity of soil water repellency. *Soil Science* 163: 780–796.

Derner, J.D., A.J. Smart, T.P. Toombs, D. Larsen, R.L. McCulley, J. Goodwin, and L.M. Roche. 2018. Soil health as a transformational change agent for US grazing lands management. *Rangeland Ecology and Management* 71: 403–408.

Dexter, A.R. 1988. Advances in characterization of soil structure. *Soil and Tillage Research* 11: 199–238.

Doerr, S.H., R.A. Shakesby, and R.P.D. Walsh. 1996. Soil water repellency variation with depth and particle size fraction in burned and unburned *Eucalyptus globulus* and *Pinus pinaster* forest terrain in the Agueda basin, Portugal. *Catena* 27: 25–47.

———. 1998. Spatial variability of soil hydrophobicity in fire-prone eucalyptus and pine forests, Portugal. *Soil Science* 163: 313–324.

———. 2000. Soil water repellency: Its causes, characteristics and hydro-geomorphological significance. *Earth-Science Reviews* 51: 33–65.

Doerr, S.H., W.H. Blake, R.A. Shakesby, F. Stagnitti, S.H. Vuurens, G.S. Humphreys, and P. Wallbrink. 2004. Heating effects on water repellency in Australian eucalypt forest soils and their value in estimating wildfire soil temperatures. *International Journal of Wildland Fire* 13: 157–163.

Doerr, S.H., R.A. Shakesby, L.H. MacDonald. 2009. Soil water repellency: A key factor in post-fire erosion. In Fire effects on soils and restoration strategies. Land reconstruction and management series, 5. eds. A. Cerdà, P.R. Robichaud. Enfield: Science Publishers: 197–224.

Durrant, S.D. 1970. Faunal remains as indicators of Neothermal climates at Hogup Cave. Appendix II. p. 241–245, University of Utah Anthropology Paper 93.

Ehrenfeld, J.G. 2003. Effects of exotic plant invasions on soil nutrient cycling processes. *Ecosystems* 6: 503–523.

Ellerbrock, R.H., H.H. Gerke, J. Bachmann, and M.O. Goebel. 2005. Composition of organic matter fractions for explaining wettability of three forest soils. *Soil Science Society America Journal* 69: 57–66.

Evans, R.D., R. Rimer, and S.P. Belnap. 2001. Exotic plant invasion alters nitrogen dynamics in an arid grassland. *Ecological Applications* 11: 1301–1310.

Eynard, A., T.E. Schumacher, M.J. Lindstrom, D.D. Malo, and R.A. Kohl. 2006. Effects of aggregate structure and organic C on wettability of Ustolls. *Soil Tillage Research* 88: 205–216.

Fahnestock, J.T., and J.K. Detling. 2002. Bison-prairie dog-plant interactions in a North American mixed-grass prairie. *Oecologia* 132: 86–95.

Ffolliott, P.F., K.N. Brooks, R. Pizarro Tapia, and P.G. Chevesich. 2013. *Soil erosion and sediment production on watershed landscapes: Processes and control.* United Nations Educational Scientific and Cultural Organization: UNESCO.

Fidanza, M.A. 2003. Combination treatments for fairy ring prove effective. *Turfgrass trends in Golfdom* 59: 62–64.

Field, J.P., D.D. Breshears, J.J. Whicker, and C.B. Zou. 2011. Interactive effects of grazing and burning on wind-and water-driven sediment fluxes: rangeland management implications. *Ecological Applications* 21: 22–32.

Food and Agriculture Organization (FAO). 1990. Production yearbook. United Nations FAO Statistics Series. Rome.

Franco, C.M.M., P.P. Michelsen, and J.M. Oades. 2000. Amelioration of water repellency: Application of slow release fertilisers to stimulate microbial breakdown of waxes. *Journal of Hydrology* 231-232: 342–351.

Fuhlendorf, S.D., R.F. Limb, D.M. Engle, and R.F. Miller. 2011. Assessment of prescribed fire as a conservation practice. In *Conservation benefits of rangeland practices: assessment, recommendations, and knowledge gaps*, 75–104. Washington, DC: USDA-NRCS.

Gifford, G.F. 1984. Vegetation allocation for meeting site requirements. In *Developing strategies for rangeland management*, 35–116. National Research Council, National Academy of Sciences. Boulder: Westview Press.

————. 1985. Cover allocation in rangeland watershed management (a review). In *Watershed Management in the Eighties, Proceedings of a Symposium*, eds. E.B. Jones and T.J. Ward, 23–31. ASCE.

Gifford, G.F., and R.H. Hawkins. 1978. Hydrologic impact of grazing on infiltration: A critical review. *Water Resources Research* 14: 305–313.

Giovannini, G., S. Lucchesi, and S. Cervelli. 1983. Water repellent substances and aggregate stability in hydrophobic soil. *Soil Science* 135: 110–113.

Glenn, N.F., and C.D. Finley. 2010. Fire and vegetation type effects on soil hydrophobicity and infiltration in the sagebrush-steppe: I. Field analysis. *Journal of Arid Environments* 74: 653–659.

Gonzales, C.L., and J.D. Dodd. 1979. Production response of native and introduced grasses to mechanical brush manipulation, seeding and fertilization. *Journal of Range Management* 32: 305–309.

Goodloe, S. 1969. Short duration grazing in Rhodesia. *Journal of Range Management* 22: 369–373.

Greene, R., P. Kinnell, and J.T. Wood. 1994. Role of plant cover and stock trampling on runoff and soil-erosion from semi-arid wooded rangelands. *Soil Research.* 32: 953–973.

Greenwood, K.L., and B.M. McKenzie. 2001. Grazing effects on soil physical properties and the consequences for pastures: A review. *Australian Journal of Experimental Agriculture* 41: 1231–1250.

Gregersen, H.M., P.F. Ffolliott, and K.N. Brooks. 2007. *Integrated watershed management: Connecting people to their land and water*. Wallingford, U.K.: CAB International.

Gutierrez-Castillo, J. 1994. *Infiltration, sediment, and erosion under grass and shrub cover in the southern high plains*. Ph.D. Dissertation, Lubbock Texas: Texas Tech University.

Hallett, P.D., and I.M. Young. 1999. Changes to water repellence of soil aggregates caused by substrate-induced microbial activity. *European Journal of Soil Science* 50: 35–40.

Hanson, C.L., A.R. Kuhlman, and J.K. Lewis. 1978. Effect of grazing intensity and range condition on hydrology of western South Dakota ranges. South Dakota State University Agr. Experiment Station. Bull. 657.

Herrick J.E., V.C. Lessard, K.E. Spaeth, P.L. Shaver, R.S. Dayton, D.A. Pyke, L. Jolley, and J.J. Goebel. 2010. National ecosystem assessments supported by scientific and local knowledge. *Frontiers in Ecology and the Environment* 8: 403–408.

Hilken T.O, and R.F. Miller 1994. Medusahead (*Taeniatherum asperum* Nevski): A Review and Annotated Bibliography. *Agricultural Experiment Station Oregon State University Station Bulletin* 644.

Holechek, J.L., R.D. Pieper, and C.H. Herbel. 2010. *Range management principles and practices*. Pearson.

Hooper, D.U., F.S. Chapin, J.J. Ewel, A. Hector, P. Inchausti, S. Lavorel, and B. Schmid. 2005. Effects of biodiversity on ecosystem functioning: A consensus of current knowledge. *Ecological Monographs* 75: 3–35.

Hubbert, K.R., H.K. Preisler, P.M. Wohlgemuth, R.C. Graham, and M.G. Narog. 2006. Prescribed burning effects on soil physical properties and soil water repellency in a steep chaparral watershed, southern California, USA. *Geoderma* 130: 284–298.

Huffman, E.L., L.H. MacDonald, and J.D. Stednick. 2001. Strength and persistence of fire-induced soil hydrophobicity under ponderosa and lodgepole pine, Colorado Front Range. *Hydrological Processes* 15: 2877–2892.

Hussain, S.B., C.M. Skau, S.M. Bashir, and M.O. Meeuwig. 1969. Infiltrometer studies of water-repellent soils on the east slope of the Sierra Nevada. In Proceedings Water-repellent soils May 6–10, 1968, eds. L.F. DeBano, L. Letey, Riverside California, University of California.

Johnson, C.W., and N.E. Gordon. 1988. Runoff and erosion from rainfall simulator plots on sagebrush rangeland. *Transactions of the American Society of Agricultural Engineers* 31: 421–427.

Johnson, M.S., J. Lehmann, T.S. Steenhuis, L. deOlivera, and E.C.M. Fernandes. 2005. Spatial and temporal variability of soil water repellency of Amazonian pastures. *Australian Journal of Soil Research* 43: 319–326.

Jungerius, P.D., and J.H. De Jong. 1989. Variability of water repellence in the dunes along the Dutch Coast. *Catena* 16: 491–497.

Keeley, J.E., and P.W. Rundel. 2005. Fire and the Miocene expansion of C4 grasslands. *Ecology Letters* 8: 683–690.

King, P.M. 1981. Comparison of methods for measuring severity of water repellence on sandy soils and assessment of some factors that affect its measurement. *Australian Journal of Soil Research* 19: 275–285.

Klatt, L.E., and D. Hein. 1978. Vegetative differences among active and abandoned towns of black-tailed prairie dogs (*Cynomys ludovicianus*). *Journal of Range Management* 31: 315–317.

Knight, R.W., W.H. Blackburn, and L.B. Merrill. 1984. Characteristics of oak mottes, Edwards Plateau, Texas. *Journal of Range Management* 37: 534–537.

Kostka, S.J. 2000. Amelioration of water repellency in highly managed soils and the enhancement of turfgrass performance through the systematic application of surfactants. *Journal of Hydrology* 231–232: 359–368.

Lane, L.J., J.R. Simanton, T.E. Hakonson, and E.M. Romney. 1987. Large plot infiltration studies in desert and semiarid rangeland areas of the southwest U.S.A. In *Infiltration development and application*, ed. Yu-Si-Fok, 365–376. Manoa: Water Research Center.

Lavee, H., P. Kutiel, M. Segev, and Y. Benyamini. 1995. Effects of surface roughness on runoff and erosion in Mediterranean ecosystems: The role of fire. *Geomorphology* 11: 227–234.

Lawton, J.H., and C.G. Jones. 1995. Linking species and ecosystems: Organisms as ecosystem engineers. In *Linking species and ecosystems*, ed. C.G. Jones and J.H. Lawton, editors, 141–150. New York: Chapman and Hall.

Linnell, L.D. 1961. Soil vegetation relationships on a chalk-flat range site in Grove County, Kansas. *Transactions Kansas Academy of Science* 64: 293–303.

Lomolino, M.V., and G.A. Smith. 2003. Terrestrial vertebrate communities at black-tailed prairie dog (*Cynomys ludovicianus*) towns. *Biological Conservation* 115: 89–100.

Lowell, B.J. 1980. Factors affecting seed germination and seedling establishment of broom snakeweed. Thesis, New Mexico State University, Las Cruces.

Ludwig, J.A., R. Bartley, A.A. Hawdon, B.N. Abbott, and D. McJannet. 2007. Patch configuration non-linearly affects sediment loss across scales in a grazed catchment in north-east Australia. *Ecosystems* 10: 839–845.

Lyford, F.P., and H.K. Qashu. 1969. Infiltration rates as affected by desert vegetation. *Water Resources Research* 5: 1373–1376.

MacDonald, L.H., and E.L. Huffman. 2004. Post-fire soil water repellency: Persistence and soil moisture thresholds. *Soil Science Society of America Journal* 68: 1729–1734.

Mack, R.N., and J.N. Thompson. 1982. Evolution in steppe with few large, hooved mammals. *The American Naturalist* 119: 757–773.

Madsen, M.D., D.G. Chandler, and J. Belnap. 2008. Spatial gradients in ecohydrologic properties within a pinyon-juniper ecosystem. *Ecohydrology* 1: 349–360.

Martinson, G.D., J.H. Cushman, and T.G. Whitham. 1990. Impact of pocket gopher disturbance on plant species diversity in a shortgrass prairie. *Oecologia* 83: 132–138.

Mataix-Solera, J., A. Cerdà, V. Arcenegui, A. Jordán, and L.M. Zavala. 2011. Fire effects on soil aggregation: A review. *Earth-Science Reviews* 109: 44–60.

Mazurak, A.P., and E.C. Conrad. 1959. Rates of water entry in three great soil groups after seven years in grasses and small grains. *Agronomy Journal* 51: 264–267.

Mazurak, A.P., W. Kriz, and R.E. Ramig. 1960. Rates of water entry into a chernozem soil as affected by age of perennial grass sods. *Agronomy Journal* 52: 35–37.

McCune, B., and M.J. Mefford. 2011. *PC-ORD. Multivariate analysis of ecological data*. Gleneden Beach: MjM Software.

Meeuwig, R.O. 1971. Infiltration and water repellency in granitic soils. In USDA Forest Service Research Note INT-111. Ogden, Utah, Intermountain Forest and Range Experiment Station.

Miller, R.F., J.C. Chambers, D.A. Pyke, F.B. Pierson, and C.J. Williams. 2013. *A review of fire effects on vegetation and soils in the Great Basin Region: Response and ecological site characteristics*. General Technical Report RMRS-GTR-308. Fort Collins: Utah.

Moffet, C.A., F.B. Pierson, P.R. Robichaud, K.E. Spaeth, and S.P. Hardegree. 2007. Modeling soil erosion on steep sagebrush rangeland before and after prescribed fire. *Catena* 71: 218–228.

Moody, J.A., and D.A. Martin. 2001. Initial hydrologic and geomorphic response following a wildfire in the Colorado Front Range. *Earth Surface Processes and Landforms* 26: 1049–1070.

Moody, J.A., D.A. Martin, and S.H. Cannon. 2008. Post-wildfire erosion response in two geologic terrains in the western USA. *Geomorphology* 95: 103–118.

Moore, E., J.E. Kinsinger, K. Pitney, and J. Sainsbury. 1979. Livestock grazing management and water quality protection EPA 910/9-79-67. US Environmental Protection Agency and USDI Bureau of Land Management.

Naeth, M.A., R.L. Rothwell, D.S. Chanasyk, and A.W. Bailey. 1990. Grazing impacts on infiltration in mixed prairie and fescue grassland ecosystems of Alberta. *Canadian Journal Soil Science* 70: 593–605.

Nash, M.S., E. Jackson, and W.G. Whitford. 2004. Effects of intense, short duration grazing on microtopography in a Chihuahuan Desert grassland. *Journal of Arid Environments* 56: 383–393.

Neary, D.G., C.C. Klopatek, L.F. DeBano, and P.F. Ffolliott. 1999. Fire effects on belowground sustainability: A review and synthesis. *Forest Ecology and Management* 122: 51–71.

Neary, D.G., K.A. Koestner, A. Youberg, and P.E. Koestner. 2012. Post-fire rill and gully formation, Schultz Fire 2010, Arizona, USA. *Geoderma* 191: 97–104.

Neff, E.L., and J.R. Wight. 1977. Overwinter soil water recharge and herbage production as influenced by contour furrowing on eastern Montana rangelands. *Journal of Range Management.* 30: 193–195.

Nielsen, M.N., A. Winding, and S. Binnerup. 2002. Microorganisms as indicators of soil health. National Environmental Research Institute, Denmark. Technical Report No. 388.

Norton, J.B., T.A. Monaco, J.M. Norton, D.A. Johnson, and T.A. Jones. 2004. Soil morphology and organic matter dynamics under cheatgrass and sagebrush-steppe plant communities. *Journal Arid Environment* 57: 445–466.

Office of Technology Assessment. 1982. Land productivity problems. In *Impacts of technology on U.S. cropland and rangeland productivity*, 23–63. Washington DC: U.S. Gov. Print. Office.

Ogle, S.M., W.A. Reiners, and K.G. Gerow. 2003. Impacts of exotic annual brome grasses (*Bromus* spp.) on ecosystem properties of northern mixed grass prairie. *American Midland Naturalist* 149: 46–58.

Ogle, S.M., D. Ojima, and W.A. Reiners. 2004. Modeling the impact of exotic annual brome grasses on soil organic carbon storage in a northern mixed-grass prairie. *Biological Invasions* 6: 365–377.

Osborn, B. 1950. *Range cover tames the raindrop. A summary of range cover evaluations, 1949.* Fort Worth: Soil Conservation Service.

Owens, L.B., R.W. Van Keuren, and W.M. Edwards. 1982. Environmental effects of a medium-fertility, 12-month pasture program: I. Hydrology and soil loss. *Journal of Environmental Quality* 11: 236–240.

Owens, L.B., W.M. Edwards, and R.W. Van Keuren. 1996. Sediment losses from a pastured watershed before and after stream fencing. *Journal of Soil and Water Conservation* 51: 90–94.

Parsons, A.J., A.D. Abrahams, and J. Wainwright. 1996. Responses of interrill runoff and erosion rates to vegetation change in southern Arizona. *Geomorphology* 14: 311–317.

Paul, E.A., and F.E. Clark. 1996. *Soil microbiology and biochemistry, second ed.* San Francisco: Academic Press.

Pearse, C.K., and S.B. Wooley. 1936. The influence of range plant cover on the rate of absorption of surface water by soils. *Journal of Forestry* 34: 844–847.

Pellant, M., P.L. Shaver, D.A. Pyke, J.E. Herrick, F.E. Busby, G. Riegel, N. Lepak, E. Kachergis, B.A. Newingham, and D. Toledo. 2019. *Interpreting Indicators of Rangeland Health, Version 5.* Tech 7 Ref 1734-6. Denver: U.S. Department of the Interior, Bureau of Land Management, National Operations 8 Center.

Pierson, F., and C. Williams. 2016. *Ecohydrologic impacts of rangeland fire on runoff and erosion: A literature synthesis.* Gen. Tech. Rep. RMRS-GTR-351. Fort Collins: U.S. Department of Agriculture, Forest Service, Rocky Mountain Research Station.

Pierson, F.B., P.R. Robichaud, and K.E. Spaeth. 2001. Spatial and temporal effects of wildfire on the hydrology of a steep rangeland watershed. *Hydrological Processes* 15: 2905–2916.

Pierson, F.B., D.H. Carlson, and K.E. Spaeth. 2002. Impacts of wildfire on soil hydrological properties of steep sagebrush-steppe rangeland. *International Journal of Wildland Fire* 11: 145–151.

Pierson, F.B., D.H. Carlson, and K.E. Spaeth. 2002a. Impacts of wildfire on soil hydrological properties of steep sagebrush-steppe rangeland. *International Journal of Wildland Fire* 11: 145–151.

Pierson, F.B., K.E. Spaeth, M.E. Weltz., and D.H. Carlson. 2002b. Hydrologic response of diverse western rangelands. *Journal of Range Management* 55: 558–570.

Pierson, F.B., P.R. Robichaud, C.A. Moffet, K.E. Spaeth, S.P. Hardegree, P.E. Clark, and C.J. Williams. 2008. Fire effects on rangeland hydrology and erosion in a steep sagebrush-dominated landscape. *Hydrological Processes* 22: 2916–2929.

Pierson, F.B., C.J. Williams, S.P. Hardegree CJ, M.A. Weltz, J.J. Stone, and P.E. Clark. 2011. Fire, plant invasions, and erosion events on western rangelands. *Rangeland Ecology and Management* 64: 439–449.

Pluhar, J., R. Knight, and R. Heitschmidt. 1987. Infiltration rates and sediment production as influenced by grazing systems in the Texas Rolling Plains. *Journal of Range Management* 40: 240–243.

Power, M.E., D. Tilman, J.A. Estes, B.A. Mente, W.J. Bond, L.S. Mills, G. Daily, J.C. Castilla, J. Lubchenco, and R.T. Paine. 1996. Challenges in the quest for keystones. *Bioscience* 46: 609–620.

Pyke, D.A. 1999. Invasive exotic plants in sagebrush ecosystems of the intermountain west. In Proceedings: Sagebrush steppe ecosystems symposium: 2000 p. 43–54.

Rauzi, F. 1960. Water-intake studies on range soils at three locations in the northern plains. *Journal of Range Management* 13: 179–184.

Rauzi, F., and C.L. Hanson. 1966. Water intake and runoff as affected by intensity of grazing. *Journal of Range Management* 19: 351–356.

Rauzi, F., and F.M. Smith. 1973. Infiltration rates: Three soils with three grazing levels in northeastern Colorado. *Journal of Range Management* 26: 126–129.

Rauzi, F., C.L. Fly, and E.J. Dyksterhuis. 1968. Water intake on midcontinental rangelands as influenced by soil and plant cover. USDA Tech. Bull. No. 1390, Washington, D.C.

Reid, K.D., B.P. Wilcox, D.D. Breshears, and L. MacDonald. 1999. Runoff and erosion in a Piñon–Juniper woodland influence of vegetation patches. *Soil Science Society of America Journal* 63: 1869–1879.

Rhodes, E.D., L.F. Locke, H.M. Taylor, and E.H. McIlvain. 1964. Water intake on a sandy range is affected by 20 years of differential cattle stocking rates. *Journal of Range Management* 17: 185–190.

Rillig, M.C. 2004. Arbuscular mycorrhizae and terrestrial ecosystem processes. *Ecology Letters* 7: 740–754.

———. 2005. A connection between fungal hydrophobins and soil water repellency. *Pedobiologia* 49: 395–399.

Rillig, M.C., and D.L. Mummey. 2006. Mycorrhizas and soil structure. *New Phytologist* 171: 41–53.

Ritsema, C.J., J.L. Nieber, L.W. Dekker, and T.S. Steenhuis. 1998a. Modeling and field evidence of finger formation and finger recurrence in a water repellent sandy soil. *Water Resources Research* 34: 555–567.

———. 1998b. Stable or unstable wetting fronts in water repellent soils—Effect of antecedent soil moisture content. *Soil and Tillage Research.* 47: 111–123.

Robinson, A.R. 1979. Sediment yield as a function of upstream erosion. In *Universal soil loss equation: past, present, and future*, Spec. Publ. 8, ed. A.E. Peterson and J.B. Swan, 7016. Madison: Soil Science Society America.

Rogers, R.D., and S.A. Schumm. 1991. The effect of sparse vegetative cover on erosion and sediment yield. *Journal of Hydrology* 123: 19–24.

Salih, M.S.A., F.K.H. Taha, and G.F. Payne. 1973. Water repellency of soils under burned sagebrush. *Journal of Range Management* 26: 330–331.

Sartz, R.S., and D.N. Tolstead. 1974. Effect of grazing on runoff from two small watersheds in southwestern Wisconsin. *Water Resources Research* 10: 354–356.

Satterlund, D.R. 1972. *Wildland watershed management.* New York: Ronald Press Co.

Savory, A. 1978. A holistic approach to ranch management using short duration grazing. In *Proceedings of the 1st International Rangelands Congress,* ed. D.N. Hyder, 555–557.

———. 1979. Range management principles underlying short duration grazing, p. 375–379. In *Beef Cattle Science Handbook (USA).* Clovis, CA: Agriservices Foundation.

———. 1988. *Holistic resource management (No. 333.7 S268h).* Washington, US: Island Press.

Savory, A., and S.D. Parsons. 1980. The Savory grazing method. *Rangelands Archives* 2: 234–237.

Schlesinger, W.H. 1977. Carbon balance in terrestrial detritus. *Annual Review of Ecology and Systematics* 8: 51–81.

Schlesinger, W.H., J.F. Reynolds, G.L. Cunningham, L.F. Huenneke, W.M. Jarrell, R.A. Virginia, and W.G. Whitford. 1990. Biological feedbacks in global desertification. *Science* 247: 1043–1048.

Schlesinger, W.H., A.D. Abrahams, A.J. Parsons, and J. Wainwright. 1999. Nutrient losses in runoff from grassland and shrubland habitats in Southern New Mexico: I. Rainfall simulation experiments. *Biogeochemistry* 45: 21–34.

Schlossberg, M.J., A.S. McNitt, and M.A. Fidanza. 2005. Development of water repellency in sand-based root zones. *International Turfgrass Society* 10: 1123–1130.

Schroedl, G.F. 1973. *The archaeological occurrence of bison in the southern plateau.* Ph.D. Dissertation, Pullman Washington: Washington State University.

Scurlock, J.M.O., and D.O. Hall. 1998. The global carbon sink: A grassland perspective. *Global Change Biology* 4: 229–233.

Shakesby, R.A. 2011. Post-wildfire soil erosion in the Mediterranean: Review and future research directions. *Earth-Science Reviews* 105: 71–100.

Shakesby, R.A., and S.H. Doerr. 2006. Wildfire as a hydrological and geomorphological agent. *Earth-Science Reviews* 74: 269–307.

Sheath, G.W., and W.T. Carlson. 1998. Impact of cattle treading on hill land: 1. Soil damage patterns and pasture status. *New Zealand Journal of Agricultural Research* 41: 271–278.

Shepherd, T.G., S. Saggar, R.H. Newman, C.W. Ross, and J.L. Dando. 2001. Tillage-induced changes to soil structure and organic carbon fractions in New Zealand soils. *Australian Journal of Soil Research* 39: 465–489.

Sommer, M.L., R.L. Barboza, R.A. Botta, E.B. Kleinfelter, M.E. Schauss and J.R. Thompson. 2007. Habitat guidelines for mule deer: California Woodland Chaparral Ecoregion. Mule Deer Working Group, Western Association of Fish and Wildlife Agencies.

Spaccini, R., A. Piccolo, C. Conte, G. Haberhauer, and H.H. Gerzabek. 2002. Increases soil organic carbon sequestration through hydrophobic protection by humic substances. *Soil Biology and Biochemistry* 34: 1839–1851.

Spaeth, K.E. 1990. *Hydrologic and ecological assessments on a discrete range site.* Ph.D. Dissertation, Lubbock, Texas: Texas Tech University.

———. 2018. Stemple creek rainfall simulation. Technical Note, USDA-NRCS Fort Worth, Texas.

Spaeth, K.E., M.A. Weltz, H.D. Fox, and F.B. Pierson. 1994. Spatial pattern analysis of sagebrush vegetation and potential influences on hydrology and erosion. In *Variability in rangeland water erosion processes,* ed. W.H. Blackburn, F.B. Pierson Jr., G.E. Schuman, and R. Zartman, 35–50. Madison: Soil Science Society of America.

Spaeth, K.E., F.B. Pierson, M.A. Weltz, and G. Hendricks, eds. 1996a. *Grazingland hydrology issues: Perspectives for the 21ˢᵗ century.* Denver: Society for Range Management.

Spaeth, K.E., F.B. Pierson, M.A. Weltz, and J.B. Awang. 1996b. Gradient analysis of infiltration and environmental variables as related to rangeland vegetation. *Transactions of the ASAE* 39: 67–77.

Stewart, C.E., D. Roosendaal, K. Denef, E. Pruessner, L.H. Comas, G. Sarath, and M. Soundararajan. 2017. Seasonal switchgrass ecotype contributions to soil organic carbon, deep soil microbial community composition and rhizodeposit uptake during an extreme drought. *Soil Biology and Biochemistry* 112: 191–203.

Swanson, S.R., and J.C. Buckhouse. 1984. Soil and nitrogen loss from Oregon lands occupied by three subspecies of big sagebrush. *Journal of Range Management* 37: 298–302.

Swenson, C.F., D. Le-Tourneau, and L.C. Erickson. 1964. Silica in Medusahead. *Weeds* 12: 16–18.

Synman, H.A., and W.L.J. Van Rensburg. 1986. Effect of slope and plant cover on run-off, soil loss and water use efficiency of natural veld. *Grassland Society South Africa* 3 (4): 153–158.

Teague, W.R., S.L. Dowhower, S.A. Baker, N. Haile, P.B. DeLaune, and D.M. Conover. 2011. Grazing management impacts on vegetation, soil biota and soil chemical, physical and hydrological properties in tall grass prairie. *Agriculture, Ecosystems and Environment* 141: 310–322.

ter Braak, C.J.F. 1987. Ordination. In *Data analysis in community ecology*, ed. R.H. Jongman, C.J.F. ter Braak, and O.F.R. van Tongeren, 91–173. New York: Cambridge University Press.

Teramura, A.H. 1980. Relationships between stand age and water repellency of chaparral soils. *Bulletin Torrey Botanical Club* 107: 42–46.

Thurow, T.L. 1991. Hydrology and erosion. In *Grazing management: An ecological perspective*, ed. R.K. Heitschmidt and J.W. Stuth, 141–159. Portland Oregon: Timber Press.

Thurow, T.L., W.H. Blackburn, and C.A. Taylor. 1986. Hydrologic characteristics of vegetation types as affected by livestock grazing systems, Edwards Plateau, Texas. *Journal of Range Management* 39: 505–509.

Thurow, T.L., W.H. Blackburn, and C.A. Taylor Jr. 1988. Infiltration and interrill erosion responses to selected livestock grazing strategies, Edwards Plateau, Texas. *Journal of Range Management* 41: 296–302.

Trimble, S.W., and A.C. Mendel. 1995. The cow as a geomorphic agent—A critical review. *Geomorphology* 13: 233–253.

Tromble, J.M. 1976. Semiarid rangeland treatment and surface runoff. *Journal of Range Management* 29: 252–255.

Turnbull, L., J. Wainwright, and R.E. Brazier. 2008. A conceptual framework for understanding semi-arid land degradation: Ecohydrological interactions across multiple-space and time scales. *Ecohydrology* 1: 23–34.

Turnbull, L., B.P. Wilcox, J. Belnap, S. Ravi, P. D'odorico, D. Childers, and T. Sankey. 2012. Understanding the role of ecohydrological feedbacks in ecosystem state change in drylands. *Ecohydrology* 5: 174–183.

Turner, R. B., C.E. Poulton, and W.L. Gould. 1963. Medusahead-a threat to Oregon rangeland. Oregon State University Agricultural Experiment Station Spec. Rep. 149. 22 p.

USDA-NRCS. 2018. National resource inventory report: rangeland 2018. https://www.nrcs.usda.gov/wps/portal/nrcs/detail/national/technical/nra/nri/results/?cid=nrcseprd1343025

van de Koppel, J., M. Rietkerk, and F.J. Weissing. 1997. Catastrophic vegetation shifts and soil degradation in terrestrial grazing systems. *Trends in Ecology and Evolution* 12: 352–356.

Vitousek, P.M., C.M. D'Antonio, L.L. Loope, M. Rejmanek, and R. Westbrook. 1997. Introduced species: A significant component of human-caused global change. *New Zealand Journal of Ecology* 21: 1–16.

Wainwright, J., A.J. Parsons, and A.D. Abrahams. 2000. Plot-scale studies of vegetation, overland flow and erosion interactions: Case studies from Arizona and New Mexico. *Hydrological Processes* 14: 2921–2943.

Warren, S.D. 1987. Soil hydrologic response to intensive rotation grazing: A state of knowledge. In Proc. of the International Conference on infiltration development and application. January 6–9, 1987. ed. Y.S. Fok, p. 488–501. Water Resources Research Center. Honolulu Hawaii: University of Hawaii at Manoa.

Warren, S.S., M.B. Nevill, W.H. Blackburn, and N.E. Garza. 1986a. Soil response to trampling under intensive rotation grazing. *Soil Science Society America Journal* 50 (5): 1336–1341.

Warren, S.D., W.H. Blackburn, and C.A. Taylor Jr. 1986b. Effects of season and stage of rotation cycle on hydrologic condition of rangeland under intensive rotation grazing. *Journal of Range Management* 39: 486–491.

Warren, S.D., T.L. Thurow, W.H. Blackburn, and N.E. Garza. 1986c. The influence of livestock trampling under intensive rotation grazing on soil hydrologic characteristics. *Journal of Range Management* 39: 491–495.

Weaver, J.E. 1954. *North American prairie*. Lincoln Nebraska: Johnsen Pub. Co.

Weltz, M.A., and W.H. Blackburn. 1995. Water budget for South Texas rangelands. *Journal of Range Management* 48: 45–52.

Weltz, M., and M.K. Wood. 1986. Short duration grazing in Central New Mexico: Effects of infiltration rates. *Journal of Range Management* 39: 365–368.

Weltz, M.A., M.K. Wood, and E.E. Parker. 1989. Flash grazing and trampling: Effects on infiltration rates and sediment yield on a selected New Mexico range site. *Journal of Arid Environments* 16: 95–100.

Weltz, M.A., M.R. Kidwell, and H.D. Fox. 1998. Influence of abiotic and biotic factors in measuring and modeling soil erosion on rangelands: State of knowledge. *Journal of Range Management* 51: 482–495.

West, N.E., F.D. Provenza, P.S. Johnson, and M.K. Owens. 1984. Vegetation change after 13 years of livestock grazing exclusion on sagebrush semidesert in west central Utah. *Journal of Range Management* 37: 262–264.

Whicker, A.D., and J.K. Detling. 1988. Ecological consequences of prairie dog disturbances. *Bioscience* 38: 778–785.

White, N.A., P.D. Hallett, D. Feeney, J.W. Palfreyman, and K. Ritz. 2000. Changes to water repellence of soil caused by the growth of white-rot fungi: Studies using a novel microcosm system. *FEMS Microbiology Letters* 184: 73–77.

Wight, J.R. 1976. Land surface modifications and their effects on range and forage watersheds. In *Proceedings of the Fifth Workshop of the United States/Australia Rangelands Panel: Watershed Management on range and forest lands*, 165–174. Logan Utah: Utah State University.

Wilcox, B.P., and D.D. Breshears. 1994. *Hydrology and ecology of pinyon-juniper woodlands: conceptual framework and field studies* (No. LA-UR-94-2782; CONF-9408157-1). New Mexico: Los Alamos National Lab.

Wilcox, B.P., and M.K. Wood. 1988. Factors influencing interrill erosion form semiarid slopes in New Mexico. *Journal of Range Management* 42: 66–70.

Wilcox, B.P., M.K. Wood, and J.A. Tromble. 1988. Factors influencing infiltrability of semiarid mountain slopes. *Journal of Range Management* 41: 197–206.

Williams, C.J., F.B. Pierson, P.R. Robichaud, and J. Boll. 2014. Hydrologic and erosion responses to wildfire along the rangeland-xeric forest continuum in the western US: A review and model of hydrologic vulnerability. *International Journal of Wildland Fire* 23: 155–172.

Williams, C.J., F.B. Pierson, K.E. Spaeth, J.R. Brown, O.Z. Al-Hamdan, M.A. Weltz, M.A. Nearing, J.E. Herrick, J. Boll, P.R. Robichaud, D.C. Goodrich, P. Heilman, D.P. Guertin, M. Hernandez, H. Wei, S.P. Hardegree, E.K. Strand, J.D. Bates, L.J. Metz, and M.H. Nichols. 2016. Incorporating hydrologic data and ecohydrologic relationships into ecological site descriptions. *Rangeland Ecology and Management* 69: 4–19.

Winter, S.L., J.F. Cully Jr., and J.S. Pontius. 2002. Vegetation of prairie dog colonies and non-colonized short-grass prairie. *Journal of Range Management* 55: 502–508.

Wood, M.K., and W.H. Blackburn. 1981. Grazing systems: Their influence on infiltration rates in the rolling plains of Texas. *Journal of Range Management* 34: 331–335.

Wood, J.C., and M. Karl Wood. 1988. Infiltration and water quality on a range site at Fort Stanton, New Mexico. *Water Resources Bulletin* 24: 317–323.

Wright, H.A., and A.W. Bailey. 1982. *Fire ecology: United States and Southern Canada*. New York: Wiley. 501 p.

Yates, C.J., D.A. Norton, and R.J. Hobbs. 2000. Grazing effects on plant cover, soil and microclimate in fragmented woodlands in South-Western Australia: Implications for restoration. *Australian Ecology* 25: 36–47.

Yensen, D.L. 1981. The 1900 invasion of alien plants in southern Idaho. *Great Basin Naturalist* 41: 176–183.

Chapter 6
Organic Matter: The Whole Truth and Nothing but the Truth

Abstract Carbon is an essential element and building block of all living organisms. The carbon cycle is the key element to soil function and ultimately soil health. Understanding the intricacies of soil health requires a rudimentary knowledge of the carbon cycle. Other nutrient cycles, nitrogen, phosphorus, etc. are closely interconnected with organic matter and its disposition in the environment. Discussions and publications on organic matter in the popular literature abound and include articles about the value of cover crops, manuring, and rebuilding soil health. Unfortunately, some popular articles overstate or exaggerate some of the facts concerning the restoration of soil organic matter in depleted soils. In this chapter, we will explore the carbon cycle with examples of losses and gains and carbon balance examples for cropland, range, and garden settings.

Questions

- What are the ranges of geomorphic and cultural soil loss per hectare (acre) per year worldwide and in cropping situations?
- Is 1 mm of soil loss sustainable in agricultural systems?
- What aspects of the ecosystem contribute the most CO_2 to the atmosphere?
- Rank the five global carbon pools.
- What is the effect of erosion on carbon balance in cropland, rangeland, and forestland?
- Can a 1% gain in soil organic matter be achieved in 1 year in cropping systems?
- During the decomposition process of organic matter, how much organic matter is sequestered in the soil, and how much is respired by microbes?
- Describe the two major processes involved in the terrestrial carbon cycle.
- Outline the contribution of soil microbes to carbon balance in the soil.
- How much soil organic matter can be restored to cropped soils in a year or a decade?
- Diagram the main attributes of the carbon cycle.
- What are the main stages in the organic matter cycle?
- What are the roles of bacteria, fungi, and mycorrhizal groups in the organic and nitrogen cycle?
- What is the single most important constituent involved in carbon sequestration?
- What is the potential enhancement of soil organic matter with raw cow manure?

© Springer Nature Switzerland AG 2020
K. E. Spaeth Jr., *Soil Health on the Farm, Ranch, and in the Garden*,
https://doi.org/10.1007/978-3-030-40398-0_6

- What effects does grazing have on soil organic matter? What are the safety factors associated with applying manure to vegetable crops commercially and in the home garden?

Introducing Carbon

On a weight basis, carbon ranks 19th in order of elemental abundance in Earth's crust (about 0.025 % of the Earth's crust); however, carbon is a component of more compounds than all the other elements combined. The element carbon (C) is the building block and foundation of all organisms on Earth. Life on our planet would not exist without the carbon cycle, renewing and recycling plant and animal matter. Chemically carbon is a nonmetal and is a component of over 10 million different compounds, thousands of which are vital to life processes. All organically derived substances contain the element carbon, and in soils, organic matter (OM) is about 58% carbon. Carbon is found throughout nature in vastly different forms: graphite is very soft, while diamond is one of the hardest substances found in nature. When carbon combines with oxygen in differing amounts, hydrocarbons are formed (coal, petroleum, and natural gas) (Fig. 6.1).

Carbon dioxide (CO_2), the most common oxide of carbon, is a trace gas [also classified as a greenhouse gas; 0.04% (410 and increasing by 0.5% per year) by volume compared to 280 ppm before the onset of the industrial revolution] of the Earth's

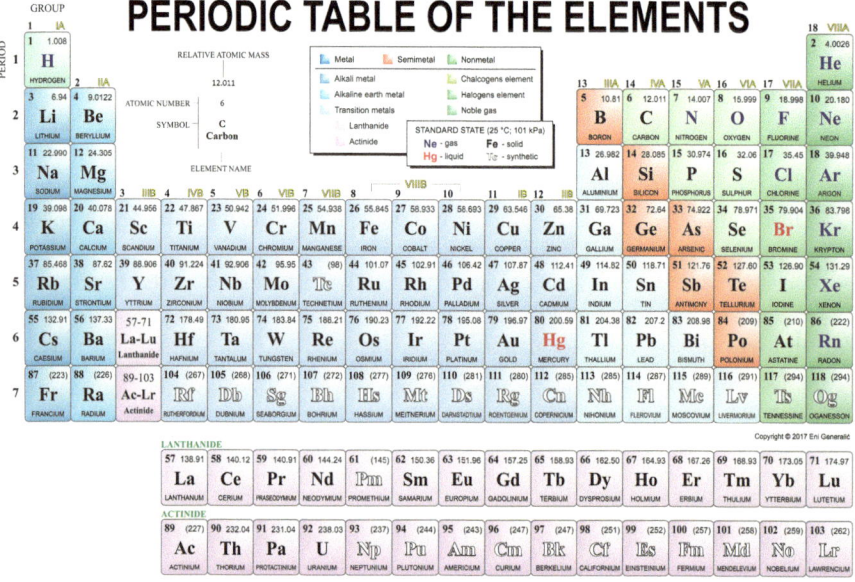

Fig. 6.1 Periodic table of the elements featuring carbon. Carbon is the foundation of all living organisms. (Courtesy of Generalic (2019))

atmosphere. The term "greenhouse gas" is derived from the concept that within a greenhouse, glass does not allow long-wave radiation to reflect out of the greenhouse. Some gases in the atmosphere possess the same properties that are analogous to glass. These gases transmit short wavelengths of light, as does glass in a greenhouse, and do hinder long-wave radiation back into space. Greenhouse gases can be natural or synthetic in origin. Natural gases include water vapor, CO_2, carbon monoxide (CO), methane (CH_4), nitrous oxide (N_2O), nitrogen oxide (NO), nitrogen dioxide (NO_2), and ozone (O_3). Synthetic compounds, which mimic and possess similar characteristics as greenhouse gases, are the chlorofluorocarbons and chlorofluorohydrocarbons (CFCs). Four basic anthropogenic components contribute to atmospheric CO_2 emissions: (1) fossil fuel consumption and transport (46%), (2) the chemical industry (20%), (3) agricultural land use changes (20%) and deforestation practices including burning, and (4) other land use changes (14%) (Lal 2003). Most methane and nitrous oxide emissions are created by bacteria in soils and from wetland ecosystems.

The burning of fossil fuels has been a major contributor to increasing levels of CO_2 in the atmosphere. Less publicized sources come from the net loss of soil organic matter (SOM) from soils around the world due to cultivation and erosion of forestland (Bolin 1977; Buringh 1984) and grasslands (Schimel et al. 1985; Burke et al. 1989; Schuman et al. 2002). Bohn (1976) states that the decline of SOM is one of the largest inputs of CO_2 to the atmosphere. Buringh (1984) estimates that about 27% (535×1015 g) of the SOC existing from prehistoric times (2014×1015 g) has been lost in the last two millennia.

Global Carbon Cycle

The carbon cycle and temporal fluxes (natural and anthropogenic) are depicted and discussed in abundance throughout the literature because it is the foundation to understanding the dynamics of how carbon is processed and compartmentalized into various pools and reservoirs. The carbon cycle involves four basic carbon reservoirs, the atmosphere, oceans, terrestrial biosphere, and fossil carbon. As seen in Fig. 6.2, natural pools and rates from heterotrophs are much greater than anthropogenic-derived carbon from burning fossil fuels and other activities, and rising CO_2 in the atmosphere is due to an imbalance between anthropogenic production and CO_2 consumption (Kirchman 2010). Global estimate of burning of fossil fuels is 4–10 Pg yr^{-1}, and open burning produces 2–3 Pg C yr^{-1} (Wiedinmyer and Neff 2007); however, soil organic carbon losses from erosion are estimated at 5.7 Pg C yr^{-1}, of which about 4 Pg C yr^{-1} is redistributed elsewhere on the landscape, 1.14 Pg C yr^{-1} is emitted to the atmosphere, and 0.57 Pg C yr^{-1} is transferred to the oceans (Lal 2003, 2008, 2010, 2018).

Müller-Nedebock and Chaplot (2015) investigated particulate organic carbon losses from runoff plots all over the world. Particulate organic carbon is organic matter that is coarse material that does not pass through a filter (0.7–0.22 um), whereas dissolved organic matter passes through. Each year, sheet erosion removes an average of 27.2 g of soil organic carbon per square meter from its original place.

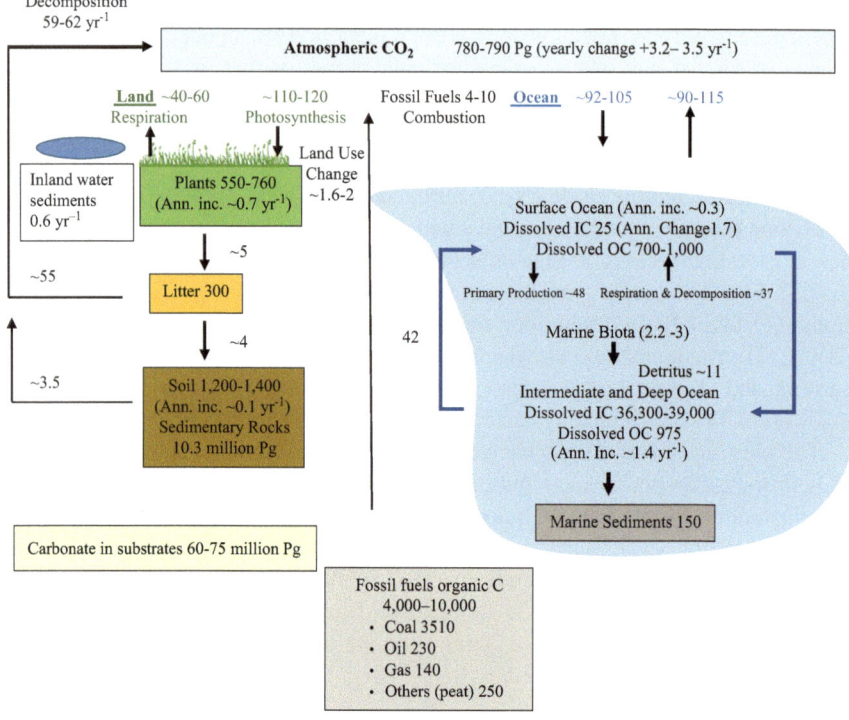

Fig. 6.2 The global carbon cycle with representative carbon pools and reservoirs that interact with Earth's atmosphere (units next to reservoir names are Pg C; Pg = 10^{15} g) (10^6 g Mg megagram (ton); 10^{12} g Tg teragram; 10^{15} g Pg petagram). The soil reservoir contains about as much of the carbon reservoir as plants and the atmosphere combined. Carbon imbalance exists from the flow of burning fossil fuels, open burning (fire); and more carbon is emitted from the soil (62 Pg) than entering the soil (59–60 Pg) due to erosion and loss of organic matter. Land use changes have been estimated to produce 1.6–2 Pg C yr^{-1}, and the largest carbon reserve is in carbonate rocks (75 million Pg). Estimates are slightly different among authors (adaptations and data from Berner 1990; Schimel 1995; Batjes 1996; Falkowski et al. 2000; Pacala and Socolow 2004; Houghton 2007; IPCC 2007; Battin et al. 2009; Haddix et al. 2011; Pan et al. 2011; Lal 2018)

The highest soil loss was observed in semiarid soils, followed by soils in tropical regions and temperate soils (Müller-Nedebock and Chaplot 2015). Sheet and rill erosion is the significant force for the displacement of soil organic carbon from its source throughout different pedoclimatic regions of the world.

Estimates of wind and water erosion amount to 75 billion metric tons per year, the majority emanating from agricultural lands (Gaia 1993; Pimentel et al. (1995a, b; Pimentel 2006). As Lal (2018) stresses: "Such a magnitude of carbon emission cannot be ignored and must be accounted for." More carbon is leaving the soil reservoir (62 Pg) than entering the soil (59–60 Pg) (Battin et al. 2009; Weil and Brady 2017). Anthropogenic carbon emissions of CO_2 are not balanced by CO_2 consumption. About half of the CO_2 emitted to the atmosphere by fossil fuel burning, open burning (fire), and terrestrial processes (mainly deforestation) is absorbed by terrestrial and marine environments; however, long-term trends are uncertain (Schimel et al. 2001).

Bicarbonates occupy the largest dissolved inorganic reservoir in the oceans and calcium carbonate (major mineral in limestone) in terrestrial ecosystems and oceans.

Focusing on the terrestrial carbon cycle, there are two major processes involved: (1) CO_2 fixation by autotrophs (organisms that produce complex organic compounds (carbohydrates, fats, and proteins) from nutrients in the environment either by using energy from light—photosynthetic process—or oxidation of reduced inorganic compounds and chemical reactions (chemosynthesis) and (2) decomposition of carbon in SOM to CO_2 by heterotrophic organisms. In the carbon cycle, the greenhouse gas, methane, is also oxidized by archaea (single-celled microorganisms, prokaryotes—no cell nucleus). Figure 6.2 displays the main parts of the carbon cycle with ranges of carbon stored and in transition on an annual basis. One important point, these natural processes are complex and variable according to large annual fluctuations in the environment, which makes it difficult to derive more exact information relating to carbon emissions and potential influence on the climate. On the average, anthropogenic activities burn about 8 Pg of carbon yearly, about 3 Pg remain in the atmosphere, and 5 Pg are sequestered in oceans and by terrestrial land plants (Kirchman 2010).

Text Box 6.1 How much O_2 is released from photosynthetic activity?
The terrestrial carbon cycle is a "balancing act" where plants take up inorganic carbon as CO_2 and synthesize organic compounds during the photosynthetic process (Fig. 6.3). Plant tissue is the source of SOM (animals are secondary sources), and rates of accumulation are in part a function of plant net primary productivity. Table 6.1 shows net primary productivity (NPP) gains of various terrestrial ecosystems with cycling of O_2 and CO_2. As a point of interest, oxygen recycled to the atmosphere for respective ecosystem types is given in terms of oxygen requirement per year for a given number of 79 kg (175 pound) individuals. The oxygen requirement information is based on the following calculations shown below.

In summary, to calculate values for terrestrial ecosystems, the following ratios apply: for each 1120 kg ha^{-1} yr^{-1} * (1000 lb ac^{-1} yr^{-1}) of aboveground net primary production, the following is produced:

- 535 kg ha^{-1} yr^{-1} (478.5 lb ac^{-1} yr^{-1}) glucose.
- 17,860 moles of O_2; 571.52 kg ha^{-1} yr^{-1} (510.4 lb ac^{-1} yr^{-1}); 4.0×105 liters O_2 yr^{-1}.
- Enough O_2 for 1.62 individuals (79 kg; 175 lb each)*.
- 17,860 moles of CO_2; 785.8 kg ha^{-1} yr^{-1} (701.75 lb ac^{-1} yr^{-1}).
- 0.78 Mg ha^{-1} (0.35 tons acre^{-1} yr^{-1}) CO_2 reduced.
- 0.3 Mg ha^{-1} (0.12 ton acre^{-1} yr^{-1}) carbon sequestered. Note: This is temporary carbon reduced via photosynthesis and is not cumulative.

An average adult uses 3 cc of O_2 per kg/body wt min^{-1}. An exercising adult uses four times that amount. On average, a 79 kg (175 pound) person would use 238 cc O_2 per min. Using the ideal gas law, a 79 kg (175 pound) individual uses a 0.23835 liter O_2 per min, or 178.895 kg O_2 yr^{-1}.

(contiuned)

Table 6.1 Net primary production (low, high, mean) of various ecosystems; CO_2 reduced during photosynthesis and recycled to ecosystem net primary productivity; and number of 79 kg (175 lb.) individuals supported per ecosystem type (Mg ha^{-1} yr^{-1}) during photosynthetic production of oxygen

Ecosystem type	Net primary productivity*			Carbon dioxide reduced			No. of 79 kg (175 lb) individuals supported per ac yr^{-1} during photosyn. O_2		
				Mg ha^{-1} yr^{-1}			No.		
	Low	High	Mean	Low	High	Mean	Low	High	Mean
Tropical rain forest	10.80	37.79	23.75	8.7	30.9	19.3	14.5	50.7	31.9
Temperate evergreen forest	6.48	26.99	14.04	5.4	22.0	11.4	8.7	36.2	18.8
Temperate deciduous forest	6.48	26.99	12.96	5.4	22.0	10.5	8.7	36.2	17.4
Boreal forest	4.32	21.59	8.64	3.6	17.7	6.9	5.8	29.0	11.6
Woodland and shrubland	2.70	12.96	7.56	2.2	10.5	6.3	3.6	17.4	10.1
Savannah	2.16	21.59	9.72	1.8	17.7	7.8	2.9	29.0	13.0
Temperate grassland	2.16	16.19	6.48	1.8	13.2	5.4	2.9	21.7	8.7
Tundra and alpine	0.11	4.32	1.51	0.0	3.6	1.3	0.1	5.8	2.0
Desert and semidesert scrub	0.11	2.70	0.97	0.0	2.2	0.9	0.1	3.6	1.3
Extreme desert, rock, sand, ice	0.00	0.11	0.03	0.0	0.0	0.0	0.0	0.1	0.0
Cultivated land	1.08	37.79	7.02	0.9	30.9	5.8	1.4	50.7	9.4
Swamp and marsh	8.64	37.79	21.59	6.9	30.9	17.7	11.6	50.7	29.0
Open ocean	0.02	4.32	1.35	0.0	3.6	1.1	0.0	5.8	1.8

Net primary productivity from Whittaker (1975)

Plants have evolved different mechanisms of fixing carbon that reduces photo-respiration. These alternate forms of carbon fixation constitute three distinct photo-synthetic pathways: the most widespread is C3 photosynthesis, C4 photosynthesis, and crassulacean acid metabolism (CAM) photosynthesis. Table 6.2 gives a brief comparison between the three photosynthetic pathways. Photosynthesis involves photochemical processes that occur in the presence of light and enzymatic pro-cesses which do not require light (Fig. 6.3). The process of diffusion brings about the exchange of CO_2 and O_2 between chloroplasts and the external air. Each of these three sub-processes is influenced by internal and external factors which can limit the yield of the overall process (Larcher 1983). Photosynthesis provides the energy and reduced carbon required for the survival of virtually all life on our planet, as well as the molecular oxygen necessary for the survival of oxygen-consuming organisms.

Table 6.2 Comparison of distinguishing characteristics of plants with different types of photosynthetic pathways

Characteristic	C3	C4	CAM
Description	Cool season plants	Warm season plants	Succulent plants
Temp. optimum (°F)	68–86 agricultural plants	95–113 hot habitats	41–59 CO_2 fixation at night
Leaf structure	Laminar mesophyll, parenchymatic bundle sheaths	Mesophyll arranged radically around chlorenchymatic bundle sheaths (Kranz anatomy)	Laminar mesophyll, large vacuole
Chloroplasts	Granal	Mesophyll: granal; bundle sheath cells: granal or agranal	Granal
Stomatal behavior	Open day, closed night	Open or closed day, closed night	Closed day, open night
Ratio of chlorophyll a/b	ca. 3:1	ca: 4:1	ca. 3:1
Light saturation point (foot candles)	3000–6000	8000–10,000	–
Low temp. Limit for CO_2 uptake (°F)	28.4–32	41–45	28.4–32
High temp. Limit for CO_2 uptake (°F)	104–122	122–140	77–86
CO_2 uptake under normal CO_2 levels, optimal temp. and water supply mg dm^2 hr^{-1}	20–45	30–80	10–15 in the dark
Primary CO_2 acceptor	RuBP	PEP	In light: RuBP In dark: PEP
Photorespiration	High	Low	Low
First product of photosynthesis	C3 acids (PGA)	C4 acids (malate, aspartate)	In light: PGA In dark: malate
Net photosynthetic capacity	Slight to high	High to very high	In light: slight In dark: medium
Light saturation reached during photosynthesis	At low to intermediate intensities	No saturation at highest intensities	At intermediate to high intensities
Photosynthesis depression by O_2	Yes	No	Yes
Maximum photosynthetic rate mg CO_2 dm^2 hr^{-1}	15–35	30–45	3–13
Maximum growth rate g dm^2 day^{-1}	1	4	0.02
Redistribution of assimilation products	Slow	Rapid	Variable
Dry matter production	Medium	High	Low

Data from Larcher (1983), Gliessman (1998)
Note: units given often do not have customary English units
ca. circa (approximately); *RUBP* pentose phosphate, ribulose-1,5 bisphosphate; *PEP* phosphoenolpyruvate; *PGA* 3-phosphoglyceric acid

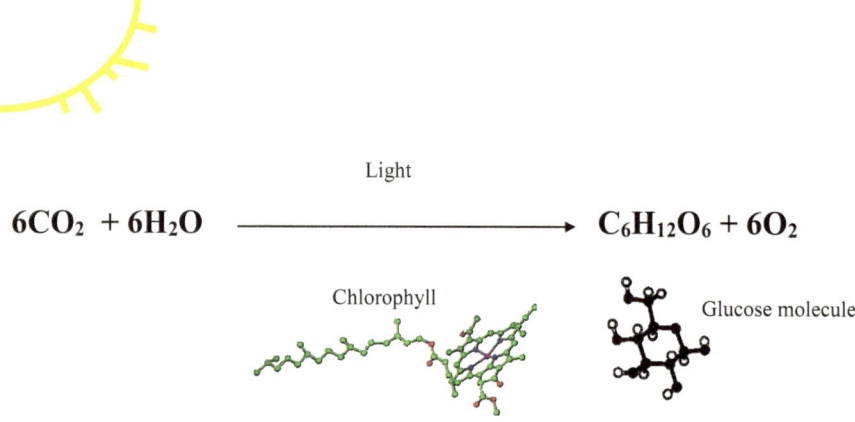

Light

$$6CO_2 + 6H_2O \longrightarrow C_6H_{12}O_6 + 6O_2$$

Chlorophyll   Glucose molecule

Fig. 6.3 Photosynthesis can be summarized by the chemical equation, where $C_6H_{12}O_6$ represents glucose, a simple sugar. Glucose is not the actual product of the carbon fixation reaction; however, the energetics of the actual carbohydrate products are approximately the same as glucose. The appearance of glucose in the equation should be regarded as a convenience, not as an absolute (Taiz and Zeiger 1998). The reduction of 6 moles of CO_2 to 1 mole of glucose results in the removal of 6 moles of H_2O and the production of 6 moles of O_2. The overall equation for photosynthesis is deceptively simple. A complex set of physical and chemical reactions must occur in a coordinated manner for the synthesis of carbohydrates. To produce a sugar molecule, such as sucrose, plants require nearly 30 distinct proteins that work within a complicated membrane structure. Carbon metabolism in plant cells begins with radiant energy; light is converted into chemical energy, which drives the synthesis of organic compounds (Fig. 6.3). On the average, for every 1 g of CO_2, a mature plant leaf manufactures 0.62 g of carbohydrate (Fig. 6.4) (Mooney 1972). Via translocation of the carbohydrate in the plant, new leaf construction results in 0.28 g dry matter. The composition of the new leaf is approximately 59% carbohydrate, 25% protein, 6% lipids, and 10% minerals (Mooney 1972)

Fossil fuels that are used to provide energy for human activity were produced by ancient photosynthetic organisms.

Although photosynthesis occurs in cells or organelles that are typically only a few microns across, the process has a profound impact on the Earth's atmosphere and climate. Each year more than 10–15% of the total atmospheric CO_2 is reduced to carbohydrate by photosynthetic organisms. It is important to note that CO_2 and O_2 are being recycled year to year. Over time (may be short or long), carbon from decaying and decomposing plant materials, wood, litter, and roots is recycled back to the atmosphere via the metabolism of the soil flora and combustion of litter and wood in fires. Terrestrial land carbon cycles and gain and loss timetables vary from within a few years to centuries.

In plants, algae, and certain types of bacteria, the photosynthetic process results in the release of molecular oxygen and the removal of CO_2 from the atmosphere that is used to synthesize carbohydrates. Conversely, terrestrial ecosystems return carbon to the atmosphere via respiration, organic matter decay and decomposition, and fire. Soils cover about 30% of the Earth's surface and account for as much respiration as ocean ecosystems. Microbes contribute to more than half of global respiration, but precise estimates are difficult in various terrestrial plant communities. In the soil, larger organisms (earthworms, nematodes, insect larvae, etc.) present a

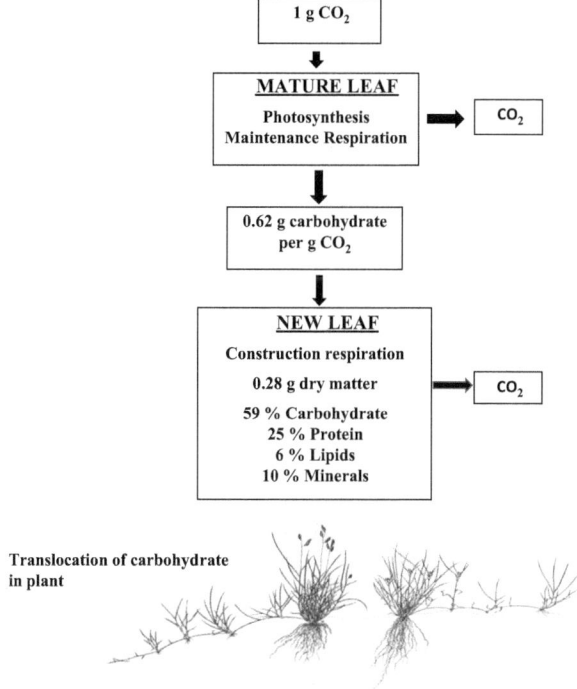

Fig. 6.4 Carbon budget for leaf development. On the average, for every 1 g of CO_2, a mature plant leaf manufactures 0.62 g of carbohydrate. Via translocation of the carbohydrate in the plant, new leaf construction results in 0.28 g dry matter. The composition of the new leaf is approximately 59% carbohydrate, 25% protein, 6% lipids, and 10% minerals. (Data from Mooney 1972)

small fraction of respiration (<5%) (Fierer et al. 2009). Saprophytic fungi (~65%) account for significantly more microbial respiration in soils than bacteria (~35%) (Joergensen and Wichern 2008). Fungi are more adapted to dry soil conditions and contribute higher overall respiration at lower temperatures (Pietikainen et al. 2005). In aquatic environments, bacteria biomass is dominant over fungi, although some fungi are indigenous and colonize large particles of fresh detritus, especially elevated plant material, e.g., marshes, etc.

Organic Matter Content in Soil

Organic Matter Losses with the Advent of Agriculture and Land Use Practices.
   The terrestrial biosphere stores about two-thirds of the global carbon in the surface meter of the soil. There are five global carbon pools (Canadell et al. 2007a).[1]

   I. Oceanic 39,000 Pg, increasing at 2.3 Pg C yr-[1]
   II. Fossil fuels 5000–10,000 Pg, mined and combusted 8 Pg C yr^{-1}

---

[1] See Fig. 6.2 for ranges by different authors.

III. Pedologic 2500 Pg @ 1 m soil depth

- 1500–1550 Pg soil organic carbon
- 950 Pg soil inorganic carbon

IV. Atmospheric 780 Pg, increasing at 4 Pg C yr^{-1}
V. Biotic pool 560–600 Pg live biomass, 60 Pg detritus material

Assuming that organized agriculture began about 10,000 years ago (Ruddiman 2003), more than 320 Pg carbon has been depleted by anthropogenic activities such as deforestation, biomass burning, cultivation, and drainage of peatlands. Of this amount, about 78 ± 12 Pg of carbon has been lost from the soil (Lal 2010). Global conversions of 1.136 billion hectares (Bha) of forest and woodlands and 0.669 Bha of savannah and grasslands have occurred between 1700 and 1990 (Foley 2005). Since the 1850s, estimates between 124 and 158 Pg of soil carbon have been depleted by various land use conversions (Canadell et al. 2007b). Lewis (2005) estimates that between 1750 and 2000, about 180 Pg carbon has been released to the atmosphere, 60% originating from the tropics.

On a global scale, intensive cultivation has resulted in a soil carbon loss of about 55 Pg carbon—25% of the original historic soil organic carbon levels (Cole et al. 1997; Lal 1998, Six et al. 2006). The cumulative loss of soil carbon from cultivated soils (55 Pg) accounts for about 7% of the current carbon in the atmosphere (Lal 1998).

In the Great Plains of the United States, organic matter depletion as a result of intense cultivation of the native grassland sod by pioneer farmers reduced soil productivity by 71% over a 28-year period (Flach et al. 1997). Many authors have reported that once cultivation begins by breaking up undisturbed native grassland and forest soils, about 20–50% of the original soil organic carbon is lost within 40–50 years (Cambell and Souster 1982; Mann 1986; Schimel 1995; Donigian et al. 1994; Rasmussen and Parton 1994; Houghton 1995) (Figs. 6.5 and 6.6). After the decline of soil organic carbon, further changes are largely a function of erosion and soil management and cropping systems (Rasmussen and Collins 1991).

Since 1750, comparative losses of soil carbon from fossil fuel combustion have released about 292 Pg of carbon (Lal 2010). In 2017, the US carbon dioxide emissions were estimated at 6.45 Pg of carbon—emissions have increased by 1.3% from 1990 to 2017 (EPA 2017) (an average annual rate increase of 0.1%). The breakdown of total fossil fuel combustion losses by economic sector is shown in Fig. 6.7 (EPA 2017).

Carbon Budgets and Balance in Terrestrial Ecosystems

In addressing soil health issues, directed management to enhance soil health requires at least a basic understanding of the cycle and balance of carbon in any particular ecosystem. Cropland settings, forest, rangeland, and gardens all have particular features concerning biomass production, plant decay rates, soil-microorganism

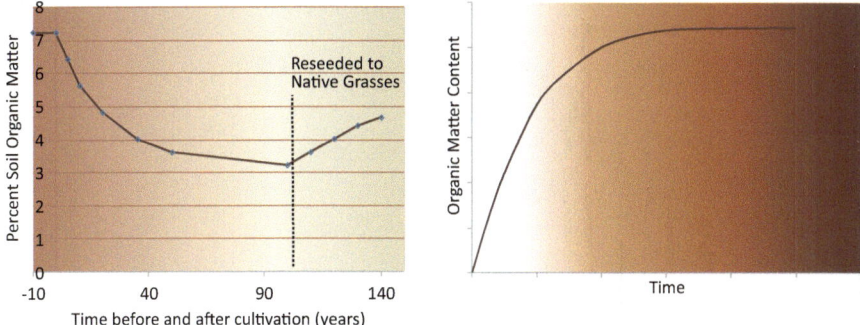

Fig. 6.5 Decline of SOM after cultivation. Native tallgrass prairie with high historic organic matter content (7.5%). Graph shows progressive loss of organic matter in the upper 25.4 cm (10 in) after cultivation. If a field is seeded back to native grasses after 100 years, organic matter would slowly increase. b) Development of organic matter over time, rate of accumulation is progressive until a constant level is reached—in equilibrium with climate, soil properties, plant community, and decomposition processes

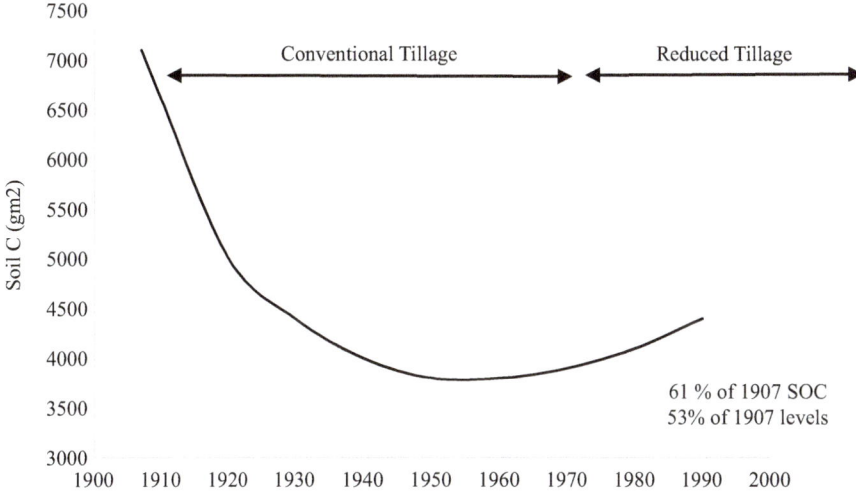

Fig. 6.6 Simulation of soil carbon changes from 1907 to 1990 for corn belt in Central United States. (Adapted from Donigian et al. 1994)

dynamics, and climatic effects—water and temperature is the principal driver in all plant communities. Organic matter gains and losses in the soil are dependent upon gains and losses of carbon. Organic matter in each land type exhibits inherent environmental inputs such as net primary productivity and litterfall, applied organic materials (manure, compost) in managed soils, respiration of living organisms (CO_2 emissions), leaching of dissolved carbon, biomass harvest (grain, stover removal, grazing), and losses from erosion.

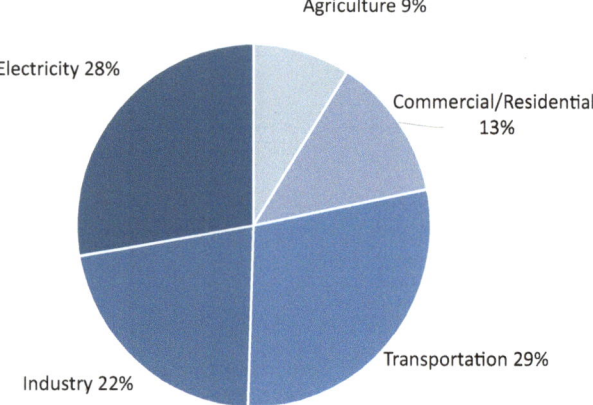

Fig. 6.7 Total US greenhouse gas emissions by economic sector (2017). *Land use, land use change, and forestry are carbon sinks and offsets approximately 11% of these greenhouse gas emissions. (Courtesy EPA 2017)

Text Box 6.2 Importance of Soil Organic Matter: Quotes

Humus plays a leading part in the storage of energy of solar origin on the surface of the earth. (Waksman 1936)

Centuries before there was any science that acquainted people with the intricacies of plant nutrition, decaying organic matter, as in manure or other forms, was recognized as an effective agent in the nourishment of plants . . . Soil organic matter is one of our most important national resources: its unwise exploitation has been devastating, and it must be given its proper rank in any conservation policy as one of the major factors affecting the levels of crop production in the future. (Albrecht 1938)

Soil organic matter is a natural product resulting from microbial activity in the inorganic and/or organic soil environment. The amount and accumulation of SOM are controlled by the composition and amounts of the plant residues, by climatic conditions, and soil texture. Other important factors are microbial activity, soil redox conditions, and other soil chemical and physical properties. (Haider and Guggenberger 2005)

In most soils, the percentage of SOM is small, but its effects on soil function are profound. This ever-changing soil component exerts a dominant influence on many soil physical, chemical, and biological properties and ecosystem functions of soil. (Weil and Brady 2017)

    The value of SOC is more than just its ability to hold water and nutrients for plant growth. Its hidden value lies in its ability to regulate the environment, especially to mitigate the greenhouse effect. (Lal 1998)

(continued)

A fertile and healthy soil is the basis for healthy plants, animals, and humans. And SOM is the very foundation for healthy and productive soils. Understanding the role of organic matter in maintaining a healthy soil is essential for developing ecologically sound agricultural practices. (SARE 2012)

The value of SOM has long been recognized. From earliest times, its level in the soil was used as a general indicator of soil productivity . . . A major factor contributing to the level of SOM is annual input of plant residues . . . Under cultivated agriculture, crop residues serve as carbon inputs and thus influence both the level and the dynamics of organic matter in the soil. (Buyanovsky and Wagner 1997)

Soil organic matter is where soil carbon is stored and is directly derived from biomass of microbial communities in the soil (bacterial, fungal, and protozoan), as well as from plant roots and detritus, and biomass-containing amendments like manure, green manures, mulches, composts, and crop residues . . . OM in its various forms greatly impacts the physical, biological, and chemical properties of the soil. OM acts as a long-term sink, and as a slow release pool for nutrients. It contributes to ion exchange capacity (nutrient storage), nutrient cycling, soil aggregation, and water holding capacity, and it provides nutrients and energy to the plant and soil microbial communities. (Moebius-Clune et al. 2016)

The weathering of minerals is an essential function in the development of soils. The second process of great significance to soil formation is the accumulation of organic matter, which tends toward an equilibrium level in well-developed soils. The level to which organic matter accumulates in the soil depends on the nature of the soil forming environment as it affects two opposing processes, namely: the addition of residues, principally from plants, and the decay of these residues by microbes and other soil-inhabiting organisms. (Hausenbuiller 1978)

Soil organic carbon is a complex of organic carbon compounds in the form of SOM. Soil organic matter includes everything in or on the soil that is of biological origin, irrespective of origin or state of decomposition (Baldock and Skjemstad 1999). It includes plant and animal remains in various states of decomposition, cells and tissues of soil organisms, and substances from plant roots and soil microbes. The ultimate product of the decomposition process is humus, an amorphous array of compounds highly resistant to further decomposition. Many organic compounds in the soil are intimately associated with inorganic soil particles. In agricultural soils, soil organic carbon content is usually less than 5 % and decreases with soil depth (Baldock and Skjemstad 1999). (citation from Xiao 2017)

Plant residues are the major source of carbon inputs in all terrestrial ecosystems. (Paustian et al. 1997)

Organic matter affects both the chemical and physical properties of the soil and its overall health. Properties influenced by organic matter include soil structure; moisture holding capacity; diversity and activity of soil organisms, both those that are beneficial and harmful to crop production; and nutrient availability. It also influences the effects of chemical amendments, fertilizers, pesticides, and herbicides. (Bot and Benites 2005)

Organic residues are carbon-containing compounds of biological origin. Decomposition is the breakdown of these complex organic materials into simpler comp (Franzluebbers 2005)

Biogeochemical Cycles

Terrestrial ecosystems are a complex of biotic and abiotic components with various cycles that are active in all environments. Energy and nutrients flow interchangeably from nonliving to living and back to nonliving components. Two kinds of cycles are represented in terrestrial environments, the gaseous (carbon, nitrogen, and oxygen cycles) and sedimentary (phosphorus and sulfur). The carbon cycle that is of primary importance in soil health is closely tied to energy flow and all the other cycles via living organisms. The carbon cycle involves the assimilation of CO_2 by plant, animal, microbial tissues, which release CO_2 by respiration.

Essentially, soil organic carbon originates from atmospheric CO_2, processed by plants during photosynthesis. In the food cycle, herbivores and secondary animal predators process this carbon, which ultimately is also returned to the soil via microbial decomposition. The amount of soil carbon that soil can eventually accumulate is a balance of carbon inputs and carbon of organic material. Soil organic matter content is dependent upon climate and climatic conditions, soil type, parent materials, physiographic influences, and how the land is used (wildlands or domestically grazed). Decomposition involves both abiotic and biotic processes. Abiotic processes include weathering, freeze/thaw cycles, alternating drying and wetting, and UV photooxidation. Microorganisms decompose SOM by cellular metabolism and extracellular excretions of exoenzymes. Although microbial biomass represents a small fraction (1–5%) of the total carbon, nitrogen, phosphorus, and sulfur pools, microorganisms are necessary for the decomposition process to proceed (Balota et al. 2003). Since microorganisms are the living component and have high surface-to-volume ratios, they respond more quickly to changes in the soil environment than chemical alteration of SOM and other soil physical and chemical properties. Microbial function and action is a key factor in diagnosing and monitoring soil health.

Temperature is an important factor in decomposition, e.g., litter is fragmented and broken down during freeze/thaw cycles and in the production of microbial exoenzymes (osmotrophs—bacteria, fungi, and archaea produced enzymes that are secreted in the environment by prokaryotic and eukaryotic cells to degrade organic polymers such as starch, cellulose and hemicellulose, lignin, chitin, and fats). Since osmotrophs do not directly ingest the coarse particular organic matter, they utilize these organic constituents as a source of energy to produce exoenzymes that degrade more complex molecules into simpler more transportable molecules needed to continue the decomposition process.

Microbial action begins on standing live vegetation as different fungal species colonize different parts of plants. When moisture is adequate, fungi can colonize the culms of grasses; some fungal species inhabit various internode and culm distances from the ground (microenvironments are more humid near the ground). Nutritional differences of the upper and lower internodes may also influence fungal flora (Hudson and Webster 1958). Fungi also can infect aboveground pine needles 5–6 months before needle fall (Burges 1963). In deciduous trees, leaf minors can damage the palisade layers of leaves, which instigates microbial attack of attached leaves. Most of the decomposition phase occurs when organic material is in contact with the soil.

A myriad of organisms (insects, earthworms, mites, millipedes, centipedes—detritivores and saprophages, etc.) fragment the debris, open plant cuticles exposing parenchyma cells to microbial invasion, and consume raw material, which, through excreta, is again acted upon by microorganisms. The physical action of these macrofeeders can expose leaf area 15 times the original leaf size (Ghilarov 1970). This initial stage of decomposition by the macrofeeders assimilates about 10% of the plant debris (mostly easily digested proteins and carbohydrates), thus allowing most of the material to pass through the gut. Plant debris, once in contact with the soil, bacteria, fungi, yeasts, actinomycetes feed on the material at rates determined by temperature and moisture—decomposition is accelerated at high temperatures (35 °C; 95 °F) provided that sufficient moisture and oxygen are available for decomposition. Oxygen can be a limiting factor in decomposition, as it is limited in oxygen-poor environments (Sierra et al. 2017).

Figure 6.8 shows some of the basic components of organic matter: dry matterwater content, elemental composition, and biologically derived components. The components in organic matter breakdown at different rates, fungal mycelium, and nonspore-forming bacteria quickly utilize carbohydrates (sugars) and simple proteins in the organic material. In viewing various soil organic carbon cycle diagrams, it is not difficult to see the complex processes that govern residue decomposition and nutrient turnover, thus, the difficulty in obtaining accurate predictions of available nutrients to subsequent crops.

Bacteria and fungi provide more than 95% of the biotic contribution to organic matter decomposition (Persson et al. 1980). Bacteria and fungi are short lived and, as they expire, are consumed by other microbial organisms. As this process continues, nutrients are immobilized in microbial tissue, and upon death, nutrients are released or mineralized—nutrients (nitrogen, phosphorus, sulfur, etc.) are again available for use by microbes and primary producers.[2] Enzymatic biochemical processes of soil microorganisms reduce decomposed organic matter into mineral compounds that may be utilized by plant roots. Mineralization is the reduction of inorganic mineral compounds. One important point that is often not recognized: for example, in a yearly cycle as plant residues are being mineralized by microorganisms and become part of the soil, a considerable amount of the original carbon (65–85% of what was part of plant litter and animal detritus) is released back into the atmosphere as CO_2 via respiration. Only about 15–35% of this remaining carbon may remain in the soil as either live biomass (~2–5%), nonliving labile carbon compounds (~3–10%), and stabilized humus (~10–30%) (Weil and Brady 2017). Stabilization and the accumulation of organic matter in the soil are regulated by the soil environment and microbial activity. In almost every aspect of our terrestrial ecosystem, especially the soil environment, organic matter is involved in the chemical and physical aspects in one way or another. The ramifications are so complex, that it is certainly beyond the scope of this book. One important soil physical

---

[2] In the ecosystem, primary producers are organisms that produce biomass from inorganic compounds (autotrophs). Autotrophs are photosynthetically active organisms (plants, certain algae, and photosynthetic bacteria, cyanobacteria and other unicellular organisms).

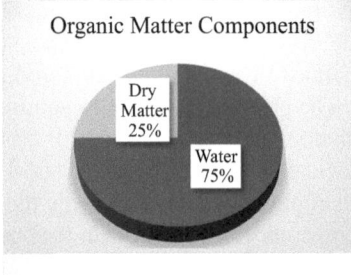

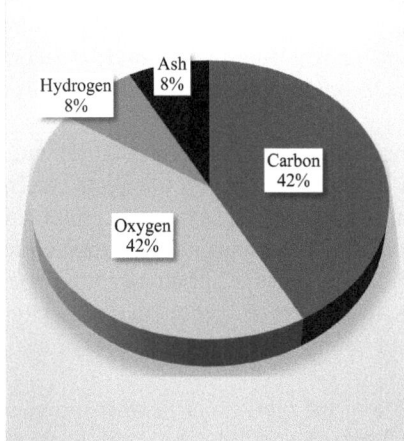

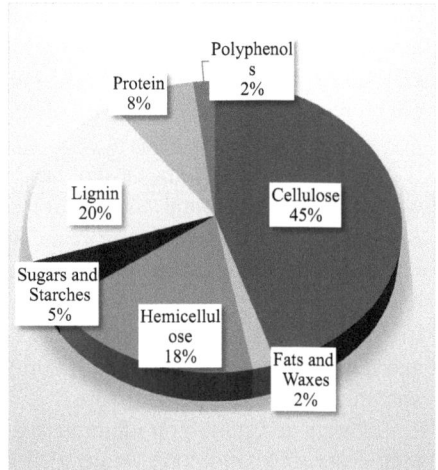

Fig. 6.8 Organic matter components

factor, soil aggregate stability, is affected by particulate organic matter (plant tissues and cell wall materials) that was in assessable to microbial action, where microaggregates form around them together with microbial exudates. These hydrophobic (water repellent) biomolecules of either plant origin or by-products of microbial synthesis (bacterial polysaccharides, fungal glycoproteins (glomalin)) are important constituents in stabilizing soil aggregates, which are an important aspect of hydrologic function and indicator of soil health. Polarity zones formed by microbial oxidation of these biomolecules allow the carbon to be stabilized and bonded to soil mineral surfaces forming stabilized aggregates. There are many other compounds produced by microbes that are associated with soil aggregate stability: waxes, humic acids, aliphatic carbon, hydrophobins (fungal proteins), fatty acids, fulvic acids, and many other extracellular enzymes and polysaccharides. Cultivation, grazing hoof impact, foot traffic, fire, and erosion can breakdown these soil aggregates.

The composition and breakdown of organic compounds vary, the ranking, from fast to slow rates are (1) sugars and simple proteins, (2) more complex proteins, (3) hemicellulose, (4) cellulose, and (5) lignin and fats (Fig. 6.8). The portion of the organic matter that is not mineralized is humus—the stable fraction of organic

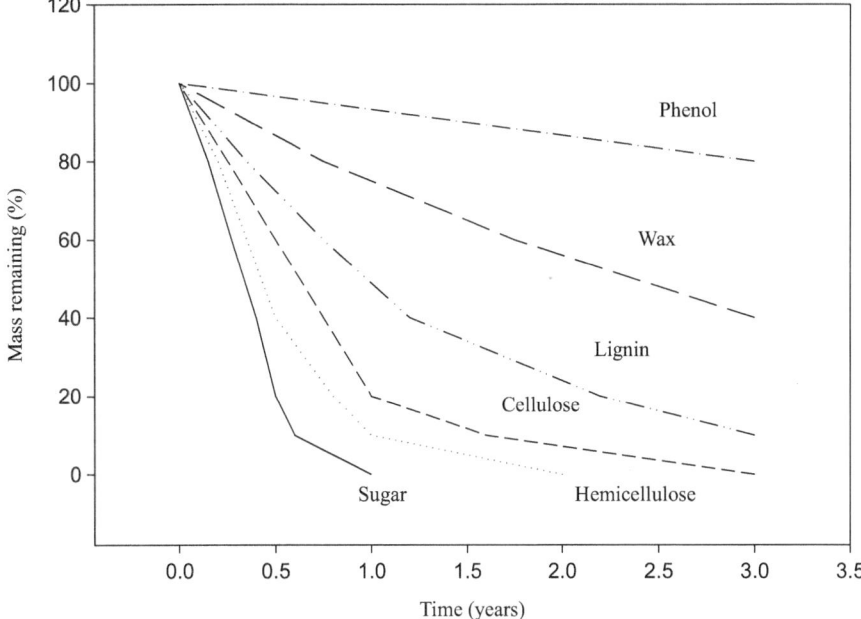

Fig. 6.9 Theoretical rates of decomposition for individual components of plant cells (source adapted from Minderman 1968; Franzluebbers 2005)

matter. There are four basic pools of SOM, each with variable turnover times: plant residues, particulate organic carbon, humus carbon, and recalcitrant organic carbon. These organic matter pools vary in chemical composition, decomposition rates, and carbon and nutrient cycles (Figs. 6.8 and 6.9). The response of these carbon pools to management is critical to soil function and health.

During the decomposition process, SOM eventually is transferred to carbon pools, as shown in Fig. 6.10 with varying turnover times (Fig. 6.11). The labile pool which turns over relatively rapidly (< 5 years) results from the addition of fresh residues such as plant roots and living organisms, while resistant residues which are physically or chemically protected are slower to turn over (20–40 years). The protected humus and charcoal components make up the stable SOM pool, which can take hundreds to thousands of years to turnover.

Although the carbon pool approach is straightforward and convenient, it perhaps is not the most accurate representation of the decomposition process (Wetterstedt 2010). The carbon pool approach is "a kind of ad-hoc theory proven to work in generating realistic figures" (Wetterstedt 2010). For example, in the century model, the SOM submodel is based on multiple compartments or pools (active, slow, passive; above- and belowground litter pools and surface microbial pools) with accompanying decomposition rates (Jenkinson 1990).

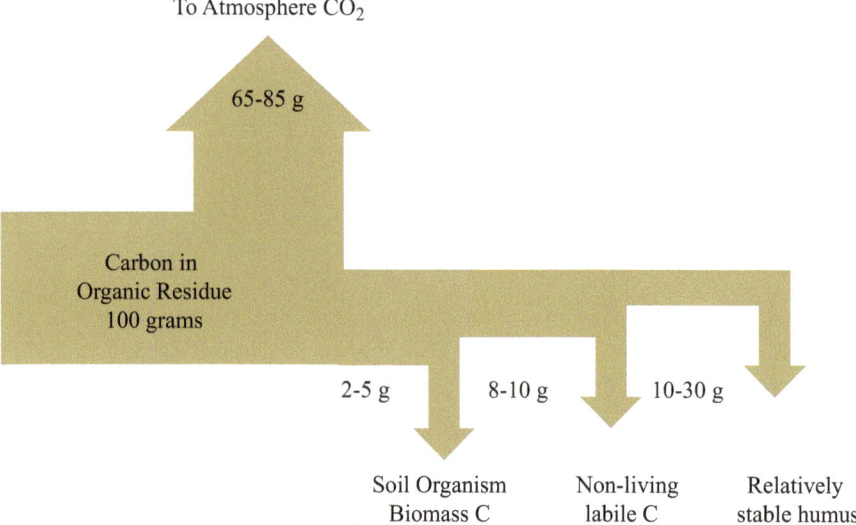

Fig. 6.10 Carbon dioxide respiration as a consequence of the breakdown of carbon-based organic residues. Example of an annual cycle where organic matter is incorporated into the soil and 65–85% of the biomass is released as CO_2. Less than one-third of the original carbon remains in the soil as microbial biomass, nonliving labile carbon, and humus. (Adapted from Weil and Brady 2017)

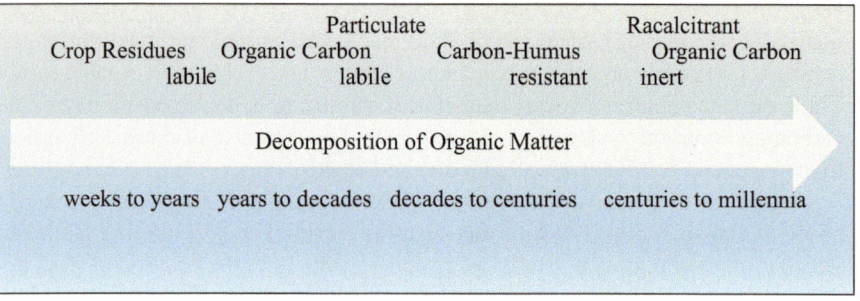

Fig. 6.11 Decomposition or turnover rates of four organic carbon pools

Conceptual Flow of Soil Organic Matter and Associated Nutrients

Several iterations of the organic matter cycle are presented because many are available in the literature. The basic patterns are similar, but each includes different details (Figs. 6.12, 6.13, 6.14, 6.15, 6.16 and 6.17). As you scan through and follow

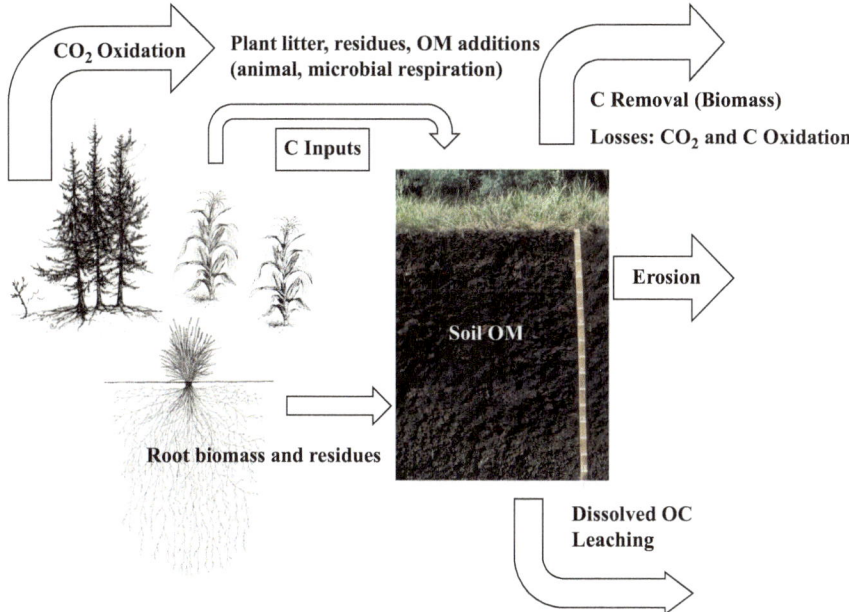

Fig. 6.12 Input and output of SOM. Balance can be sustainable or unsustainable depending on disturbances (natural and anthropogenic) and land use and management. Management in terrestrial plant communities, especially those concerned with soil health, should focus on inputs and minimize outputs (adapted from Weil and Brady 2017)

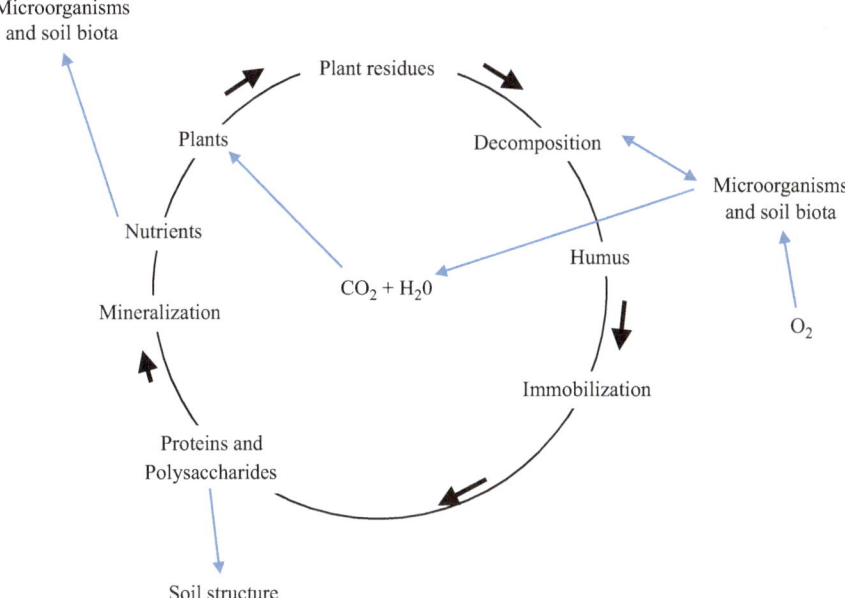

Fig. 6.13 Simple representation of organic matter cycle (adapted from Bot and Benites 2005)

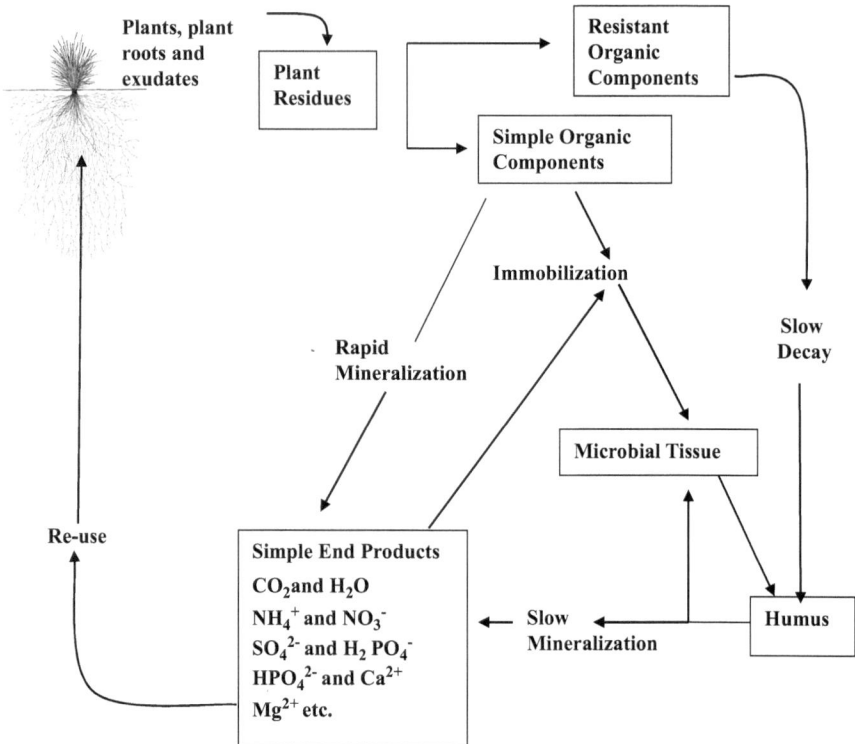

Fig. 6.14 Representation of the organic matter cycle in soils (adapted from Hausenbuiller 1978)

the pathways of the models presented here, I believe you will gain a better idea of the complexity, but also notice that more emphasis is now given to the importance of soil microorganisms.

In Liang et al. (2017) (Fig. 6.17), they created a "conceptual scheme" that shows related pathways and effects of fungi and bacteria growth, metabolism, and death in a terrestrial carbon cycle. The primary inputs of carbon are achieved via two pathways, in vivo turnover (inside the living organism) and ex vivo (outside the organism) modification driven by microbial catabolism (the breakdown of complex molecules and compounds to form simpler ones, together with the release of energy) and/or anabolism (creation of other molecules that catabolism breaks down). The authors use the microbial carbon pump concept as a sequestration system during in vivo turnover resulting in anabolism-induced necromass, the persistent carbon pool. The authors stress that much more work is needed to understand microbial contribution to the terrestrial carbon cycle.

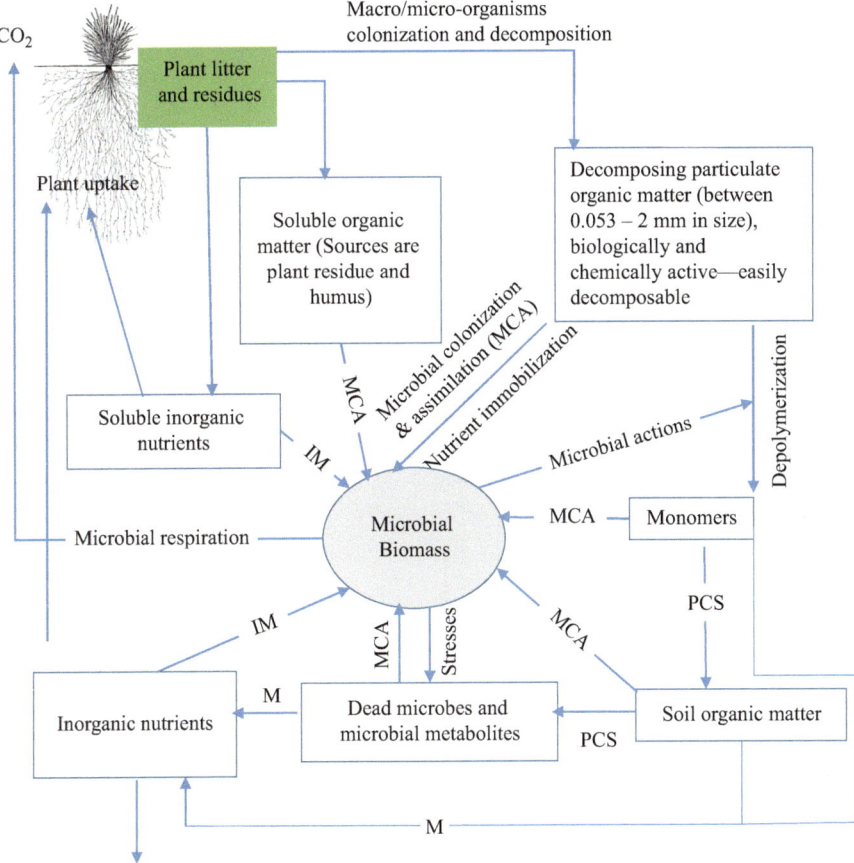

Losses from erosion, leaching, gaseous emission, and erosion

Fig. 6.15 Organic matter cycle with plant litter and residue decomposition. (MCA microbial colonization and assimilation, PCS phytochemical stabilization, M mineralization, IM immobilization (adapted from Singh and Rengel 2007)

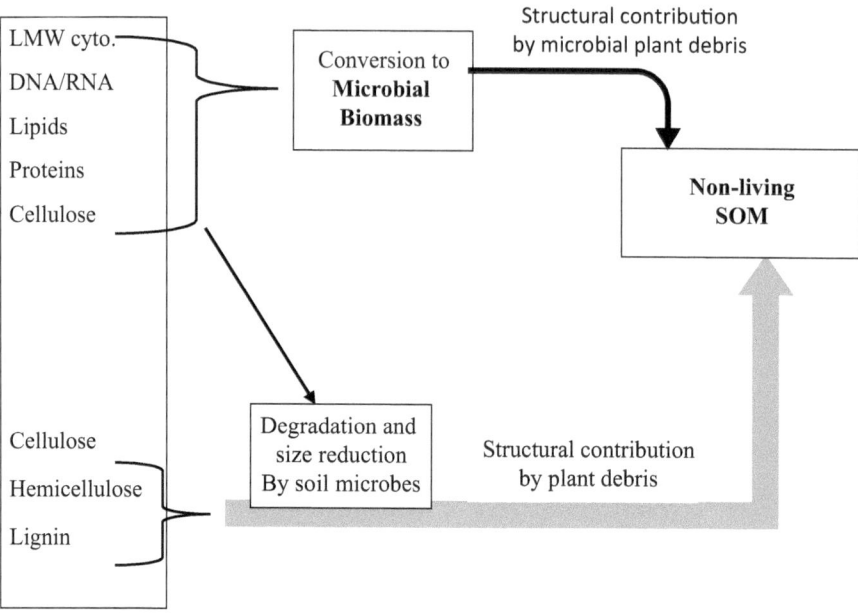

Fig. 6.16 Flow of carbon from degradation of plant litter in soil. Carbon is supplied by plant material; however, the molecular form of the carbon originated from microbial synthesis of LMW (low molecular weight cytosolic compounds) and DNA/RNA nucleic acids. (adapted from Miltner et al. 2012). Miltner et al. (2012) state that SOM should not solely be classified as humic substances, but rather a "complex mixture of plant and microbial residues at various stages of decay" of which, 50% of the bacterial biomass is mineralized, 10% remains in living microbial populations, and 40% is transferred to the nonliving SOM pool

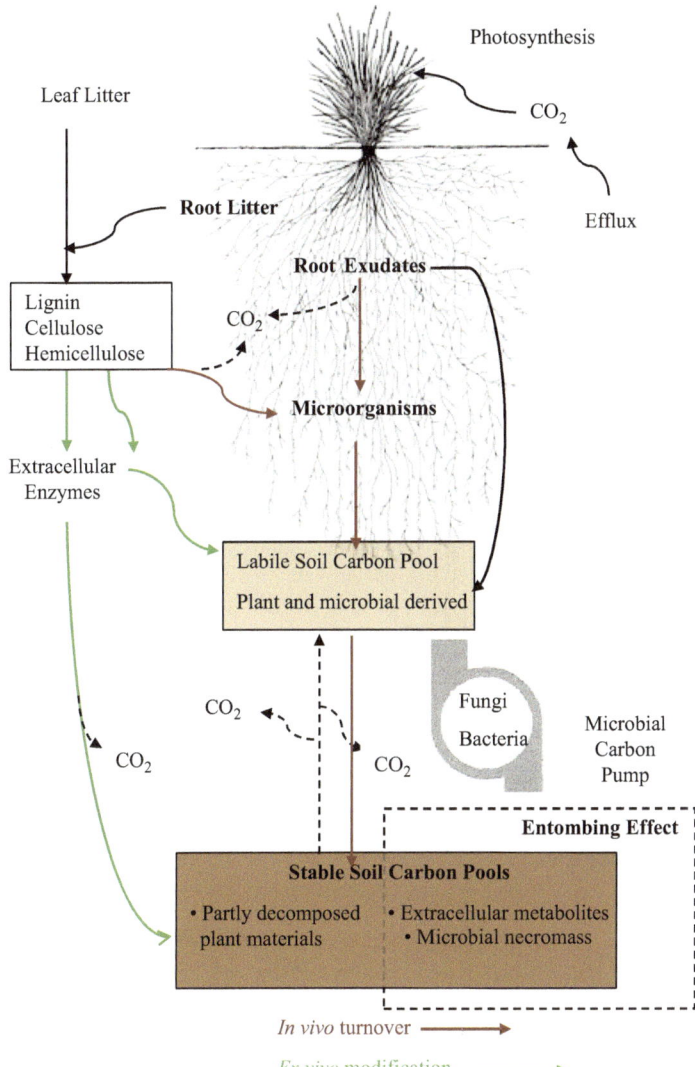

Fig. 6.17 Microbial metabolic processes associated with carbon cycling. (Adapted from Liang et al. 2014)

Text Box 6.3 Common Terms Used in Describing the Organic Soil Carbon Cycle

Decomposition

Decomposition is the physical and chemical breakdown of dead biological materials. Decomposition is responsible for the formation of SOM, and transformations result in organism-available energy and nutrient sources (Evans et al. 2017).

Decomposition is the process whereby dead organic material is broken down to its constituent parts, ultimately to CO_2 and inorganic ions such as ammonium, calcium (Ca^{2+}), potassium (K^+), and other elements originally assimilated by the organisms now decomposing. At various points along this path, the decomposing organic matter is consumed and metabolized by other organisms, which themselves will die and decompose. Decomposition is thus a dynamic recursive process that involves a complex array of physical, chemical, and biological interactions that complete the biogeochemical nutrient cycles. Rates of decomposition vary tremendously as a function of the structure and chemical composition of the organic matter being decomposed (substrate quality), the abiotic environment in which decomposition is occurring (temperature, moisture, and aeration status in particular), and the degree to which the decomposing substrate is exposed to heterotrophs, both microbial and faunal. (Robertson and Paul 2000)

When plant residues are returned to the soil, various organic compounds undergo decomposition. Decomposition is a biological process that includes the physical breakdown and biochemical transformation of complex organic molecules of dead material into simpler organic and inorganic molecules (Juma 1998)... Decomposition of organic matter is largely a biological process that occurs naturally. Its speed is determined by three major factors: soil organisms, the physical environment, and the quality of the organic matter (Brussaard 1994). In the decomposition process, different products are released: CO_2, energy, water, plant nutrients, and resynthesized organic carbon compounds. Successive decomposition of dead material and modified organic matter results in the formation of a more complex organic matter called humus (Juma 1998). This process is called humification. Humus affects soil properties. As it slowly decomposes, it colors the soil darker; increases soil aggregation and aggregate stability; increases the CEC (the ability to attract and retain nutrients); and contributes nitrogen, phosphorus and other nutrients. (Bot and Benites 2005)

Decomposition is the process by which organic matter in (or on top of) the soil is converted into progressively smaller pieces and eventually inorganic compounds. Organic matter contains carbon and hydrogen as well as several other elements such as oxygen, nitrogen, phosphorus, etc., and can originate from plants or animals. The major input into the soil is from plants, in the form of aboveground litter, i.e., leaf litter, twigs, and stems, or belowground material, i.e., root litter, exudates, and mycorrhizal hyphae. There are both abiotic and biotic processes involved in decomposition. Abiotic processes include mechanic forces acting on the litter, caused by, e.g., freezing/thawing and drying/wetting cycles. Part of the effect of bioturbation, e.g., by earthworms, is also mechanical. Light, i.e., UV photooxidation, plays an important part by having both direct and indirect effects on litter and OM (as well as DOC, DIC, POC, SOM) in terrestrial and marine ecosystems. Bacterial and fungal decomposers are more than 95 % of the biotic part of organic matter decomposition (Persson et al. 1980). Decomposers rely on water as a medium for transport of substrate (Marschner and Kalbitz 2003). To break down macromolecules into pieces

(continued)

small enough for ingestion, decomposers produce exoenzymes that diffuse through the water films to the substrate. Different kinds of litter require different types of enzymes to break them down. For example, degradation of lignin, an abundant structural part of wood, requires special lignolytic enzymes. (In Wetterstedt 2010)

Mineralization

Mineralization is a biological process where organic substances are converted to inorganic substances by soil microorganisms with end products that include CO_2, H_2O, and nutrients.

Mineralization of SOM is a microbial-mediated transformation of organically bound elements such as C, N, S, and P into inorganic compounds such as CO_2, CH_4, NH_4^+, NO_3^-, SO_4^{-2}, H_2S, and HPO_4^{-2}. The mineralization process affects different SOM pools, above and below ground as primary resources, organic compounds, and dissolved organic carbon. (Al-Kaisi and Lal 2017)

Mineralization (the release of nitrogen from organic matter) is an important factor to consider when planning nitrogen applications for crops grown on soils containing 1 % or more organic matter. Many factors affect nitrogen release from organic matter, but under average conditions, a silt loam will release about 2.25 % of the nitrogen in organic matter during an average growing season. A clay or clay loam will release an average of 1.8 % of its total organic matter nitrogen, while a sandy loam will release up to 5 % of its total organic matter nitrogen. (Thorup 1984)

Bot and Benites (2005): Soil organisms, including micro-organisms, use SOM as food. As they break down the organic matter, any excess nutrients (nitrogen, phosphorus, and sulfur) are released into the soil in forms that plants can use. This release process is called mineralization. The waste products produced by micro-organisms are also SOM. This waste material is less decomposable than the original plant and animal material, but it can be used by a large number of organisms. By breaking down carbon structures and rebuilding new ones or storing the carbon into their biomass, soil biota plays the most important role in nutrient cycling processes and, thus, in the ability of a soil to provide the crop with sufficient nutrients to harvest a healthy product. The organic matter content, especially the more stable humus, increases the capacity to store water and store (sequester) carbon from the atmosphere.

Mineralization is the biological process where organic compounds in organic matter are chemically converted by the microorganisms in the soil to simpler organic compounds, other organic compounds, or mineralized nutrients. Bacteria and fungi are responsible for most of the mineralization of organic matter in soils. Microorganisms release enzymes that oxidize the organic compounds in organic matter. The oxidation reaction releases energy and carbon, which micro-organisms need to live. The end product of mineralization is nutrients in the mineral form. Plants require nutrients to be in mineral form to take them up from the soil. Therefore, all nutrients in organic matter must undergo mineralization before they can be used again by living organisms. For example, consider a protein molecule containing carbon nitrogen, phosphorus, and sulfur. When microorganisms mineralize the protein molecule, it may undergo several changes to simpler organic molecules before the carbon is converted to CO_2, the nitrogen to ammonium, the phosphorus to phosphate and the sulfur to sulfate (http://www.soilhealth.com/soil-health/biology/organic.htm).

Drobnik (1960): Two processes, mineralization, and synthesis of SOM are involved in the turnover of organic substances. Several stages of mineralization occur: the first stage is direct transformation of the organic matter is called primary oxidation, which is characterized by increases in oxygen consumption by microorganisms caused by

(continued)

the addition of organic matter. As the organic matter is oxidized by the primary process of oxidation (mineralized), it is no longer a substrate for the synthesis of the organic matter as it does not persist in its original form. The organic matter now is converted into various metabolites and to microorganisms. The second stage, the intermediate stage of the secondary oxidation involves a portion of the organic material that is converted in microorganism's cellular material and an increase of endogenous respiration. With the increase of endogenous respiration and oxidation of more decomposable metabolites (e.g., polysaccharides), and deceased microbial cells, oxygen consumption remains high. This last stage is referred to as residual oxidation.

Johnson et al. (2005): Mineralization is the process by which microbes decompose organic nitrogen from manure, organic matter, and crop residues to ammonium. Because it is a biological process, rates of mineralization vary with soil temperature, moisture, and the amount of oxygen in the soil (aeration).

$$R\ NH_2 \rightarrow NH_3 \rightarrow NH_4^+$$
$$organic N \quad ammonia \quad ammonium$$

Mineralization readily occurs in warm (20–35 °C; 68-95 °F), well-aerated, and moist soils.

In New York State, approximately 27.3–36.3 kg (60–80 lbs) of nitrogen per acre is mineralized on average from SOM each year.

Mineralization: Carbon-nitrogen ratios (C:N) are important in carbon and nitrogen cycles, as materials with ratios less than 30:1 (30 parts carbon to 1 part nitrogen) (see C:N discussion) have sufficient nitrogen for the microorganisms involved in decomposition (mineralization) as well as providing some excess inorganic nitrogen to plants.

Immobilization

Johnson et al. (2005): Immobilization in the organic matter cycle involves the incorporation of mineralized inorganic compounds into organic molecules within living cells of organisms.

Immobilization is the reverse of mineralization. All living things require nitrogen; therefore, microorganisms in the soil compete with crops for nitrogen. Immobilization refers to the process in which nitrate and ammonium are taken up by soil organisms and therefore become unavailable to crops.

$$NH_4^+ \qquad and\,/\,or\,NO_3^- \rightarrow R\ NH_2$$
$$ammonium \quad nitrate \qquad organic N$$

Incorporation of materials with a high carbon to nitrogen ratio (e.g., sawdust, straw, etc.) will increase biological activity and cause a greater demand for nitrogen and thus result in nitrogen immobilization

Immobilization only temporarily locks up nitrogen. When the microorganisms die, the organic nitrogen contained in their cells is converted by mineralization and nitrification to plant available nitrate.

(contiuned)

Immobilization: when the C:N ratio of organic materials exceeds 30:1, micro-organisms will utilize nitrogen from the soil reserve (nitrates or ammonium) to decompose the organic materials (nitrogen deficit). Nitrogen is immobilized or unavailable to plants until the nutrients are released after the death of the microorganisms.

Carbon Sequestration

The fixation of atmospheric carbon in the organic matter of terrestrial eco-systems and its long-term preservation in the SOM with minimal risk of immediate emission is referred to as soil carbon sequestration (Semenov et al. 2008).

Carbon sequestration refers to the process of transferring CO_2 from the atmosphere into the soil. Once carbon is transferred to the soil, carbon can be stored for decades or longer. This sequestering process may be accomplished by 1) increasing crop yields through the use of management practices such as fertilization, irrigation, and grazing management, and 2) reducing decomposition of existing or new soil organic matter. Soil carbon sequestration helps offset emissions from combustion of fossil fuels and other human-induced activities. (Silveira et al. 2012)

Carbon sequestration is the process by which atmospheric carbon dioxide is taken up by trees, grasses, and other plants through photosynthesis and stored as carbon in biomass (trunks, branches, foliage, and roots) and soils. The sink of carbon sequestration in forests and wood products helps to offset sources of carbon dioxide to the atmosphere, such as deforestation, forest fires, and fossil fuel emissions.

Sustainable forestry practices can increase the ability of forests to sequester atmospheric carbon while enhancing other ecosystem services, such as improved soil and water quality. Planting new trees and improving forest health through thinning and prescribed burning are some of the ways to increase forest carbon in the long run. Harvesting and regenerating forests can also result in net carbon sequestration in wood products and new forest growth (USDA-USFS 2019).

Carbon Sequestration Potentials and Benefits of Conservation Management

Land management practices have varying effects on carbon balance and sequestration (additions or maintenance of carbon from plant biomass). By implementing conservation practices (see Table 6.9), the benefits of maintaining soil health, and where needed, restoration of soil organic matter, include improved water holding capacity, functional nutrient cycling, increased potential of plant production, management of soil-plant related diseases and insect pests, enhanced soil structure, and reduced soil erosion (Derner and Schuman 2007; Biardeau et al. 2016). The primary benefits of soil health to society are increased security of agricultural products and maintaining environmental resilience. Amundson et al. (2015) state that many conventional farming practices, such as intensive tillage of cropland, threaten long-term soil health. In Chapters 1 and 4, farming practices and cover crop benefits were explored—there is no doubt that reduced tillage and no-till farming practices result in improved hydrology and less erosion. Biardeau et al. (2016) state that "despite the numerous agronomic and ecological benefits, there is not enough sustained adoption of soil health practices by farmers in the US . . . With rapid urbanization, increasing demand for food, and shifting climate conditions, a critical public policy question is how to improve and maintain cropland soil health." In a report to USDA-Natural Resources Conservation Service, the Goldman School of Public Policy points out that the agency does not have the resources to help all farmers throughout the United States with overcoming the challenges of implementing soil health practices (Biardeau et al. 2016). They state that a small percentage of cropland acres in the United States are benefitting from the two main USDA-NRCS conservation cost-sharing programs [Environmental Quality Incentives Program (EQIP) and the Conservation Technical Assistance Program (CTA)]. The associated numbers of farm acres nationwide benefitting from EQIP and CTA are 2.7 and 5.1%, respectively (Biardeau et al. 2016).

What are the potentials for restoring carbon in the terrestrial biosphere? Plant biomass production where CO_2 is utilized during photosynthesis is considered the most efficient way to extract CO_2 from the atmosphere; however, microbial degradation then re-releases greenhouse gases back to the atmosphere (Schmidt et al. 2019). Biochar production is being investigated as a means to convert organic carbon into solid biochar, liquid bio-oil, and gaseous pyrogas carbonaceous products; however, the efficiency of the thermal conversion process is only 30–50% efficient (Schmidt et al. 2019).

Table 6.3 shows the potentials for sequestering carbon for land use activities (Canadell and Raupach 2008; Lal 2010; USFS 2015; Chambers et al. 2016; Zomer et al. 2017).

Lal (2010) estimates that terrestrial carbon pools in cropland, rangeland, forests, and wetlands can act as a sink for sequestering atmospheric CO_2 (as much as 50 ppm

Table 6.3 Yearly potential carbon storage in terrestrial biomes (United States and globally)

Land use activities	United States (Pg yr^{-1})	Globally (Pg yr^{-1})
Afforestation, agroforestry, natural forest succession, peatlands		1.2–1.4
Natural forest plantings (plantations) Improved forest management USFS-National Forest System	0.0317–0.05	0.2–0.5 0.08
Rangelands improved management	0.0054–0.016	
Pastureland grazing management Pastureland fertility management Pastureland manure management Pastureland improved species	0.0046–0.019 0.0015–0.0031 0.0036–0.009 0.0008–0.0023	
Total grazingland intensification and improvement	0.016–0.0504 (avg. 0.033)	0.3–0.5
Desertification control		0.2–0.7
Management of salic soils		0.3–0.7
Cropland conservation and cultural practices	0.144–0.432 (avg. 0.288)	0.4–1.85
Total potential	0.2–0.48	2.55–4.96 (avg. 3.8)

of CO_2 for 100–150 years). In the United States, the sink capacity of sequestering additional carbon ranges from 0.2 to 0.48 when all land uses are tallied (Table 6.3). Forests (1.2–1.4 Pg C yr^{-1}) and cropland (0.4–1.2 Pg C yr^{-1}) have the largest potentials for sequestering carbon, although grazinglands (range and pasturelands) can contribute up to 10% of the sink capacity (Lal 2010). On a global perspective, rangelands occupy about half of the world's land area, 10% of the terrestrial biomass, and 10–30% of the soil organic carbon (Schlesinger 1997). An average estimate of globally sequestered soil carbon on rangelands is 0.5 Pg C yr^{-1} (Schlesinger 1997; Scurlock and Hall 1998).

On a global basis, Earth's terrestrial ecosystems could store up to 3.8 Pg C yr^{-1}. Lal (2010) points out that costs and co-benefits need to be closely studied, but the outcomes of recarbonization of terrestrial biomes—enhancing the sequestration of atmospheric CO_2—are great: increase soil health and global food security, watershed health and water quality, and biodiversity (above and below ground). "Regardless of climate change, it is important to realize that the basic needs (e.g., food, feed, fiber, fuel) of Earth's 10 billion inhabitants by 2100 cannot be met without restoring the services of the terrestrial ecosystems, especially soil quality, for which recarbonization of the biosphere and enhancement of the SOC pool are essential prerequisites" Lal (2010).

At both the biome and more refined landscape scales, plant communities containing specific plant life and growth forms display inherent productivity levels, above and below ground, and therefore have different capabilities concerning inherent and potential sequestration of soil organic carbon (Table 6.4). Also, the capacity of science to model consequences of land cover changes and effects on climate is dependent upon terrestrial plant community carbon dynamics. Plant productivity and subsequent carbon balance in plant communities are dependent upon

Table 6.4 Primary productivity and biomass potentials for major ecosystems

Ecosystem	Net primary productivity					
	Range		Avg.	Range		Avg.
	kg ha^{-1} yr^{-1}			lbs ac^{-1} yr^{-1}		
Tropical rain forest	10,000	35,000	22,000	8922	31,227	19,628
Tropical seasonal forest	10,000	25,000	16,000	8922	22,305	14,275
Temperate evergreen forest	6000	25,000	13,000	5353	22,305	11,599
Temperate deciduous forest	6000	25,000	12,000	5353	22,305	10,706
Boreal forest	4000	20,000	8000	3569	17,844	7138
Woodlands and shrublands	2500	12,000	7000	2231	10,706	6245
Savannahs	2000	20,000	9000	1784	17,844	8030
Temperate grasslands	2000	15,000	6000	1784	13,383	5353
Tundra and alpine	100	4000	1400	89	3569	1249
Desert and semidesert	100	2500	900	89	2231	803
Extreme desert	0	100	30	0	89	27
Cultivated land	1000	35,000	6500	892	31,227	5799
Marsh and swamplands	8000	35,000	20,000	7138	31,227	17,844
Lakes and streams	1000	15,000	2500	892	13,383	2231

Data Whittaker and Likens (1973), Jobbagy and Jackson (2000)

Table 6.5 Organic and inorganic carbon mass in upper 1 m of global soils

Soil order	Global area 10^3 km^2	Global carbon in upper 1 m			Total % of global soil
		Organic	Inorganic	Total	
		Pg			
Aridisols	15,699	59	456	515	20.6
Entisols	21,137	90	263	353	14.2
Gelisols	11,260	316	7	323	12.9
Mollisols	9005	121	116	237	9.5
Inceptisols	12,863	190	34	224	9.0
Alfisols	12,620	158	43	201	8.0
Histosols	1526	179	0	180	7.2
Ultisols	11,052	137	0	137	5.5

Note: inorganic matter is largely calcium carbonate in soils of arid regions. About 60–90% of carbon is in upper 1 m of soil profile although stored carbon is also significant below 1 m in histosols and gelisols. Data from (Eswaran et al. 2000; Weil and Brady 2017)

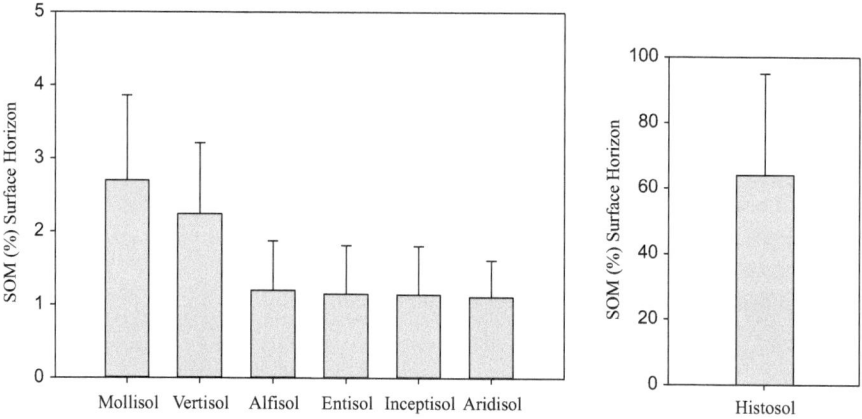

Fig. 6.18 Average SOM content for selected soil orders on non-federal rangeland (USDA-NRCS soil data from National Resource Inventory on-site study 2003-2012).)

temperature, timing, and amount of precipitation, both of which enhance or constrain the microbial process of decomposition. In arid plant communities, productivity and decomposition of organic matter are constrained and tenuous because of low annual precipitation, higher summer temperatures, and high occurrence of drought, wherein subhumid ecosystems, plant growth, and decomposition are more constant. Soil texture also plays an integral role in baseline carbon contents. Jabbagy et al. (2000) found that precipitation and climate were the best predictors of soil organic carbon in the upper 20 cm soil layer; however, clay content was the best predictor in deeper soil layers. Higher clay content may be responsible for slower soil organic carbon cycling due to clay particles and noncrystalline minerals that act to stabilize organic matter (Trumbore 2000). The soil orders, Histosols, Mollisols, and Vertisols, contain the highest soil organic carbon contents (Fig. 6.18); however, from a global perspective they do not comprise the highest global carbon percentages in the upper 1 m of soil (Table 6.5). Some generalizations can be made in regards to SOC and specific soil orders (U.S classification). Aridisols (20.6% of global soils) generally contain the lowest amounts of organic matter; whereas, histosols (7.2% of the global soils) have the highest amounts (Table 6.5) (Weil and Brady 2017). Forest soils in humid tropical regions (Oxisols and Ultisols) have similar amounts of organic carbon as Alfisols and Spodisols in humid temperate regions. Three soil orders (Histosols, Inceptisols, and Gelisols) contain about one-half of the total global soil organic carbon. Weil and Brady (2017) suggest that organic matter accumulations can be several meters thick in Gelisols, this soil order may contain more organic carbon than all the other soil orders combined. Fig. 6.18 shows average soil order SOM percentages of the surface soil horizon from a sample of over 23,000 non-federal rangeland National Resource Inventory points in the U.S.

Soil Loss Effects on Carbon Budget

Our terrestrial landscapes of today would not exist if it were not for the erosive forces of nature. As Ruhe (1969) states: "most landforms are products of erosion." Accelerated erosion has been an environmental concern ever since widespread cultivation began. The rate of natural erosion from geomorphic processes has now been surpassed by human-associated activities such as agriculture, construction, and mining. Wilkinson and McElroy (2007) state that "humans are now the most important geomorphic agent on the planet's surface, natural and anthropogenic processes serve to modify quite different parts of Earth's landscape." Soil erosion by water (inter-rill/rill erosion) is a four-stage process (Lal 2004): (1) detachment, (2) splash, (3) transport and redistribution over the landscape, and (4) deposition and burial at depressional and aquatic sites.

Soil formation begins with mechanical weathering accompanied by biological and chemical actions of parent material or from sedimentary deposits transported by wind or water. Jenny (1941) identified temperature, moisture, topographical relief, geological substrate, biological activity, and time as the main factors associated with physical and chemical weathering. Soil formation rates are higher in humid areas compared to semiarid and arid regions. As mechanical weathering progresses, parent material or bedrock is slowly transformed into smaller fragments from temperature fluctuations, hydrologic forces, erosivity of wind and water, and the colonization of vascular and non-vascular plants and animals. Chemical weathering is augmented by the composition, age, and exposed surface area of the parent materials and by the following actions: oxidation (reaction of oxygen in water with mineral iron), hydrolysis (chemical breakdown of parent material when combined with water), dissolution (chemical weathering in rocks containing either magnesium carbonate or calcium carbonate that dissolve in water by acidic solutions), carbonation (formation of carbonic acid from CO_2 dissolved in water), hydration (absorption and expansion of water in the mineral structure), and dehydration (removal of water from rock or mineral structure). As weathering continues, inorganic nutrients begin to influence soil fertility and eventually plant production.

What is the perceived balance of soil formation versus soil erosion? Soil deposition by wind into the global atmosphere is estimated at 1.0–2.1 Gt yr^{-1}, and the Sahara contributes 40–60% of this material (Maher et al. 2010). Much of this material is deposited in the North Atlantic, but rates of 0.0008 and 0.015 mm yr^{-1} are deposited throughout Europe; in the Great Plains of the United States, estimates of soil deposition are 0.2 mm yr^{-1} (Verheijen et al. 2009). In riverine systems, about 83% of the current sediment flux worldwide originates from the highest 10% of the Earth's surface elevations (Wilkinson and McElroy 2007).

On a global scale, weathering contributes substantially more to soil formation than dust deposition. Wilkinson and McElroy (2007) estimate that geologic fluxes have varied over time but are about equal to 0.1 mm yr^{-1} for our most recent geologic epoch. Land use changes leading to erosion, nutrient loss, salinization, and decline in SOM have a substantial impact on the global carbon cycle and role in emission of greenhouse gases (Lal 2018).

The decline of soil organic carbon from soils following land use changes in temperate regions (estimated of 20% or more in first 2–5 years) is rapid after immediate disturbance (Kirschbaum et al. 2008; Olson et al. 2012). These rates are accelerated (up to 70%) in tropical environments (Guillaume et al. 2015).

Soil erosion is a natural phenomenon that has occurred over geological time scales; however, this eroded soil is often deposited and becomes part of soil formation in other locations. In the United States, the most extreme rates of erosion are confined to Cordilleran regions greater than ~2.5 km in elevation (> 135 m/m.y.), and the average erosion rate is estimated at 21 meters per million years (m/m.y.) or 0.021 mm yr^{-1} (Summerfield and Hulton 1994). In examining geologic rates of erosion from stratigraphic data of sediment deposits, erosion rates for loess and glacial till areas of Iowa, United States, were within the bounds of 0.8–1.9 Mg ha^{-1} yr^{-1} (0.35–0.85 t ac^{-1} yr^{-1}) (Ruhe and Daniels 1965; Walker 1966). In the Fort Sage Mountains of northeastern California (16,000 years BP), average erosion rates were 0.95–1.6 Mg ha^{-1} yr^{-1} (0.42–0.71 t ac^{-1} yr-1) (Granger et al. 1996). On US croplands, average estimated erosion rates are about 600 m/m.y. (0.6 mm yr^{-1}) (USDA NRCS 2000) (Table 6.6).

Some of the highest soil erosion rates exist in China, and Yue et al. (2016) estimates that between 1995 and 2015, water erosion was estimated at 180 ± 80 Tg C yr^{-1} Although China has made significant efforts to reduce soil loss with the planting of trees and shrubs, erosion from mass wasting and gully erosion are still a major problem[3] (Fig. 6.19)

---

[3] Author observations from Inner Mongolia, July 2018. The predominant planting was pine trees planted on the contour throughout many areas of Inner Mongolia. Grasses and shrubs are also planted. Unfortunately, once major gullies are initiated, vegetation plantings have little effect once this process starts. These processes were initiated centuries ago from farming on uplands and steep slopes.

Table 6.6 Erosion (wind and water)-induced displacement of soil organic carbon (SOC)

Country and erosion source	SOC displaced	Reference
Australia water/wind	0.3–1.0 Pg C yr⁻¹	Chappell et al. (2014), Chappell and Baldock (2016)
China wind	75 Tg C yr⁻¹	Yan et al. (2005)
China water	180 ± 80 Tg C yr⁻¹	Yue et al. (2016)
Asia, Africa, and South America	30–40 tons ha yr⁻¹	Barrow (1991), Pimentel et al. (1995a, b)
United States and Europe	12.3–13.4 tons ha yr⁻¹	Lal (2003)
US (California) water	1.4–8.0 g C m² yr⁻¹	Yoo et al. (2005)
Global water	5.7 Pg C yr⁻¹	Lal (2003)
US and Iowa loess and glacial till	0.8–1.9 Mg ha⁻¹ yr⁻¹	Norton (1986)
US croplands, avg. estimated	0.6 mm yr⁻¹	USDA NRCS (2000)
Developed agriculture	1.2 mm yr⁻¹	Pimentel et al. (1995a, b)
Worldwide	0.1 mm yr⁻¹	Wilkinson and McElroy (2007)

Fig. 6.19 Near Linzheyu, China, Yellow River, and Great Wall, 250 km south of Hohhot, Inner Mongolia, China. (Photo by author). Severe accelerated gully erosion, especially in upper right corner. Pine trees planted on the contour in the distant upslope from the Great Wall

Open Burning Effects on Carbon Balance

Open biomass burning is an important biogeochemical pathway where carbon, oxygen, nitrogen, hydrogen, other trace gases, and particulates from terrestrial ecosystems are transferred to the atmosphere (Table 6.7a and 6.7b). Fire in terrestrial ecosystems includes prescribed burning (managed burns), wildfire (Figs. 6.20 and 6.21), agricultural burning, and trash burning (Wiedinmyer et al. 2006; Wiedinmyer and Neff 2007). In the United States, the trend of acres burned by wildfire has diminished and has remained fairly constant since 2001 (between 5 and 10 million acres per year).

Table 6.7a Fire emission components from wildfire

Emissions	Mass g per kg fuel burned	(%) of total emission
Carbon dioxide (CO_2)	1564.8	71.44
Carbon monoxide (CO)	120.9	5.52
Organic carbon	5.2	0.24
Elemental carbon	0.4	0.02
Particulate matter pd < 2.5 μ	10.3	0.47
Particulate matter 2.5 μ < pd < 10 μ	1.9	0.09
Particulate matter pd > 10 μ	3.8	0.17
Nitric oxide (N_2O)	8.5	0.39
Methane (CH_4)	5.9	0.27
Non-methane hydrocarbon	4.3	0.20
Volatile organic compounds	5.2	0.24
Water	459.2	20.97

Data from NRC (2004)
pd particle diameter μ

Table 6.7b Global Warming Potential and Atmospheric Lifetime for Major Greenhouse Gases

Greenhouse gas	Chemical formula	Global Warming Potential (GWP), 100-year time span	Atmospheric lifetime (years)
Carbon dioxide	CO_2	1	100
Methane	CH_4	25	12
Nitrous oxide	N_2O	298	114
Chlorofluorocarbon-12 (CFC-12)	CCl_2F_2	10,900	100
Hydrofluorocarbon-23 (HFC-23)	CHF_3	14,800	270
Sulfur hexafluoride	SF_6	22,800	3200
Nitrogen trifluoride	NF_3	17,200	740

Two characteristics of atmospheric gases determine the potency of their greenhouse effect: (1) Global Warming Potential (GWP) is a measure of the radiative effect of each unit of gas over a specified period of time relative to the radiative effect of carbon dioxide (CO_2). High GWP gases have a greater warming potential compared to an equal amount of CO_2. (2) Atmospheric lifetime related to the residence time of the gas in the atmosphere before natural chemical decomposition reactions in the environment. A gas with a long lifetime can exercise more warming potential and influence than a gas with a short atmospheric lifetime (assuming the GWPs are equal). Fourth Assessment Report (Intergovernmental Panel on Climate Change IPCC 2013)

Fig. 6.20 Andrea Booher, FEMA; John McColgan Sula Montana fire, 2000

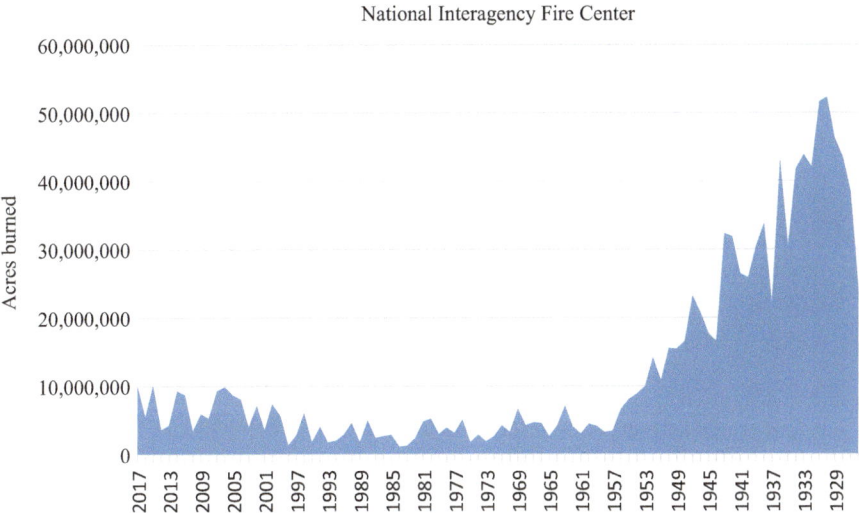

Fig. 6.21 Yearly trend of acres burned in the United States. Data from National Interagency Fire Center

There are many different compounds released by terrestrial fire (Urbanski et al. 2009); however, the principal greenhouse gases are CO_2 [the largest contribution from combustion of plant fuels (Table 6.7a and 6.7b)], methane (CH_4), nitrous oxide (N_2O), photochemically reactive compounds (e.g., carbon monoxide (CO), non-methane volatile organic carbon (NMVOC), nitrogen oxides (NOx), and fine and coarse particulate matter (PM). Note that methane and nitrous oxide have 25 and 298 times the Global Warming Potential (GWP) compared to CO_2, respectively (Table 6.7b).

Throughout the world, fires burn about 3–4 million km², and atmospheric carbon emissions are estimated at 2–3 Pg C yr⁻¹ (Wiedinmyer and Neff 2007). Wiedinmyer and Neff (2007) state: "average annual CO_2 emissions from fires in the lower 48 (LOWER48 U.S.) states from 2002–2006 are estimated to be 213 (± 50 std. dev.) Tg CO_2 yr⁻¹ and 80 (± 89 std. dev.) Tg CO_2 yr⁻¹ in Alaska. These estimates have significant inter annual and spatial variability. Needleleaf forests (conifers) in the Southeastern US and the Western US are the dominant source regions for US fire CO_2 emissions. Very high emission years typically coincide with droughts, and climatic variability is a major driver of the high inter annual and spatial variation in fire emissions. The amount of CO_2 emitted from fires in the US is equivalent to 4–6% of anthropogenic emissions at the continental scale and, at the state level, fire emissions of CO_2 can, in some cases, exceed annual emissions of CO_2 from fossil fuel usage". Although, CO_2 from open burning (a pulsed release 2–3 Pg C yr⁻¹) is about 6% of the total carbon budget of global plant respiration (50–59 Pg C yr⁻¹), and 30–48% compared to fossil fuel consumption (6–10 Pg C yr⁻¹), fire represents a large and highly variable part of the U.S. carbon budget. One important point, regrowth of vegetation from fire does offset a portion of the CO_2 emitted, as plants uptake CO_2 during photosynthesis. Changes in fire severity and frequency can lead to chemical composition changes in the atmosphere and the earth's climate system" (Lapina et al. 2006; Simpson et al. 2006; Wiedinmyer and Neff 2007; Urbanski et al. 2009).

Contribution of Plant Residues to Soil Organic Matter and Fertility

Cropland

Plant residues are an integral part of managing soil erosion and maintaining soil health (chemical, physical, and biological properties), including organic matter and nutrient pools. Decomposed or partially decomposed plant residues in cropland settings include the material from previous crops, cover crops, and the current crop. In garden environments, plant residues and mulches, fertilizers, compost, and manure are vital to healthy gardens. On a worldwide basis, cereal crop residues account for 75% of the estimated 3.8 billion tons of residue per year, 10% from sugar crops, 8% from legumes, 5% from tubers, and 3% from oil crops (Lal 2005). Nutrients and potential organic matter returned to the soil varies because the organic carbon cycle is complex and is subject to the type of plant matter, inherent soil chemistry and physical properties, climate—available moisture and temperature—biotic organisms present, and past and current management. Examples of plant residue decomposition and carbon balance are given in following text boxes.

Grazingland

On grazinglands, plant residues are important in protecting the soil surface from erosion, cycling of nutrients, and maintaining organic matter. In this chapter, several text boxes provide examples of residue decomposition and organic matter accumulation. On pasturelands, forage species that have been introduced after previous cropping or forest practices may receive nutrient inputs such as fertilizer and manure (depending upon land user management practices) to boost production; however, on rangelands in the U.S., agronomic inputs such as fertilizer, liming, and manure applications are not typically used. However, fertilizer use on native rangelands has been studied, but response varies with plant species and location (Rauzi et al. 1968). Often on native rangelands, fertilizer applications favor introduced grasses, herbaceous forbs, and/or weedy forb species over grasses (Kipple and Retzer 1959; Huffine and Elder 1960; Cosper and Thomas 1961; McKenzie et al. 2003). On annual rangelands in California, fertilizer applications were not overly beneficial because of limited rainfall (Hoagland et al. 1952). In Pacific Northwest cool season ranges, introduced grass species, e.g., intermediate wheatgrass (*Thinopyrum intermedium*), tall oatgrass (*Arrhenatherum elatius*), and timothy (*Phleum pratense*), are often more responsive to fertilization if present on native rangelands (Pumphrey and Hart 1973). On rangeland, surplus nutrients from fertilizer or manure applications (over and above normal livestock stocking rates) often favor non-native species at the expense of native plants, which could alter native biodiversity and species composition (McKenzie et al. 2003).

Grazing and Haying Effects on Soil Organic Carbon

On rangelands, carbon sequestration dynamics are quite complex, and estimation of rates and amounts is systematically more difficult than for cultivated croplands (Schuman et al. 2002) because of more heterogeneous edaphic characteristics, wide daily temperature fluctuations, intermittent precipitation, diverse vegetation life and growth forms—productivity, root-shoot ratios, and herbivore use—and imposed disturbance and management practices. Range and pasturelands make up about 10% of the overall terrestrial plant capacity, and management practices such as grazing affect soil organic carbon as plant community production and composition are dynamic throughout the season.

Rangeland plant communities can be diverse and include grasses (bunch- and/or sod-forming spp.), forbs, shrubs, and trees; plants with different root morphologies, and mixtures of C3 (cool season spp.), C4 (warm season spp.), and CAM (succulents) plants. Included in the mix of complex enviornmental interactions are perturbations such as fire, herbivory, and a propensity to drought. Schuman et al. (2005) point out that the combination of severe drought and heavy grazing can result in significant losses of soil organic carbon that was previously stored under normal production levels—rangelands then shift from sequestering carbon to releasing CO_2 to the atmosphere (Balogh et al. 2005). However, proper grazing management can avoid and/or alleviate serious soil disturbance factors. In addition, soil microbial shifts also occur during drought, which can result in losses of soil carbon (Ingram et al. 2004).

Vegetation composition changes in response to grazing can alter soil carbon in the soil (Hewins et al. 2018). There is inconclusive evidence about how grazing affects the distribution and maintenance of soil carbon in different rangeland ecosystems. Since rangelands are so diverse with respect to vegetation composition, climate, and soils, deriving a general conclusion about grazing effects and systems on carbon sequestration is unlikely—dynamics at the ecological site level will determine specific responses.

Numerous studies show varying effects of grazing on carbon balance:

• Soil organic content was reduced on grazed native grasslands compared to ungrazed grasslands (Bauer et al. 1987).
• Heavily grazed fescue grasslands in Alberta, Canada [0.2 ha per animal unit month (AUM^{-1})], had less soil organic matter compared to grazing at 0.8 ha per AUM^{-1} (Johnston et al. 1971).
• Sheep grazing on native *Stipa-Bouteloua* prairie at 2.5 ha AUM^{-1} showed increased soil carbon compared to grazing at 1.7 ha AUM^{-1} (Smoliak 1986).
• In northern mixed-grass prairie, North Dakota, moderate grazed treatments contained 17% less soil carbon compared to ungrazed exclosures. Heavily grazed treatments did not result in less soil carbon compared to the exclosure. The authors surmised that an increase in blue grama (*Bouteloua gracilis*), a grass species with a heavy dense root system, may be responsible for maintaining soil carbon levels equal to the exclosure (Frank et al. 1995).
• In meadow bromegrass (*Bromus riparius*) pastures in Alberta, Canada, "heavy and medium grazing intensities produced 83 and 90% as much above ground dry matter and 87 and 90% above ground carbon as the light intensity . . . heavy grazing reduced the contribution of vegetative dry matter in vitro digestible organic matter, carbon and nitrogen to the residual 41, 50, 36, and 52% of that for light grazing . . . estimated fecal carbon inputs were 68, 51, and 42% of all carbon inputs for heavy, medium, and light grazing, respectively" (Baron et al. 2002).
• Derner and Schuman (2007) stated: "although there was no statistical relationship between change in soil carbon with longevity of the grazing management practice in native rangelands of the North American Great Plains, the general trend seems to suggest a decrease in carbon sequestration with longevity of the grazing management practice across stocking rates".
• In northern mixed-grass prairie, Schuman et al. (1999) found that light or heavy grazing resulted in higher carbon sequestration compared to nongrazed exclosures. Eighty-one years of moderate and heavy grazing in northern mixed-grass prairie showed increases of 19 and 34% of soil carbon at 0-5 cm and 5-15 cmsoil depth (Wienhold et al. 2001). Manley et al. (1995) found that short-duration rotational grazing, rotationally deferred grazing, and continuous season-long grazing at heavy stocking rates did not affect carbon sequestration rates.
• In shortgrass steppe in northeastern Colorado, Derner et al. (1997) found increased soil carbon storage in grazed (1983 g m^2; 0.41 lb ft^2) compared to ungrazed areas (1321 g m^2; 0.27 lb ft^2) at 0–15 cm (0–5.9 in) soil depth, while no differences were found at the 15–30 cm (5.9–11.8 in) soil depth.
• During a 12-year period in "coastal" Bermuda grass (*Cynodon dactylon*)/tall fescue (*Lolium arundinaceum*) paddocks, annual soil organic carbon change at

0–90 cm soil depth was as follows: low grazing pressure (1.17 Mg C ha^{-1} year^{-1}; 0.52 t C ac^{-1} yr^{-1}) was greater than unharvested grass (0.64 Mg C ha^{-1} year^{-1}; 0.29 t C ac^{-1} yr^{-1}) but nearly equal to high grazing pressure (0.51 Mg C ha^{-1} year^{-1}; 0.23 t C ac^{-1} yr^{-1}). Hayed paddocks showed the lowest annual rate of soil organic carbon change (0.22 Mg C ha^{-1} year^{-1}; 0.1 t C ac^{-1} yr^{-1}) (Franzluebbers and Stuedemann 2009).

- During a 5-year evaluation of soil organic carbon sequestration in Bermuda grass pasture, sequestration in the surface 6 cm was 1.4 Mg C ha^{-1} year^{-1} (0.62 t C ac^{-1} yr^{-1}) when grazed in summer by cattle, 0.65 Mg C ha^{-1} year^{-1} (0.29 t C ac^{-1} yr^{-1}) when ungrazed, and 0.29 Mg C ha^{-1} year^{-1} (0.13 t C ac^{-1} yr^{-1}) when hayed (Franzluebbers et al. 2001).
- Hewins et al. (2018) evaluated 108 pairs of long-term grazed and ungrazed study sites in dry mixed-grass prairie, central parkland, foothills fescue, and montane and upper foothills representing six distinct climate subregions across 5.7 M ha^{-1} (14 M ac^{-1}) of Alberta, Canada. Their findings found that moderate grazing increased soil organic carbon by 12% in the upper 15 cm (5.9 in) of soil, and soil organic carbon concentrations in deeper mineral soil layers were associated more with regional climate and increase from dry to mesic subregions. They concluded that "longterm livestock grazing may enhance soil organic carbon concentrations in shallow mineral soil and affirm that climate rather than grazing is the key modulator of soil carbon storage across northern grasslands."
- Mountain meadows grazed for 1–3 months by sheep and cattle in the Medicine Bow National Forest in Wyoming showed higher soil organic carbon in grazed compared to ungrazed treatments at 0–7.5 cm (0–3 in) soil depth (Povirk 1999).
- "Proper grazing management has been estimated to increase soil carbon storage on US rangelands from 0.1 to 0.3 Mg C ha^{-1}year^{-1} (0.02–0.13 t C ac^{-1} yr^{-1}) and new grasslands have been shown to store as much as 0.6 Mg C ha^{-1}year^{-1} (0.26 t C ac^{-1} yr^{-1})" (Schuman et al. 2002).

Carbon Balance Examples in Cropland, Rangeland, Forest, and Garden Settings

Text boxes are presented in this discussion to demonstrate carbon turnover in cropland, rangeland, forest, and garden settings:

- Effect of corn stover removal on soil carbon
- Carbon balance with corn residues and rye cover crop
- Carbon balance, in tallgrass prairie
- Carbon balance on grassland soil (mollisols)
- Standing crop turnover on a saline desert (saltbush, *Atriplex confertifolia*) site
- Comparison of annual net primary productivity and soil carbon in a grassland and woody savannah
- Carbon and nitrogen turnover in a ponderosa pine (*Pinus ponderosa*) forest
- Carbon balance in an Indiana deciduous forest
- Carbon balance in a vegetable garden setting with compost application

Text Box 6.4 What Is the Effect on Soil Carbon with Stover Removal? (Adapted from Sawyer and Mallarino 2007)

Removing corn stover not only reduces carbon return to soil but also nitrogen. Corn stover harvest removes carbon that potentially could be recycled and incorporated into SOM pools. However, decomposition of crop residue by soil microbes with associated large carbon loss as CO_2 is not commonly recognized. Figure 6.22 shows decomposition rates and cumulative CO_2 loss to the atmosphere at 1–8 years. In the long term, 85% of the original corn stover biomass is lost as atmospheric CO_2.

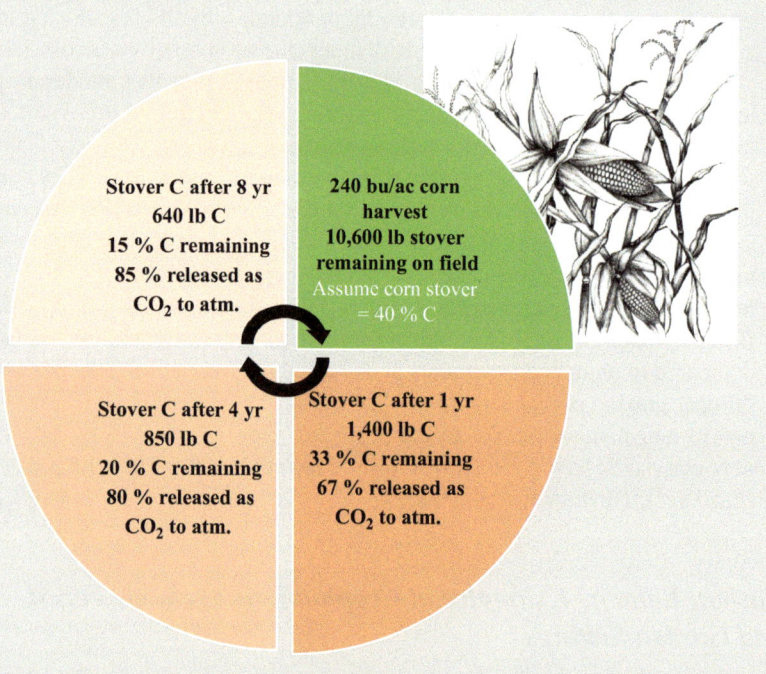

Fig. 6.22 Effect on soil carbon with corn stover removal over time

Text Box 6.5 Example of Carbon Balance in Iowa Corn Field. Units in parentheses are kg ha^{-1} (Fig. 6.23)

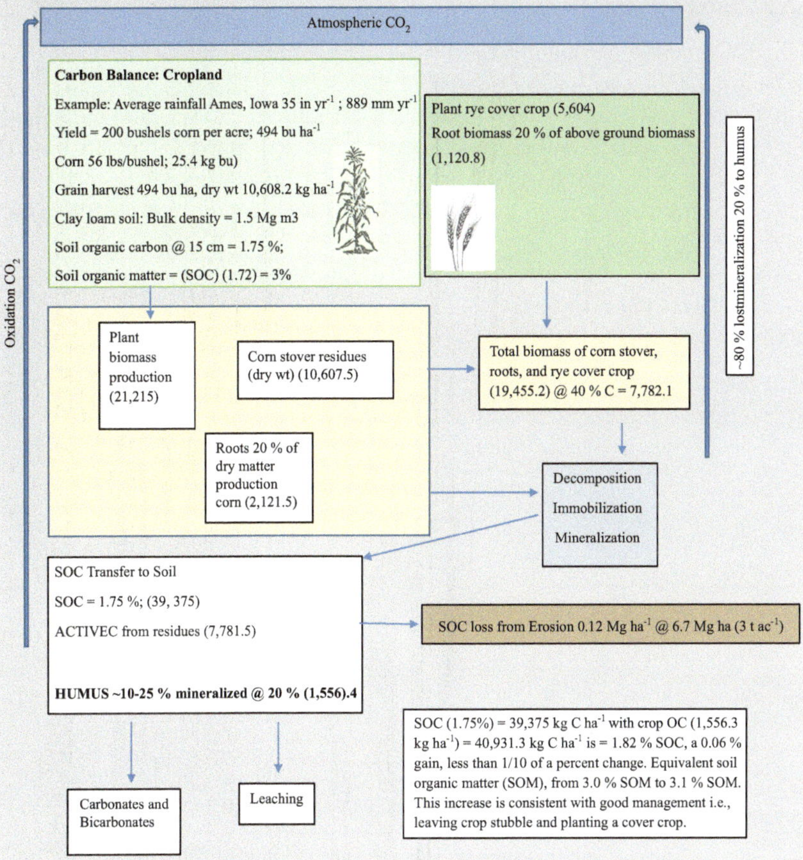

Atmospheric CO_2

Oxidation CO_2

Carbon Balance: Cropland

Example: Average rainfall Ames, Iowa 35 in yr^{-1} ; 889 mm yr^{-1}

Yield = 200 bushels corn per acre; 494 bu ha^{-1}

Corn 56 lbs/bushel; 25.4 kg bu)

Grain harvest 494 bu ha, dry wt 10,608.2 kg ha^{-1}

Clay loam soil: Bulk density = 1.5 Mg m3

Soil organic carbon @ 15 cm = 1.75 %;

Soil organic matter = (SOC) (1.72) = 3%

Plant rye cover crop (5,604)

Root biomass 20 % of above ground biomass

(1,120.8)

~80 % lost/mineralization 20 % to humus

Plant biomass production (21,215)

Corn stover residues (dry wt) (10,607.5)

Total biomass of corn stover, roots, and rye cover crop (19,455.2) @ 40 % C = 7,782.1

Roots 20 % of dry matter production corn (2,121.5)

Decomposition
Immobilization
Mineralization

SOC Transfer to Soil

SOC = 1.75 %; (39, 375)

ACTIVEC from residues (7,781.5)

SOC loss from Erosion 0.12 Mg ha^{-1} @ 6.7 Mg ha (3 t ac^{-1})

HUMUS ~10-25 % mineralized @ 20 % (1,556).4

SOC (1.75%) = 39,375 kg C ha^{-1} with crop OC (1,556.3 kg ha^{-1}) = 40,931.3 kg C ha^{-1} is = 1.82 % SOC, a 0.06 % gain, less than 1/10 of a percent change. Equivalent soil organic matter (SOM), from 3.0 % SOM to 3.1 % SOM. This increase is consistent with good management i.e., leaving crop stubble and planting a cover crop.

Carbonates and Bicarbonates

Leaching

Fig. 6.23 Carbon balance in Iowa corn field. Units in parentheses are kg ha^{-1}

Weight of soil kg ha^{-1} @ 00.15 m soil depth
(100 m) (100 m) (0.15 m) = 1500 m^3 ha^{-1}
Weight of soil = (1500 m^3 ha^{-1}) (bulk density 1.5 Mg m^3) = 2250 Mg ha^{-1}
  = 2250,000 kg ha^{-1}
SOC = 1.75%; therefore, total wt. of SOC per ha = (2250,000 kg ha^{-1})
  (0.0175% SOC) = 39,375 kg C ha^{-1}

(continued)

SOC Carbon to nitrogen ratio = 12:1; 39,375/12 = 3281.3 kg N ha^{-1}
SOC Carbon to phosphorus ratio = 50:1; 39,375.00/50 = 787.5 kg P ha^{-1}
Yield = (200 bu ac^{-1}) (56 lbs bu) = 11,200 lbs ac^{-1} (12,555.5 kg ha^{-1})
Estimated grain to corn stover residue = 1:1 ratio; therefore, corn stover pro-
   duced = 12,553,5 kg grain ha^{-1} (11,200 lbs grain ac^{-1}). Dry wt. of
   grain = (12,553.5) (0.845) = 10,608.2 kg ha^{-1} (9464 lbs ac^{-1})
Ratio of grain to stover ~1; therefore, dry corn stover on field = 10,608.2 kg ha^{-1}
   (9464 lbs ac^{-1})
Root mass of corn crop is ~20% of aboveground prod. = (10,608.2)
   (0.20) = 2121.5 kg ha^{-1} (1892.8 lbs ac^{-1})
Total above- and belowground biomass = 10,608.2 kg ha^{-1} + 2121.5 kg ha^{-1} =
   12,729 kg ha^{-1} (11,357.5 lbs ac^{-1})

Plant rye cover crop, aboveground production = 5604 kg ha−1 (5000 lbs ac−1)
Root biomass, rye cover crop = 20% of aboveground biomass = 1120.9 kg ha^{-1}
   (1000 lbs ac^{-1})
Total biomass of corn stover (10,608.2 kg ha^{-1}), roots (2121.5 kg ha^{-1}), and
   cover rye cover crop aboveground production (5604 kg ha-1) roots
   (1120.9) = 19,455.2 kg ha^{-1} (17,357.9 lbs ac^{-1})
Carbon content of biomass ~40%; therefore, 19,455.2 kg ha^{-1} (0.4) =
   7782.1 kg ha^{-1} (6943.2 lbs ac^{-1})
Wt. of SOC and crop OC = 39,375 kg ha^{-1} + 7782.1 kg ha^{-1} = 47,157.1 kg h
   a^{-1} (42,073.5 lbs ac^{-1})
Percentage carbon retained in decomposed residue considered as humus
   ~10–25%. (7782.1 kg ha^{-1}) (0.20) = 1556.4 kg ha^{-1} (1388.6 lbs ac^{-1})
SOC (39,375 kg C ha^{-1}) + Crop SOC (1556.4 kg C ha^{-1}) = 40,931.4 kg C ha^{-1}
   (36,519 lbs C ac^{-1})
Increase of SOC (1.75%) = 39,375 kg C ha^{-1} with crop OC
   (1556.4 kg ha^{-1}) = 40,931.4 kg C ha^{-1} is = 1.82% SOC, a 0.07% gain, less

(contiuned)

than 1/10 of a percent change. Equivalent SOM from 3.0% SOM to 3.1% SOM. This increase is consistent with good management, i.e., leaving crop stubble and planting a cover crop.

Potential erosion and SOC loss: assume 6725 kg ha^{-1} soil loss (3 tons ac^{-1}), equivalent SOC loss = 0.12 Mg ha^{-1}

Note: see Duiker (2018). https://extension.psu.edu/can-i-increase-soil-organic-matter-by-1-this-year

Text Box 6.6 Content and Turnover of Organic Dry Matter in the Tallgrass Prairie

Standing crop, litter, and root values are pounds per acre. No grazing of large herbivores, only from insects and rodents. During the growing season, the vegetation grows rapidly, and at the same time, parts of the shoots and roots die off and/or are eaten. The annual increase in organic matter fluctuates between large positive and negative values, but the long-term average is approximately a zero gain (Data from Kucera et al. 1967) (Fig. 6.24).

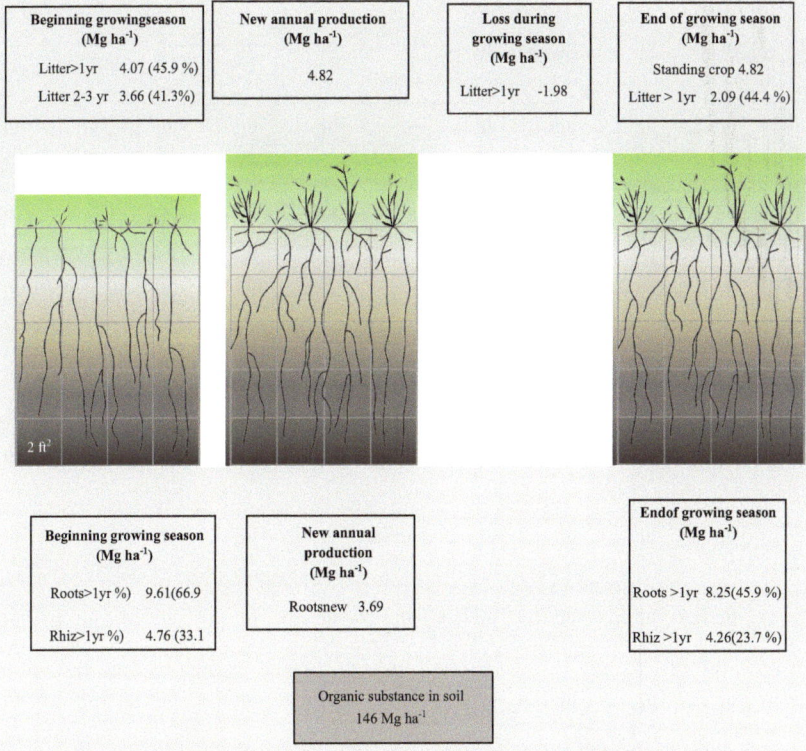

Fig. 6.24 Content and turnover of organic dry matter in the tallgrass prairie

Text Box 6.7 Plant Litter Carbon Dynamics in a Grassland Soil (Mollisol) (20 cm Soil Depth)

In diagram, carbon pools (kg C m^{-2}) and annual allocations (kg C m^{-2} yr^{-1}). At 20 cm soil depth, total carbon content is 10.4 kg C m^{-2} (104 Mg ha^{-1}; 46.4 t ac^{-1}); 84% of carbon respired. Adapted from Schlesinger (1977) (Fig. 6.25).

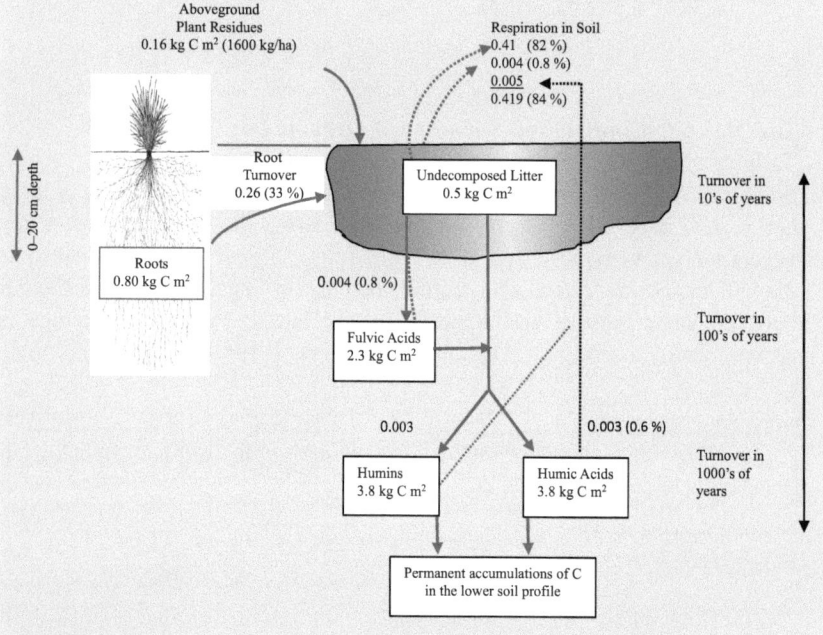

Fig. 6.25 Plant litter carbon dynamics in a grassland soil (mollisol) (20 cm soil depth)

Text Box 6.8 Content and Turnover of Standing Crop Production on a Saline Desert Site with Shadscale Saltbush (*Atriplex confertifolia*)
Standing crop, litter, and root values are pounds per acre. The amount of litter fluctuates considerably from year to year. During very dry years, this fluctuation can be larger than the production of dry matter (due to drought damage to plants). During the productive phase of the stand and during the colonization phase of saltbush, net primary productivity is larger, and there is a surplus of organic matter. As the stand becomes mature, the annual increase in biomass slows significantly, and the ratio of green to non-green tissues is smaller. Yield from photosynthesis is used for renewing leaves and for respiration of the root system. Carbon is sequestered in the woody tissue with little additional storage as the stand becomes mature (Data from Larcher 1983) (Fig. 6.26).

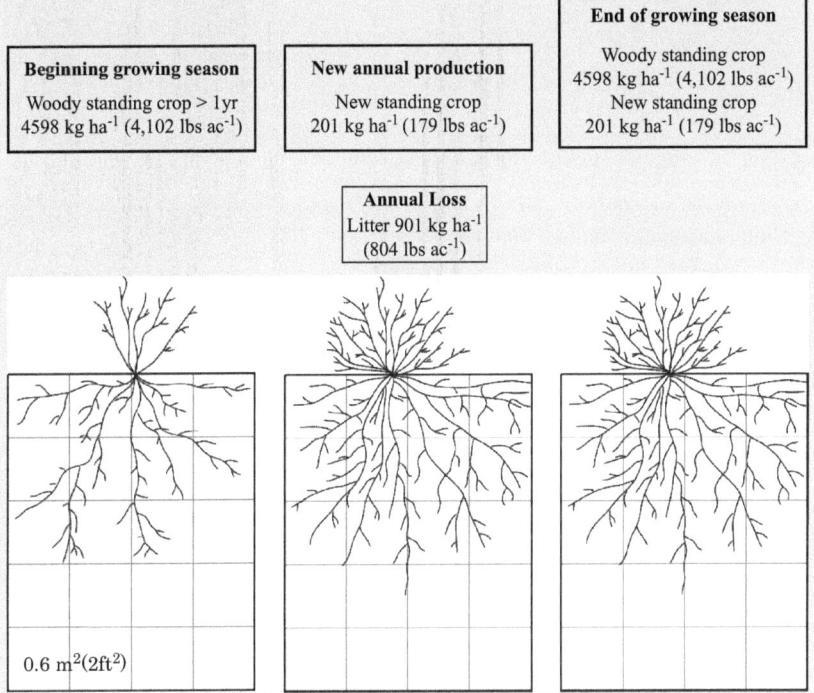

Fig. 6.26 Content and turnover of standing crop production on a saline desert site with shadscale saltbush (*Atriplex confertifolia*)

Text Box 6.9 Comparison between a grazed grassland community and woody savannah parkland (mesquite, *Prosopis glandulosa* dominated overstory) in sandy loam uplands at the La Copita Research Area in southern Texas (Data from Archer et al. 2001) (Fig. 6.27)

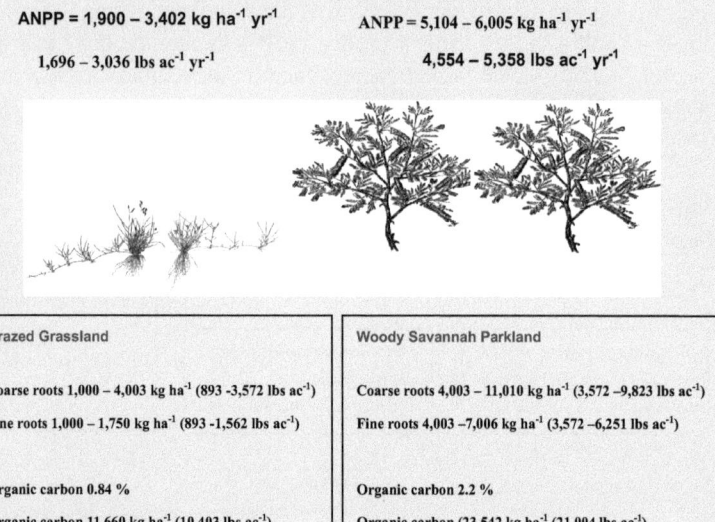

Fig. 6.27 Comparison between a grazed grassland community and woody savannah parkland (mesquite, *Prosopis glandulosa* dominated overstory) in sandy loam uplands at the La Copita Research Area in southern Texas (Archer et al. 2001). Maximum and minimum values represent annual cycle for roots. A range of values is given for annual net primary productivity (ANPP)

Text Box 6.10 Decomposition of Carbon and Nitrogen in Ponderosa pine (*Pinus ponderosa*) Forest in Sandy Alfisol (Data from Hicks Pries et al. 2017) (Fig. 6.28)

Aboveground tree litterfall and belowground fine root litter carbon and nitrogen levels in the surface soil horizon were evaluated after 10 years. After the 10-year period, root litter mineralized slower than needle litter. About 44% of root litter carbon and 22–28% of pine needle litter remained in the soil surface. Conversely, about 10–15% of root nitrogen and 5–6% of litter nitrogen remained. Downward movement of litter-derived carbon was minimal after 10 years.

(continued)

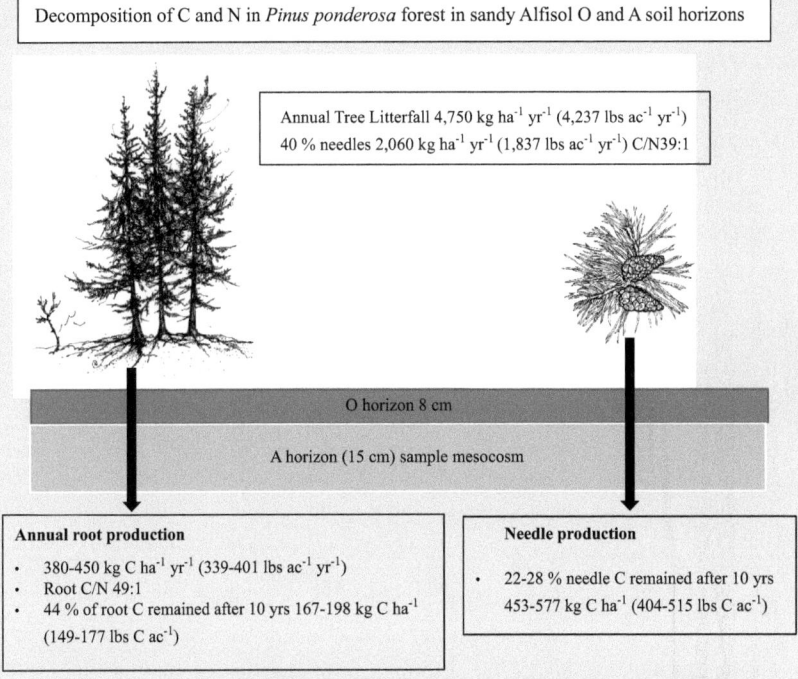

Fig. 6.28 Decomposition of carbon and nitrogen in ponderosa pine (*Pinus ponderosa*) forest in sandy alfisol (data from Hicks Pries et al. 2017)

How much did SOC increase in 10 years? Given the sandy alfisol soil, assume SOC was 1% and bulk density 1.6 Mg m^{-3}. The wt. of 1 ha soil at 0.15 m = 2400 Mg ha^{-1}, or 2,400,000 kg ha^{-1}. After 10 years.

Root carbon = 198 kg C ha^{-1} (176.6 lb ac^{-1})
Pine needle C = 577 kg C ha^{-1} (514.8 kg C yr^{-1})
Total = 775 kg C ha^{-1} (691.4 kg C yr^{-1})

If SOC = 1%, then (2400,000) (0.01) = 24,000 kg C ha^{-1} (21412.3 lb C ac^{-1}).

After 10 years, carbon levels remaining were 775 kg C ha^{-1} + 24,000 kg C ha^{-1} = 24,775 kg C ha^{-1} (22,103.7 lb C ac^{-1}).

SOC would only increase by about 0.03% after 10 years. Given that erosion is minimal, during some years there is a positive buildup of organic matter, other years a negative loss; however, the long-term average is constant or zero (Larcher 2003; Wolf and Snyder. 2003).

Text Box 6.11 Example of plant residue breakdown with carbon and nitrogen interactions in an Indiana deciduous forest. Plant residue, carbon, and nitrogen values from Welch et al. (2007); organic carbon model adapted from Weil and Brady (2017) (Table 6.8 and Fig. 6.29)

The carbon-nitrogen (C:N) ratio of the organic material is critical to the amount of nitrogen required by microbes for decomposition. If the C:N ratio is above 25:1, soil microbes will mine soluble nitrogen from the soil. In the maple-beech to oak-hickory forest, the C:N ratio was 32:1; therefore, the example shows that the required nitrogen was not supplied by the aboveground plant litter and therefore was obtained from the soil. In addition, the breakdown of organic material can be slowed if sufficient nitrogen is available in the litter or the soil solution.

Table 6.8 Annual litter contributions from ground-layer plants and tree biomass. Forest stand is 60–80-year-old maple-beech to oak-hickory forest transition zone, south-central Indiana

Variable	Ground layer	Tree leaves	Tree seeds	Total
Biomass g m^2 yr^{-1}	26.52	460.77	30.70	517.99
Macronutrients g nutrient m^2 yr^{-1}				
C	11.12	210.60	14.6	236.32
Total nitrogen	0.68	6.66		7.34
C:N ratio	16.35	31.62		32.19
P	0.04	0.16		0.20
Ca	0.33	3.59		3.92
K	0.80	0.46		1.26
Mg	0.09	0.36		0.45
Micronutrients g nutrient m^2 yr^{-1}				
Al	0.02	0.77		0.79
Fe	0.01	0.40		0.41
Mn	0.01	0.65		0.66

Data from Welch et al. (2007)

(contiuned)

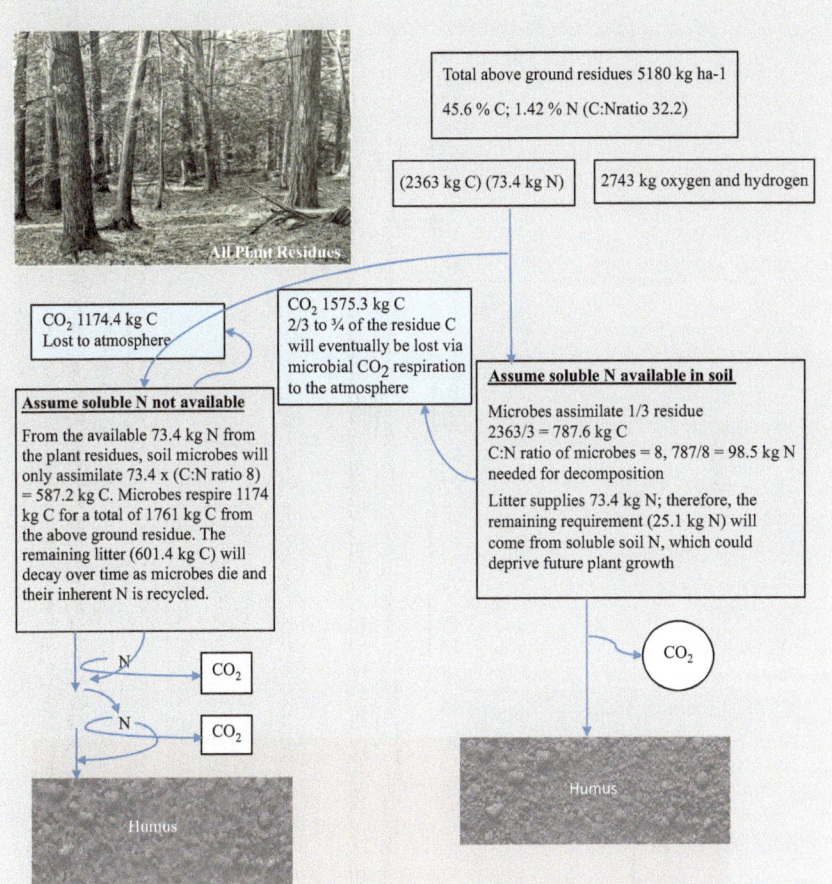

Fig. 6.29 Example of plant residue breakdown with carbon and nitrogen interactions in an Indiana deciduous forest. Plant residue, carbon, and nitrogen values from Welch et al. 2007; organic carbon model adapted from Weil and Brady 2017

Text Box 6.12 Example of Compost Application in Vegetable Garden with Carbon and Nitrogen Interactions.

On a 7.6 × 7.6 m (25 ft × 25 ft) garden space, apply 10 cm (4 in) compost. Compost weighs between 453 and 735 kg per 0.028 m³ (1000–1600 lbs ft³). A 10 cm layer over the entire garden area ~4560 kg. The C:N ratio of the compost example = 43.8/1. In applying this compost mixture, a total of 83.2 kg of N (183 lbs) is needed by microbes to decompose the compost. The compost supplies 45.6 kg N (100.3 lbs N); therefore, 37.6 kg (82.7 lbs) is needed from the soluble soil nitrogen pool. Most likely, this will create a nitrogen deficit for growing garden vegetables; therefore, additional nitrogen is needed from organic and/or inorganic fertilizers.

Table 6.9 Average carbon and nitrogen contents and C:N ratios for commonly available organic materials[4]

Organic materials	Carbon (%)	Nitrogen (%)	C:N ratio
Chicken manure (fresh)	15	1.8	8:1
Topsoil (high organic matter)	56	4.9	11:1
Alfalfa hay	40	3.0	13:1
Alfalfa pellets	40	2.7	15:1
Kitchen veg. Wastes	40	2.7	15:1
Finished garden compost	30	0.5–2.0	15:1
Bluegrass from fertilized lawn	42	2.2	20:1
Cow manure fresh	16	0.6–4.2	20:1
Decomposed barnyard manure	41	1–2.0	20:1
Leaf litter	48	1.4	34:1
Horse manure (fresh)	27	0.7	38:1
Sphagnum peat moss	43	0.9	48:1
Bermuda grass hay	44	1.1	49:1
Timothy grass hay	52	0.9	58:1
Corn stalks	40	0.7	57:1
Wheat straw	38	0.5	80:1
Paper	39	0.3	120:1
Sawdust (hardwood)	46	0.1	400:1
Sawdust (spruce)	50	0.05	600:1

Table 6.10 Calculating compost mixtures using the weighted average technique. For example, the example compost pile contains discarded Bermuda grass hay, some snippets of left over alfalfa stems, goat and llama manure, and soil (author farm example)

	C:N ratio	Parts in mix (%)	(C:N ratio) × (parts)
Spoiled Bermuda hay	49	85	4165
Alfalfa stems	13	5	65
Manure	20	5	100
Soil	11	5	55
Sum		100	4385
C:N ratio 4385/100 = 43.8			

(contiuned)

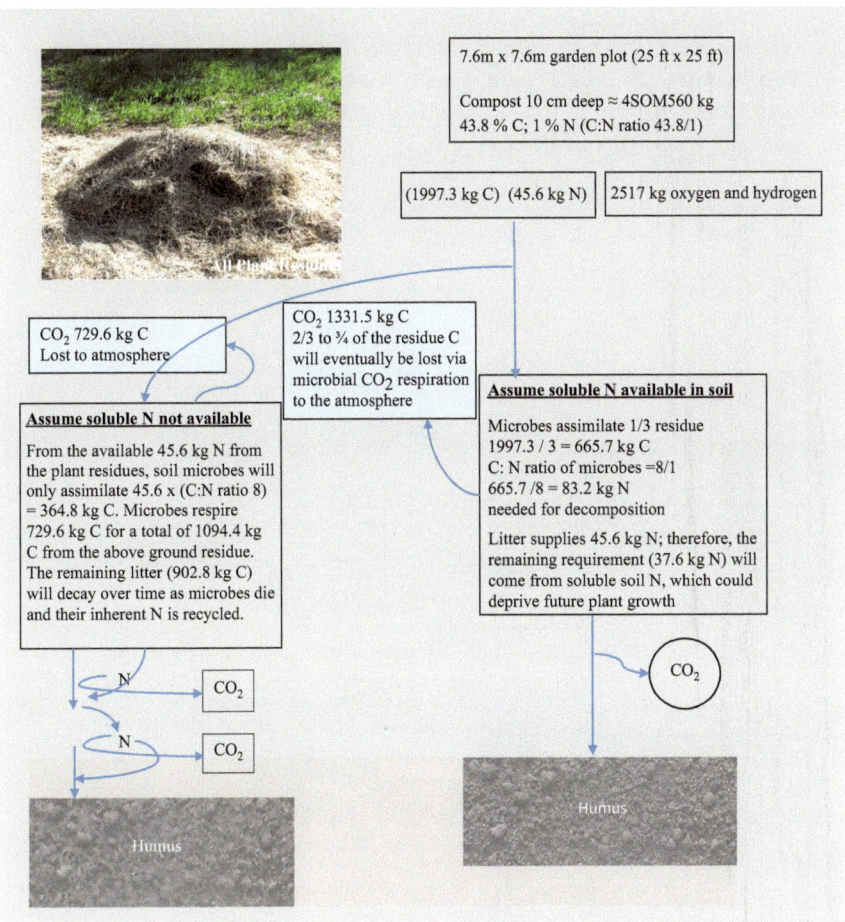

Fig. 6.30 Fate of carbon and nitrogen during composting

Text Box 6.13 Carbon balance and flow in a terrestrial forest landscape, with stream and lake components (adapted from Whittaker 1975). Location at Hubbard Brook New Hampshire, stream from Bear Brook, and lake from Mirror Lake (Fig. 6.31)

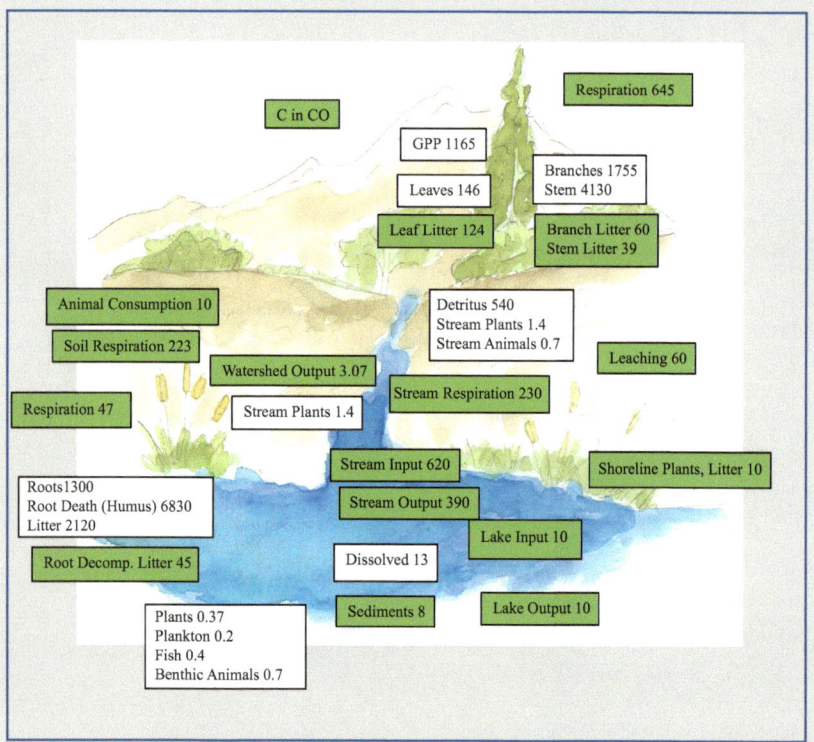

Fig. 6.31 Carbon balance and flow in a terrestrial forest landscape, with stream and lake components (adapted from Whittaker 1975). Location at Hubbard Brook New Hampshire, stream from Bear Brook, and lake from Mirror Lake. Units of measure in boxes are grams C m^2. Numbers of green boxes represent carbon transfer (grams C m^2 yr^{-1}). GPP gross primary productivity

Manure Applications and Carbon Balance

In any farming operation, it is extremely important to understand how various cropping practices and grazing affect soil organic carbon storage and balance. What are the dynamics surrounding the use of manures for augmenting soil organic carbon and nutrients? Some important questions are:

1. How much soil organic carbon is added with varying amounts of manure?
2. How much nitrogen and phosphorus is added to the soil with manure applications?
3. What is the normal range of soil organic matter loss without erosion losses?

4. How much manure is needed to maintain soil organic matter levels in crop fields?
5. What is the effect of manure applications on mycorrhizae in the soil?
6. What is the effect of manure on soil microbial in general?
7. In newly established vegetable gardens, can applications of manure supply the needed nutrients to plants?
8. What are the risks to human health by pathogens on produce from manure contamination?

Animal manures are a valuable commodity and have many benefits in amending soil health. Manure is more economical to use if producers can obtain a supply from nearby dairy or feedlots; otherwise, transporting manure is an expensive enterprise. Typically, recommended manure application rates are about 11.2–22.4 Mg ha^{-1} yr^{-1} (5–10 tons ac^{-1} yr^{-1}) (for corn) and 4.4–6.7 Mg ha^{-1} yr^{-1} (2–3 tons ac^{-1} yr^{-1}) for grain crops. If manure is available, application rates can be increased two to three times the average recommended rates (Thorup 1984). In Text Box 6.13, the example uses a high rate 44.83 Mg ha^{-1} (20 tons ac^{-1}) to demonstrate how slowly organic carbon buildup occurs.

Manure helps to modify soil structure, supplies nutrients to plants, gives positive influence on soil microbia, enhances disease suppression, and helps maintain and/or augment soil organic carbon (Stirling et al. 2016). Nutrient contents and dry matter percentages after decomposition vary widely and are highly dependent on the class of livestock, age and stage of growth, feeds and feeding practices, amount and kinds of bedding materials, time of year—ammonia volatilizes as temperatures increase [under aerobic conditions, between 9 and 44% of the nitrogen in manure can be volatilized as ammonia (Kirchmann and Witter 1989)]—and how the manure is handled and stored. Livestock excrete 50–90% of fed nutrients, and the amount is dependent on the animal species, growth stage, and quality of the feed source and supplements.

There is no doubt that applications of fertilizers increase crop yields, but what is the effect on soil microbia? Studies with treatments of organic manures and chemical fertilizers (NPK) show varying results. Zhong et al. (2010) found that gram-negative bacteria and actinobacterial PLFA's populations were greatest in organic manure with NPK treatments followed by organic manure alone and lowest in singular nitrogen treatments. In long-term continuous wheat, cattle manure applications promoted the growth of bacteria, but not fungi (Parham et al. 2003). Chemical fertilizers (NPK) enriched K-strategist bacterial populations, whereas manure enhanced both r- and K-strategists. Overall, bacterial species richness and evenness were enhanced by applications of manure, nitrogen, and phosphorus (Parham et al. 2003). The diversity of soil organisms is immense and unknown to many people; soil microbes comprise the vast unseen majority of genetic biodiversity on Earth. Estimates of 8.7 million microbe species have been upgraded to about 1 trillion species; 99.9% are undiscovered. As many as 10^{10}–10^{11} bacteria; 6000–50,000 bacterial species; and up to 650 feet of fungal hyphae can be found in 1 gram of soil. DNA-based methods for detecting and quantifying soil microorganisms are being used to assay microbes in distinct niches in the soil matrix: inside soil aggregates and soil pores, on decomposing organic matter, and on the surface of soil organisms—including pests and pathogens. DNA sequencing identified 9885 species in a rhizosphere sample of wheat, the majority of species were bacteria from the phyla *Proteobacteria*

(300 genera), *Actinobacteria*, and *Firmicutes* (Stirling et al. 2016). Fertilizers do affect soil biology, and combinations of manure and NPK seem to optimize plant growth and enhance disease resistance (Stirling et al. 2016). Overapplication of soil nutrients can have a negative impact on beneficial organisms associated with the suppression of soilborne diseases (Stirling et al. 2016). Excessive applications of nitrogen depress decomposer fungi, arbuscular mycorrhizae, nematode-trapping fungi, and large omnivorous and predatory nematodes causing a shift toward bacterial-dominant populations in the soil. In addition to fertilizers, pesticides (insecticides, fungicides, and nematocides) applied to the soil affect many organisms besides the target pests. Broad-spectrum biocides, especially soil fumigants, are the most damaging to soil organisms (Stirling et al. 2016). Soil bacteria, fungi, and algae respond differently to the pesticides' biochemical action and concentration in the soil (Moorman 1989). He explains that total microbial populations are often not affected or minimally by pesticide applications; however, individual species or groups such as cellulose decomposers may be greatly affected. "Changes in microbial populations due to pesticide applications are no more severe than changes caused by natural environmental stresses" (Moorman 2018). Ortho- and meta-cleavage pathways in bacteria are responsible for breaking down the aromatic fraction (benzene rings, phenols, and anilines) of pesticide molecules (Moorman 2018).

Text Box 6.14 Carbon balance with manure application
Clay loam
- Soil Organic Carbon (1.6%)
Soil Organic Matter (2.0%)
Bulk Density 1.4 Mg m^3
Wt. of soil @ 0.15 m = 2,128,000 kg ha^{-1} (1,898,601.6 lbs ac^{-1})
At 3% SOM = 63,844.3 kg ha^{-1} (56,958 lbs ac^{-1}) SOM

Photo Agromet-Pilmet Tyten 10 Plus. (Courtesy Adikpl 2010, Wikipedia)

(continued)

Apply 44.83 Mg ha^{-1} solid dairy manure (20 t ac^{-1})

25% dry matter = 11,209 kg ha^{-1} (10,000 lbs ac^{-1})
Percent soil organic matter in manure = 64.43% (Larney et al. 2005)
Dry weight of soil organic matter = 7222.3 kg ha^{-1} (6443.3 lbs ac^{-1})
Percent ash in manure = 35.57% (Larney et al. 2005)
Weight of ash in manure = 3986.7 kg ha^{-1} (3556.6 lbs ac^{-1})
Ash is a primary component of *manure* and is *defined* as "the mass fraction of
inorganic residue remaining after dry oxidation at 575 °C ± 25 °C relative
to the oven-dried mass" Sakirkin et al. (2010).
Assume 75% of total decomposed manure returns as CO_2 to the atmosphere
and 25% remains as OM (7222.3 kg ha^{-1}) (0.25) = 1805.6 kg ha^{-1}(1610.8
lbs ac^{-1}) (see Schlesinger 1977; Sawyer and Mallarino 2007; Weil and
Brady 2017; Duiker 2018).
Soil organic matter at 3% = 63,844.3 kg ha^{-1} (56,958 lbs ac^{-1}) + soil organic
matter from manure (= 1805.6 kg ha^{-1}(1610.8 lbs ac^{-1}) = 65,649.9 kg ha^{-1}
(58,568.9 lbs ac^{-1}).
Adding 1805.6 kg ha^{-1} (1610.8 lbs ac^{-1}) OM from manure increased SOM
from 3% to 3.085%, an increase of 0.085%.

Manure is highly variable among different classes of livestock. In addition, animal
diets, animal bedding materials, and handing further alter its composition. In the
above example, soil organic matter was 3%, and after addition of 44.83 Mg ha^{-1}
solid dairy manure (20 t ac^{-1}), soil organic matter increased to 3.085%. The net
increase in soil organic matter was 0.085%. As Magdoff and Van Es (2009) point
out,0 annual decomposition of organic matter in cropping situations with cultivation
can result in a perpetual loss of soil organic matter. However, crop stubbles, cover
crops, and additions of manure can help balance the loss of soil organic matter;
however, yearly increases are minimal (Weil and Brady 2017; Duiker 2018).

Quick Facts Regarding Manure? (Tables 6.11 and 6.12)

Manure is highly variable with respect to dry matter and nutrient contents (Figs. 6.32
and 6.33). Other factors such as embedded litter will also affect the dynamics of
manure as it decomposes and releases available nutrients. For example, in dairy
operations, sand bedding/manure vs. organic bedding/manure will not have the
same results in increasing soil organic matter.

Table 6.11 Concentrations of water, total carbon, total nitrogen, inorganic nitrogen, total
phosphorus, and carbon/nitrogen ratio of beef cattle manure. Manure stockpiled in open area
for 3 years

Age of manure	Water (%)	Total C (%)	Total N (%)	Inorg. N (%)	Total P (%)	C/N ratio
Fresh	65.1	10.75	0.565	0.125	0.160	19.7
Stockpiled	57.15	10.6	0.660	0.190	0.225	15.85

Data from Larney et al. (2006)

Table 6.12 Average N, P, and K, timing of release, effect on pH, and notes for commonly used organic fertilizers

Material	Total N (%)	Total P (%)	Total K (%)	Nutrient release	Effect on pH	Notes
Cattle manure, dry	0.5–2	0.5–0.7	0.5–2.0	Slow	Slightly acidic	May contain soluble salts*
Cattle manure, fresh	0.6–2.5	0.3	0.5	Slow	Slightly acidic	May contain soluble salts*
Poultry manure, dry	1.5–4.4	2.1	2.6	Medium to rapid		Can burn plants,* strong odor, decomposed has less burn risk
Sheep manure, dry	3.5	0.6	1	Slow	Slightly acidic	*
Hog manure, dry	2.1	0.8	1.2	Slow	Slightly acidic	*
Horse manure, fresh	0.7	0.3	0.6	Slow	Slightly acidic	Weed seeds
Swine manure, dry	2.25	2	1	Slow	Slightly acidic	*
Swine manure, fresh	0.6	0.5	0.4	Slow	Slightly acidic	*
Rabbit manure, dry	2.25	1	1	Slow	Slightly acidic	
Blood meal	12	0–1	0–1	Medium to rapid	Acidic	$, can burn plants
Cottonseed meal	6–7	0.4–2	1–1.5	Slow to medium	Acidic	Pesticide residues and GMO
Soybean meal	7	2	1	Slow, 1–4 months		$, can be GMO
Fish meal	10	3–6	2–3	Slow, 1–4 months	Acidic	$, odor
Feather meal	10–14	0	0	Slow, 6–9 months		$, odor
Fish emulsion	4–5	1–2	1–2	Slow		Micronutrients, odor
Bone meal, steamed	1–6	15–30	0	Slow to medium	Alkaline	$, can burn
Hoof and horn meal	13	0	0	Slow to medium	Alkaline	$
Bat guano, high N	10	3	1	Slow, 4+ months		$, stimulates microbes
Bat guano, high P	3	10	1	Slow, 4+ months		$, stimulates microbes
Green manure in crop	2.5	0.2	2.1	Medium		Nutrient content decreases with age of plant
Alfalfa meal, pellets	2–3.6	2	1	Slow, 1–4 months		Micronutrients
Seaweed	1.7	0.75	5	Slow		Micronutrients
Kelp	1–1.5	0.5–1	5–10	Medium		$, zinc, iron

(continued)

Table 6.12 (continued)

Wood ashes	0	1–2	5–10	Medium	Alkaline	Use sparingly and with care
Compost	0.5–3.5	0.5–2	0.5–2	Slow to very slow	Acidic to alkaline	pH depends on materials, gradual increase in P and K
Rock phosphate	0	30	0	Slow to medium for sedimentary RP; apply 4 to 5 weeks before planting	Acidic	A mined resource; igneous RP deposits are less soluble than sedimentary RP and are not very effective when applied directly to the soil
Greensand	0	1	6	Slow release, does not burn plants, beneficial to soil microbes, good soil conditioner to loosen heavy and tight soils	Slightly acidic to slightly alkaline, usually has little effect on soil pH	Mined from ocean deposits, "glauconite," a sedimentary rock

Data Spaeth (2018); Sources: Zibilske (1998); USDA NRCS (2010); NRAES (1992, 1999)
*Growing scientific evidence shows that pathogens, antimicrobials, and hormones may exist in commercially produced livestock and poultry manure across the United States. $ = cost consideration

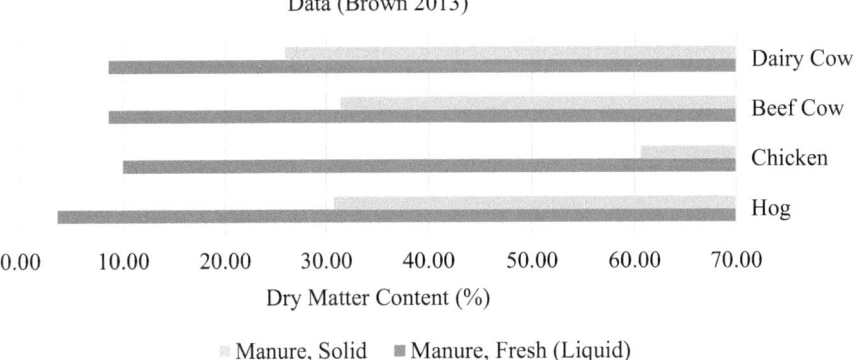

Fig. 6.32 Carbon balance with manure application

1. Cow manure contains near equal (balanced) amounts of nitrogen, phosphorus, and potassium.
2. There are two basic sources of N in manure (ammonium and organic N (Fig. 6.34).

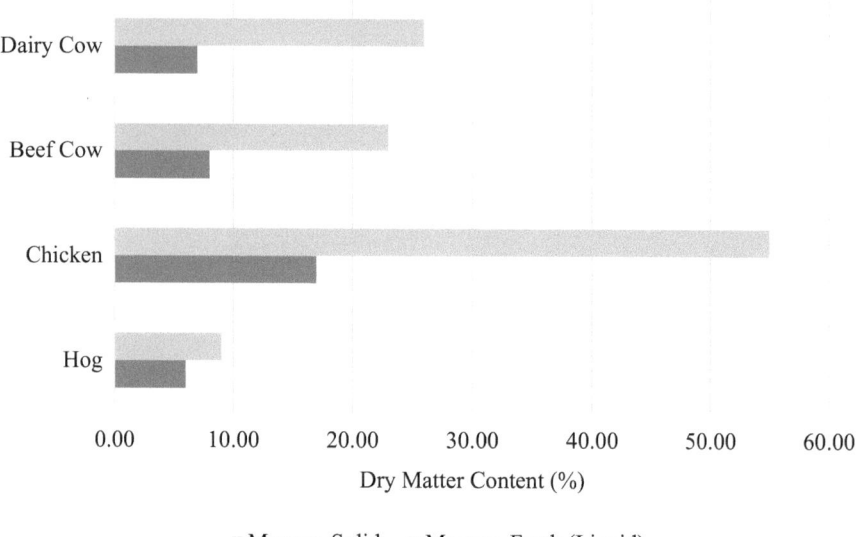

Fig. 6.33 Percent dry matter content of different animal manures in raw liquid form and solid (data from Magdoff and Van Es 2009). In dairy operations that use very little bedding and sweep alleys are flushed with water, then material is mechanically removed are indicative of liquid manure handling systems; whereas, other operations may use organic bedding materials. In situations where manure is semisolid (half-solid and liquid), dry matter contents may range from 12 to 20%

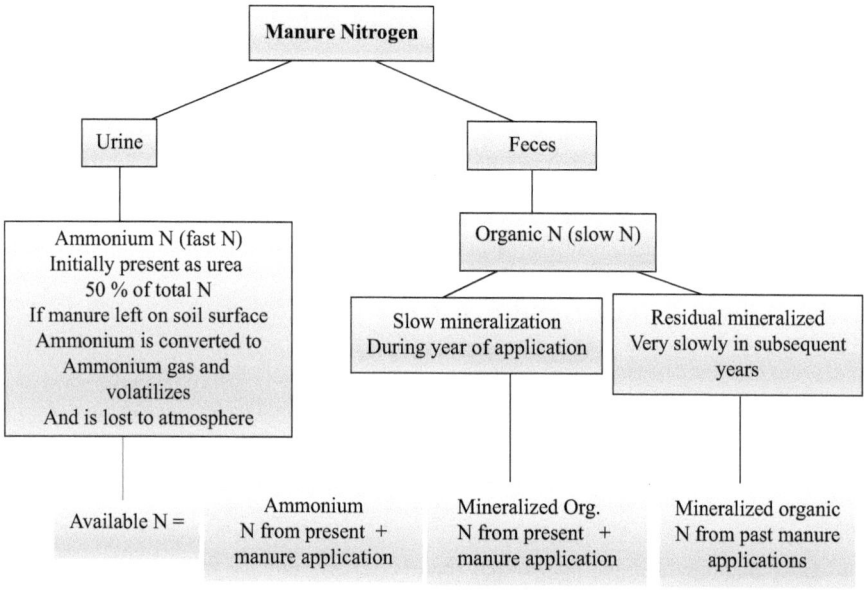

Fig. 6.34 Two sources of nitrogen in manure (ammonium and organic N) (adapted from Klausner 1997)

3. The carbon nitrogen (C:N ratio) of cow manure is about 20:1. Manures that contain high amounts of high carbon bedding material, e.g., straw, sawdust, wood chips, will have considerable high C:N ratio and will use the nitrogen in the manure to decompose the high carbon organic material. The optimum C:N ratio of m*anure* or *organic* material for crop production is under 20:1.

4. Manure and manure-based compost can harbor pathogenic bacteria (*Escherichia coli*—especially *E. coli* O157:H7, enterohemorrhagic *Escherichia coli* (EHEC) , *Salmonella, Campylobacter* bacteria, and *Giardia* or *Cryptosporidium* protozoa that are harmful to humans, even when properly composted. For example, the risk of food contamination on vegetables is a serious concern, and the USDA-National Organic Program (NOP) has specific rules regarding application of raw manures in vegetable production. The NOP states that edible parts of vegetables that might contact the soil (either directly or by raindrop splash or irrigation) where manure has been applied must wait at least 120 days before harvest. That rule makes it difficult as many vegetable crops mature in less than 120 days. There are always risks with fresh produce (contamination can also come from harvest and handling, and bird and wildlife droppings), although with proper manure handling practices, washing produce before use, and cooking, the health benefits are much greater than the risk of food-borne illness. For example, millions of bags of salad greens are sold in the United States without incident; however, outbreaks of food-borne illness do occur and are reported to the public. Contamination on vegetables causing human illness is usually from two culprits (*Salmonella enterica* and *Escherichia coli*). These pathogens are killed by cooking; however, if they are present in the soil, they can occur both inside and outside the plant. For example, during harvesting of lettuce, workers cut and core the lettuce with a knife, which can then pick up pathogens from infected soil. Lettuce produces a milky latex or sap (spinach and Swiss chard do not), which can trap these pathogens in the plant tissue. Washing and cleansing during processing do not affect pathogens that are living in the internal tissues of the plant. For example, in tomatoes, if Salmonella comes in contact with flowers, stems, or fruits, it can infiltrate and colonize inside plant tissues, where washing the outside cannot remove it. Tomato plants can also be contaminated by *Salmonella* uptake via hydroponic solutions (Guo et al. 2002). Gurtler et al. (2013) found that EHEC can translocate and persist in leek shoots for up to 22 days in the presence of arbuscular mycorrhizal fungi and 8 days in its absence. Even more insidious, in the leek shoot, *Salmonella* may internalize and persist longer than EHEC. Salmonella is much more problematic as it persists for much longer periods in the soil and is associated with a large number of animal species, including birds and flies (including the common housefly) (Davies and Wray 1996; Mian et al. 2002). Contaminated manure and detection of *Salmonella* on vegetables have been correlated with warm, moist summer conditions (average daily temperature, 20 °C), whereas no contamination on vegetables was observed with manure applications in late fall with average daily temperature of <10 °C with repeated free/thaw cycles (Natvig et al. 2002).

Some interesting facts about *Salmonella*:

- Can survive 10–15 days in septic systems (Parker and Mee 1982) (*E. coli* has a negative growth rate Gordon et al. 2002), brackish lagoons (Winfield and Groisman 2003), and aquatic environments (Chao et al. 1987) and persist in soil and sediments for at least a year (Davies and Wray 1996; Thomason et al. 1977).
- *Salmonella* has enhanced survival capability compared to *E. coli*. Its long-term survival in the environment can result in transmission to new hosts (Winfield and Groisman 2003).
- *Salmonella* can infect a large number of animal species, and the source and identification of environmental contamination in the environment may not be possible (Winfield and Groisman 2003).
- *Salmonella* has shown a fourfold greater adherence to soil mineral particles than *E. coli* (Stenstrom 1989).
- *Salmonella* is more resistant to death by microbial predators and/or competing organisms than *E. coli* (Wright 1989).
- *Salmonella* survival is significantly higher in estuarine water at temperatures below 10 °C than *E. coli* (Rhodes and Kator 1988).
- *Salmonella* has a 1000-fold higher level of vegetational attachment to the plant compared to *E. coli* (Barak et al. 2002).
- *Salmonella* and *E. coli* have different rates of survivability in the environment and life traits; therefore, the use of using the presence of *E. coli* as an indicator of *Salmonella* may not be appropriate (Barak et al. 2002).
- *Salmonella* can persist in bathrooms and toilets for several weeks following a bacterium-induced illness. In addition, *Salmonella* was also present in air samples following flushing for up to 50 days (Barker and Bloomfield 2000).
- Infected eggshells used in compost or buried in the garden can infect garden soil.
- The incorporation of organic materials (compost, decomposed manure, cover crops) can stimulate non-pathogenic antagonist microorganisms and suppress pathogenic microorganisms (Stirling et al. 2016).
- Moving this to Manure section of Organic Matter Fact Sheet
- Moving this to Manure section in Organic Matter Fact Sheet There can be potential problems associated with nutrient leaching from long-term applications of manure. Excess accumulation of nitrogen and phosphorus can occur. Organic phosphorus can be leached and transported by water runoff to streams, rivers, and lakes, causing pollution and degraded water quality. Too much of a good thing can have negative results.

Manure Applications and Nitrogen Dynamics

There are four main available sources of nitrogen for field applications: (1) soil organic matter; (2) organic residues (animal and green manure from cover crops, compost, and incorporation of other annual or perennial sods); (3) legumes fixation;

and (4) inorganic nitrogen fertilizer. Nitrogen dynamics are complex and highly variable as losses can be immediate depending upon the status of microbial populations and moisture (Table 6.13).

Manure applications can provide plant nutrients to crops over several years (Eghball et al. 2000, 2002). Part of the nitrogen from some organic fertilizers (manure and compost) is often not immediately available during the season of application because of the slow release of organically bound nitrogen (Table 6.14).

Nitrogen availability is dependent on environmental factors such as temperature, soil moisture, soil microorganisms, and the characteristics of the organic matter. Release of nitrogen from organic sources involves the biological process of microbial decomposition. Since biological processes are involved with the breakdown and mineralization of organic residues, nitrogen release from organic materials during the growing season is difficult to predict. Nitrogen available from manure includes NH_4-N and NO_3-N and some organic mineralized N following application;

Table 6.13 Estimated ammonia-N losses from manure application and time to incorporation

Manure application method	% Remaining
Injected during growing season	100
Incorporated within 1 day	65
Incorporated within 2 days	53
Incorporated within 3 days	41
Incorporated within 4 days	29
Incorporated within 5 days	17
No conservation or injected in fall	0

Data from Cornell Agronomy Fact Sheet 4 (2005)

Table 6.14 Average time needed and amount of nitrogen released by common organic materials

Organic material	Time needed (weeks)	Avg. % nitrogen in material	Avg. % nitrogen released
Urea*	12	46	91–96
Blood meal	12–26	11–12.6	56–75
Feedlot cow manure (raw)	26	0.5–1.5	30–40
Dairy manure (composted)	10–19	0.5–2.0	5.5–20
Poultry manure (decomposed)	12–26	1.8	37–53
Alfalfa pellets	12	3.6	41–52
Yard trimmings (composted woody and grassy materials)	6–19	1.7	4.75
Pelleted organic fish fertilizer	17–19	9.7	77
Rabbit manure	17–19	3.0	63
Rabbit manure (composted)	6–19	1.8	15–19

Sources: Pratt et al. (1973); Klausner et al. (1994); Brady and Weil (1996); Eghball et al. (2000); Gale et al. (2006). *Note:* The percentage of nitrogen in organic materials varies among studies.
*Urea is a naturally occurring organic compound, but it is also chemically synthesized

the rate of mineralization depends on the nitrogen components in the manure. The amount of N is available from manure during the first, second, third, and fourth years after application. For example, on average, only about 35% of mineralized nitrogen is available from raw feedlot manure during the first year, or year of application. The rate subsequently decreases to 10% the second year, 5% the third year, and 2% the fourth year (Eghball et al. 2002). Mineralization of composted cow manure is slower on average than that of raw manure: 20% the first year, 8% the second year, 4% the third year, and 1% the fourth year (Table 6.15). In the application of raw cow manure as a nitrogen source, adjustments must be made to meet target applications because manure is still undergoing decomposition and nitrogen is still undergoing mineralization. As a comparison, for synthetic mineral fertilizers such as ammonium nitrate (33% nitrogen), about 50%–80% (average 60%) of the nitrate is readily available as it dissolves in soil water. On average, phosphorus and potassium from some common organic materials such as compost and manure are more readily available the first year of application (Table 6.15).

Table 6.15 Percentages of available N, P, and K released by microbial mineralization for various organic materials

Organic material	Avg. total % N, dry wt.	Avg. % release N, year 1	Avg. % release N, year 2	Avg. % release N, year 3	Avg. % release N, year 4
Feedlot cow manure	0.5–2.5	35	10	5	2
Composted cow manure	0.5–2	20	8	4	1
Poultry floor litter	4.4	50	15	8	3
Poultry manure		90	0.01	–	–
Fresh dairy manure	0.5–2.1	21	9	3	3
Swine manure	2.1	82	3	1	–

Organic material	Available P %, year 1
Garden compost	30–51
Beef feedlot manure	80–85
Composted feedlot manure	73
Manures in general	>70
	Available K %, year 1
Dairy compost	73–90
Higher-N animal manures	>98

Sources: Pratt et al. (1973); Klausner et al. (1994); Brady and Weil (1996); Eghball et al. (2000, 2002); Gale et al. (2006). *Note*: These rates are averages and are somewhat variable because of climate and the types of feed given to the animals. The percentage of nitrogen in organic materials varies among studies

Organic Matter Fact Sheet

Global Carbon

- The carbon cycle involves four basic carbon reservoirs, the atmosphere, oceans, terrestrial biosphere, and fossil carbon.
- The decline of SOM is one of the largest inputs of CO_2 to the atmosphere, and about 27% (535 × 1015 g) of the SOC existing from prehistoric times (2014 × 1015 g) has been lost in the last two millennia.
- Anthropogenic carbon emissions of CO_2 are not balanced by CO_2 consumption. About half of the CO_2 emitted to the atmosphere by fossil fuel burning, open burning (fire), and terrestrial processes (mainly deforestation) is absorbed by terrestrial and marine environments; however, long-term trends are uncertain (Schimel et al. 2001).
- More carbon is leaving the soil reservoir (62 Pg) than entering the soil (59–60 Pg) (Battin et al. 2009; Weil and Brady 2017).
- Bicarbonates occupy the largest dissolved inorganic reservoir in the oceans and calcium carbonate (major mineral in limestone) in terrestrial and ocean ecosystems.
- Nonliving soil organic matter constitutes the largest pool of terrestrial organic carbon (Jobbagy and Jackson 2000).
- "Globally, soil contains a large amount of carbon—twice that in the atmosphere and more than carbon in vegetation and 2the atmosphere combined" (Liang et al. 2017).
- Long-term carbon storage in terrestrial ecosystems occurs primarily when plant biomass is stabilized in soils as SOM (Liang et al. 2017).
- "SOM is thus one of the key factors controlling the CO_2 concentration in the atmosphere and proper management of this resource can help mitigate global climate change and maintain or even enhance global food security" (Liang et al. 2017 citing Lal 2004).
- "Soil carbon sequestration can be viewed as a key mitigation strategy for rising CO2 concentration in the atmosphere but should also be recognized for its important role in improving the fertility and quality of soil, especially at a moment in history when agricultural lands have been seriously degraded by widespread, unsustainable management" (Franzluebbers 2012).
- "Soil organic matter enhances soil fertility and therefore increases net primary production and photosynthetic CO_2 fixation by plants" (Miltner et al. 2012).
- The terrestrial carbon cycle is a "balancing act" where plants take up inorganic carbon as CO_2 and synthesize organic compounds during the photosynthetic process.
- Fire represents a large and highly variable part of the US carbon budget, and the amount of CO_2 emitted from fires in the United States is equivalent to 4–6% of anthropogenic emissions at the continental scale. At a state level, fire emissions of CO_2 can, in some cases, exceed annual CO_2 emissions from fossil fuel usage.

- Deforestation is a major contributor to climate change—20% of anthropogenic emissions of CO_2 emanating from tropical forests (NIACS 2007). Tropical forests sequester 46% of the world's living terrestrial carbon pool and 11% of the global soil carbon pool (Brown and Lugo 2018).
- Grasslands contain 10–30% of the world's soil organic carbon and is sequestered by plants through photosynthesis and carbon loss by decomposition of organic matter. Carbon input from grasslands is mediated by plant life forms, climate, temperature, and precipitation regimes (Hewins et al. 2018).
- "Proper grazing management has been estimated to increase soil carbon storage on US rangelands from 0.1 to 0.3 Mg C ha^{-1}year^{-1} and new grasslands have been shown to store as much as 0.6 Mg C ha^{-1}year^{-1}. Grazing lands are estimated to contain 10–30 % of the world's soil organic carbon" (Schuman et al. 2002).
- Terrestrial carbon pools in cropland, rangeland, forests, and wetlands can act as a sink for sequestering atmospheric CO_2 (as much as 50 ppm of CO_2 for 100–150 years). In the United States, the sink capacity of sequestering additional carbon ranges from 0.2 to 0.48 Pg C yr^{-1} when all land uses are tallied. Forests (0.03–0.05 Pg C yr^{-1}) and cropland (0.144–0.432 Pg C yr^{-1}) have the largest potentials for sequestering carbon, although grazinglands (range and pasturelands) can contribute up to 10% of the sink capacity (Lal 2010).

Soils

- Carbon is the 15th most abundant element in the Earth's crust and 4th most abundant in the universe.
- Soil organic matter (SOM) is an integration of biologic and mineral components.
- Additions of organic matter to the soil may not supply the nutrient needs of vegetable plants in newly established gardens. Soil organic matter and nutrients accumulate over time. Additional nutrients from soil amendments may be needed to supply macronutrients for plants in new garden soils.
- In average soil, about 50% of the mass is carbon (C).
- SOM on the average contains 5% nitrogen (N).
- Decomposed soil organic matter, or humus, contains about 58% carbon, 4.8–5% N, 1.2% P, and 0.8% S and an array of micronutrients.
- Organic compounds have a tenuous hold on soil N (90–95%), which is unavailable to plants.
- Plants can uptake soluble organic compounds (mostly soluble proteins and amino acids), which comprise about 0.3–1.5% of the total organic N in soils.
- SOM is the primary source of nitrogen, phosphorus, and sulfur. Carbon ratios are 12:1 C:N, 50:1 C:P, and 70:1 C:S.
- Potential SOM content and the end point in the soil are dependent upon the soil parent materials, soil environment, climate, and native vegetation.
- Forests and cropland have the largest potentials for sequestering carbon, although grazinglands (range- and pasturelands) can contribute up to 10% of the sink

capacity. Currently, grazinglands contain 10–30% of the world's soil organic carbon.

- SOM accumulates over long periods (centuries) and eventually reaches an equilibrium level (levels off) that is correlated with the particular environment.
- The decline of soil health is due to many factors; however, the loss of soil organic carbon from tillage practices over time is the most significant factor as it affects the chemical, physical, and biological properties of soils.
- Cultivation of native grasslands to cropland results in rapid depreciation of soil organic matter, up to 40% loss in five years (Davidson and Ackerman 1993).
- In non-till cropping systems, soil organic carbon may be close to a steady-state level.
- On the average, one tillage operation with a one-way plow buries about 60% of surface crop residue.
- The decline of soil organic carbon can be slowed, and a degree of restoration is possible with cover crops and manure applications; however, the reality is that restoration of soil organic carbon may be limited, especially in more dry environments.
- Once optimum soil organic carbon levels are reached (equilibrium), the annual increase in SOM fluctuates between positive and negative values, but the long-term average is approximately a zero gain (Larcher 1983).
- Organic matter development is greater (1.5–3 times) in the surface soil layer for grassland plant communities compared to forests.
- During soil development, nutrients in the soil are derived from specific minerals in the parent material. As soils develop and become older, nutrients are supplied by the organic fraction of the soil.
- On the average, after 1 year, most of the carbon from soil organic matter added to the soil returns to the atmosphere as carbon dioxide (CO_2). About 25–33% remains in the soil (~5% in organism biomass, ~20% humic substances, and ~5% nonhumic fractions).
- SOM buildup in the soil is a slow process.
- Decomposition of organic matter is a complex process and is affected by certain carbon dynamics such as the makeup of carbon compounds in plant cells, linkages to other nutrient cycles (N, P, and S), and diversity of microorganisms.
- The principal elements of SOM are carbon, hydrogen, oxygen, nitrogen, phosphorus, potassium, and sulfur.
- SOM enhances soil physical properties such as soil structure, aggregate stability, infiltration capacity, porosity, water holding capacity, granulation, and friability—ease of working the surface.
- At 5% soil organic matter, water field capacity is two times greater compared to an identical soil with 1% soil organic matter.
- SOM is linked directly to soil health and the productivity of a soil.
- SOM generally decreases with soil depth.
- SOM varies in content in soils; soils in desert and arid regions contain on the average less than 1.5% organic matter. Soils developed with prairie grasses generally contain more organic matter (3–5%) than forested soils.

- The average carbon-to-nitrogen (C/N) ratio of undisturbed soils is relatively stable at 12:1.
- When native soils are first cultivated, about 20–40% of the SOM is lost over a 5–20 year period.
- Soil erosion has and is a major contributor to carbon losses, which ultimately affects hydrologic function and biotic integrity.
- Carbon storage on rangelands can reduce soil productivity in source areas and potentially increase it in depositional areas where carbon is redistributed.
- SOM affects chemical processes in the soil, increases cation exchange capacity (CEC), and acts as a buffer to maintain soil pH.
- Organic residues on the soil surface decompose more slowly than if incorporated into the soil because aboveground material is beyond the reach of soil microorganisms.
- Organic matter content is dependent upon plant productivity. Grasslands develop higher organic matter content than forest soils. As soil organic matter increases, plant productivity increases.
- Plant nutrients are retained by soil organic matter; thus nutrients do not leach readily into subsurface soil layers (horizons).
- Soil pathogens are suppressed to a large extent in natural ecosystems and undisturbed soils. Non-pathogenic and pathogenic microorganisms coexist.
- The incorporation of organic materials (compost, decomposed manure, cover crops) can stimulate non-pathogenic antagonist microorganisms and suppress pathogenic microorganisms.
- Hydrophilic and hydrophobic aspects of SOM and inherent minerals adsorb pesticides and other toxic compounds in the soil.
- For each 1% organic matter, on the average there is 435.5 kg (1000 pounds) of N, 45.35 kg (100 pounds) of P, and K in the upper 15 cm (6 inches) of soil.
- In plant ecosystems, the primary source of SOM is derived from plant tissues; animal waste products are secondary.
- SOM sustains and regulates nutrient cycling in soils.
- About 60–90% of carbon is in upper 1 m of soil profile, although stored carbon is also significant below 1 m in histosols and gelisols (Weil and Brady 2017).

Microbial Roles in SOM Formation

- SOM enhances microbial biodiversity in soils.
- One gram of soil can contain up to 1010–1011 bacteria, 6000–50,000 bacteria species, and up to 200 m of fungal hyphae.
- Microorganisms in the soil mineralize nitrogen, phosphorus, and sulfur as they decompose organic matter. Nutrients are released slowly, over years.
- Soil pathogens are suppressed to a large extent in natural ecosystems and undisturbed soils. Non-pathogenic and pathogenic microorganisms coexist.
- SOM enhances microbial biodiversity in soils.

- Organic residues on the soil surface decompose more slowly than if incorporated into the soil because aboveground material is beyond the reach of soil microorganisms.
- Soil organic matter and soil microorganisms function together to improve soil physical properties that are related to greater soil resilience and resistance to erosion. As organic matter is added to the soil, biological activity increases, soil porosity is greater, and more stable soil aggregates or formed, which is related to greater infiltration capacity.
- "Microorganisms have two critical, contrasting roles in controlling terrestrial carbon fluxes: promoting release of carbon to the atmosphere through their catabolic activities, but also preventing release by stabilizing carbon into a form that is not easily decomposed" (Liang et al. 2017 citing Schimel and Schaeffer 2012).
- Miltner et al. (2012): "Although most soil carbon ultimately derives from plant material (Kögel-Knabner 2002), a large proportion may pass through microbial biomass before being transformed to SOM. Microbes grow on plant residues and utilise plant-derived carbon to build their biomass, and after cell death, part of this carbon is transformed into nonliving SOM (Kindler et al. 2006, 2009; Miltner et al. 2009)."
- The living part of soil organic matter includes a wide variety of microorganisms, such as bacteria, viruses, fungi, protozoa, and algae. It also includes plant roots and the insects, earthworms, and larger animals, such as moles, woodchucks, and rabbits that spend some of their time in the soil. The living portion represents about 15% of the total soil organic matter (Magdoff and van Es 2009).
- Microbes contribute to more than half of global respiration, but precise estimates are difficult in various terrestrial plant communities.
- Bacteria and fungi provide more than 95% of the biotic contribution to organic matter decomposition (Persson et al. 1980).
- "Fungal and bacterial necromass are the primary carbon-containing constituents contributing to the stable soil organic matter (SOM) pool" (Liang et al. 2017 citing Kindler et al. 2006; Schweigert et al. 2015).
- "When bacteria degrade plant residues, they use low-molecular-weight compounds, nucleic acids, lipids, proteins, and carbohydrates from the plant biomass to build their own biomass. If this biomass is then incorporated into non-living SOM, this portion is structurally derived from microbial biomass components even though the carbon ultimately originates from plant residues" (Miltner and Bombach 2012).
- "Microbial communities in forests are better adapted to degrading complex carbon compounds than microorganisms in grassland. Yet grassland microorganisms degrade grass litter more effectively than forest litter, while microorganisms in forests do not preferentially degrade forest litter" (Liang et al. 2017 citing Waldrop and Firestone 2004; Strickland et al. 2009).
- "Microbes grow on plant residues and utilise plant-derived carbon to build their biomass, and after cell death, part of this carbon is transformed into nonliving SOM" (Miltner et al. 2012 citing Kindler et al. 2006, 2009; Miltner et al. 2009).

- "Microbial biomass is considered to be turned over much faster than plant residues (Kästner 2000; Schink 1999), and therefore, the microbe-derived carbon input to SOM formation may be much higher than might be expected from its small pool size" (Miltner et al. 2012).
- "The molecular imprint of SOM by molecules and fragments derived from microbial biomass is presumably much more important than previously considered" (Miltner et al. 2012).
- Soil organic matter (SOM) is composed of the "living" (microorganisms), the "dead" (fresh residues), and the "very dead" (humus) fractions. The "very dead" or humus is the long-term SOM fraction that is thousands of years old and is resistant to decomposition.
- Soil organic matter comprises the active (35%) and the passive (65%) pools. The active SOM pool is live and dead plant and animal matter and is the food and energy source for microbes. The passive SOM pool can be high in lignin and is resistant to decomposition by microbes.

Manure

- The nutrient composition of manure varies considerably among livestock classes (see Table 6.12).
- Typically, recommended manure application rates are about 11.2–$22.4\,Mg\,ha^{-1}\,yr^{-1}$ (5–10 tons $ac^{-1}\,yr^{-1}$) (for corn) and 4.4–6.7 Mg $ha^{-1}\,yr^{-1}$ (2–3 tons $ac^{-1}\,yr^{-1}$) for grain crops. If manure is available, application rates can be increased two to three times the average recommended rates (Thorup 1984).
- Nutrient release is variable over time (see Table 6.15).
- Average water, carbon, nitrogen, and phosphorus contents of fresh and stockpiled beef cattle manure (Data from Larney et al. 2006).

Age of manure	Water (%)	Total C (%)	Total N (%)	Inorg. N (%)	Total P (%)	C/N ratio
Fresh	65.1	10.75	0.565	0.125	0.160	19.7
Stockpiled	57.15	10.6	0.660	0.190	0.225	15.85

- As with growing cover crops to increase soil organic carbon, adding manure enhances microbial populations in the soil; however, soil carbon increases are minimal for singular applications (see Text Box 6.14).
- Percent ash in beef cattle manure = 35.57% (Larney et al. 2005).
- On the average, about 75% of the N, 80% P, and 90% K ingested by animals pass through the digestive system as manure and urine (Weil and Brady 2017).
- In any given year, the USDA estimates that about 5% of US cropland receives manure applications (Weil and Brady 2017).

- Manure applications on cropland do have environmental concerns both locally and globally. Local concerns are air quality (odors from ammonia and sulfurous gases), nutrients in runoff, and fecal pathogens (see discussion on soilborne pathogens in this chapter). Global implications are related to CO_2, NO_x, NH_3, and CH_4 emissions from decomposition (Weil and Brady 2017).
- "There is a significant trend to using more natural organic nutrient sources such as manure, compost, and cover crops. Organic methods tend to be more appealing to home gardeners since they are more able to manage the addition of organic materials in their garden settings. However, in large-scale production agriculture, agricultural scientists and producers point out that to produce abundant food, both organic and synthetic fertilizers have specific and important roles" (Spaeth 2018)
- There can be potential problems associated with nutrient leaching from long-term applications of manure. Excess accumulation of nitrogen and phosphorus can occur. Organic phosphorus can be leached and transported by water runoff to streams, rivers, and lakes, causing pollution and degraded water quality.

References

Albrecht, W.A. 1938. Loss of soil organic matter and its restoration. In *Soils and men, yearbook of agriculture 1938*, ed. H.G. Knight et al., 347–376. Washington D.C.: United States Department of Agriculture.

Al-Kaisi, M.M., and R. Lal. 2017. Conservation agriculture systems to mitigate climate variability effects of soil health. In *Soil health and intensification of agroecosystems*, ed. M.M. Alkaisi and B. Lowery. London, United Kingdom: Academic Press.

Amundson, R., A.A. Berhe, J.W. Hopmans, C. Olson, A.E. Sztein, and D.L. Sparks. 2015. Soil and human security in the 21st century. *Science* 348 (6235).

Archer, S., T.W. Boutton, and K.A. Hibbard. 2001. Trees in grasslands: Biogeochemical consequences of woody plant expansion. In *Global biogeochemical cycles in the climate systems*, ed. D. Schulze, M. Heimann, S. Harrison, E. Holland, J. Lloyd, I.C. Prentice, and D. Schimel, 115–130. San Diego: Academic Press.

Baldock, J.A., and J.O. Skjemstad. 1999. Organic soil C/soil organic matter. In *Soil analysis: An interpretation manual*, ed. K.I. Prveril, L.A. Sparrow, and D.J. Reuter, 159–170. Collingwood: CSIRO Publishing.

Balogh, J., S. Czóbel, S. Fóti, Z. Nagy, O. Szirmai, E. Péli, and Z. Zuba. 2005. The influence of drought on carbon balance in loess grassland. *Cereal Research Communications* 33: 149–152.

Balota, E.L., A. Colozzi-Filho, D.S. Andrade, and R.P. Dick. 2003. Microbial biomass in soils under and crop rotation systems. *Biological Fertility Soils* 38: 15–20.

Barak, J.D., L.C. Whitehand, and A.O. Charkowski. 2002. Differences in attachment of *Salmonella enterica* serovars and Escherichia coli O157: H7 to alfalfa sprouts. *Applied Environmental Microbiology* 68: 4758–4763.

Barker, J., and S.F. Bloomfield. 2000. Survival of Salmonella in bathrooms and toilets in domestic homes following salmonellosis. *Journal Applied Microbiology* 89: 137–144.

Baron, V., S.E. Mapfumo, A.C. Dick, M.A. Naeth, E.K. Okine, and D.S. Chanasyk. 2002. Grazing intensity impacts on pasture carbon and nitrogen flow. *Journal of Range Management*: 55: 535–541.

Barrow, C.J. 1991. *Land degradation*. Cambridge: Cambridge Univ. Press.

Batjes, N.H. 1996. Total carbon and nitrogen in the soils of the world. *European Journal of Soil Science* 47: 151–163.

Battin, T.J., S. Luyssaert, L.A. Kaplan, A.K. Aufdenkampe, A. Richter, and L.J. Tranvik. 2009. The boundless carbon cycle. *Nature Geoscience* 2: 598–600.

Bauer, A., C.V. Cole, and A.L. Black. 1987. Soil property comparisons in virgin grasslands between grazed and nongrazed management systems. *Soil Science Society of America Journal* 51 (1): 176–182.

Berner, R.A. 1990. Atmospheric carbon dioxide levels over Phanerozoic time. *Science* 249: 1382.

Biardeau, L., R. Crebbin-Coates, R. Keerati, S. Litke, and H. Rodríguez. 2016. Soil health and carbon sequestration in US croplands: A policy analysis. *University of California Berkeley* 9-11 (21): 46.

Bohn, H.L. 1976. Estimate of organic carbon in world soils. *Soil Science Society of America Journal* 40: 468–470.

Bolin, B. 1977. Changes in land biota and their importance in the carbon cycle. *Science* 196: 613–615.

Bot, A., and J. Benites. 2005. *The importance of soil organic matter. Key to drought-resistant soil and sustained food production*. Rome: FAO Land and Plant Nutrition Management Service. Food and Agriculture Organization of the United Nations.

Brady, N.C., and R.R. Weil. 1996. *The nature and properties of soils*. 11th ed. Upper Saddle River: Prentice Hall.

Brown, C. 2013. Available nutrients and value for manure from various livestock types. Ontario Ministry of Agriculture and Food and the Ministry of Rural Affairs. Fact Sheet 13-043.

Brown, S., and A.E. Lugo. 2018. The storage and production of organic matter in tropical forests and their role in the global carbon cycle. *Biotopica* 14: 161–187.

Brussaard, L. 1994. Interrelationships between biological activities, soil properties and soil management. In *Soil resilience and sustainable land use*, ed. D.J. Greenland and I. Szabolcs, 309–329. Wallingford: CAB International.

Burges, A. 1963. The microbiology of a podzol profile. In *Soil organisms*, ed. J. Doeksen and J. van der Drift, 151–157. Amsterdam: North Holland Publ. Co.

Buringh, P. 1984. Organic carbon in soils of the world. In *The role of terrestrial vegetation in the global carbon cycle: Measurement by remote sensing*, ed. G.M. Woodward. New York: Wiley and Sons.

Burke, I.C., C.M. Yonker, W.J. Parton, C.V. Cole, D.S. Schimel, and K. Flach. 1989. Texture, climate, and cultivation effects on soil organic matter content in US grassland soils. *Soil Science Society of America Journal* 53: 800–805.

Buyanovsky, G.A., and G.H. Wagner. 1997. Crop residue input ot soil organic matter in Sanborn field. In *Soil organic matter in in temperate ecosystems: Long-term experiments in North America*, ed. E.A. Paul, 73–84. Boca Raton: CRC Press.

Cambell, C.A., and W. Souster. 1982. Loss of organic matter and potentially mineralizable N from Saskatchewan soils due to cropping. *Canadian Journal of Soil Science* 62: 651–656.

Canadell, J.G., and M.R. Raupach. 2008. Managing forests for climate change mitigation. *Science* 320: 1456–1457.

Canadell, J.G., M.U.F. Kirschbaum, W.A. Kurz, M.J. Sanz, B. Schlamadinger, and Y. Yamagata. 2007a. Factoring out natural and indirect human effects on terrestrial carbon sources and sinks. *Environmental Science and Policy* 10: 370–384.

Canadell, J.G., D.E. Pataki, R. Gifford, R.A. Houghton, Y. Luo, M.R. Raupach, and W. Steffen. 2007b. Saturation of the terrestrial carbon sink. In *Terrestrial ecosystems in a changing world*, 59–78. Berlin: Springer.

Chambers, A., R. Lal, and K. Paustian. 2016. Soil carbon sequestration potential of US croplands and grasslands: Implementing the 4 per thousand initiative. *Journal of Soil and Water Conservation* 71: 68A–74A.

Chao, W., R. Ding, and R. Chen. 1987. Survival of pathogenic bacteria in environmental microcosms. *Chinese Journal Microbial Immunology (Taipei)* 20: 339–348.

Chappell, A., and J.A. Baldock. 2016. Wind erosion reduces soil organic carbon sequestration falsely indicating ineffective management practices. *Aeolian Research* 22: 107–116.

Chappell, A., N. Webb, R. Rossel, and E. Bui. 2014. Australian net (1950s-1990) soil organic carbon erosion: Implications for CO_2 emission and land-atmosphere modeling. *Biogeosciences* 11: 5235–5244.

Cole, C.V., J. Duxbury, J. Freney, O. Heinemeyer, K. Minami, A. Mosier, K. Paustian, N. Rosenberg, N. Sampson, D. Sauerbeck, and Q. Zhao. 1997. Global estimates of potential mitigation of greenhouse gas emissions by agriculture. *Nutrient Cycling in Agroecosystems* 49: 221–228.

Cornell Agronomy Fact Sheet 4. 2005. Cornell University Cooperative Extension. http://nmsp. cals.cornell.edu/publications/factsheets/factsheet4.pdf

Cosper, H.R., and J.R. Thomas. 1961. Influence of supplemental water and fertilizer on production and chemical composition of native forage. *Journal of Range Management* 15: 292–297.

Davidson, E.A., and I.L. Ackerman. 1993. Changes in soil carbon inventories following cultivation of previously untilled soils. *Biogeochemistry* 20: 161–193.

Davies, R.H., and C. Wray. 1996. Seasonal variations in the isolation of *Salmonella typhimurium*, *Salmonella enteritidis*, *Bacillus cereus* and *Clostridium perfringens* from environmental samples. *Journal of Veterinary Medicine Series* B43: 119–127.

Derner, J.D., and G.E. Schuman. 2007. Carbon sequestration and rangelands: a synthesis of land management and precipitation effects. *Journal of soil and water conservation* 62: 7--85.

Derner, J.D., D.D. Briske, and T.W. Boutton. 1997. Does grazing mediate soil carbon and nitrogen accumulation beneath C4, perennial grasses along an environmental gradient? *Plant and Soil* 191: 47–156.

Donigian, A.S., T.O. Barnwell, R.B. Jackson, A.S. Patwardhan, and K.B. Weinrich. 1994. *Assessment of alternative management practices and policies affecting soil carbon in agroecosystems of the central United States (No. PB-94-189420/XAB)*. California: AQUA TERRA Consultants, Mountain View.

Drobnik, J. 1960. Primary oxidation of organic matter in the soil. *Plant and Soil* 12 (3): 212–222.

Duiker, S.W. 2018. Can I increase soil organic matter by 1% this year? University Park, PA. Penn. State Extension. https://extension.psu.edu/can-i-increase-soil-organic-matter-by-1-this-year

Egball, B., B.J. Weinhold, J.E. Gilley, and R.A. Eigenberg. 2002. Mineralization of manure nutrients. *Journal of Soil and Water Conservation* 57: 470–473.

Eghball, B. 2000. Nitrogen mineralization from field-applied beef cattle feedlot manure or compost. *Soil Science Society of America Journal* 64: 2024–2030.

EPA. 2017. https://www.epa.gov/sites/production/files/2019-04/documents/us-ghg-inventory-2019-chapter-2-trends.pdf

Eswaran, H., P.F. Reich, J.M. Kimble, F.H. Beinroth, E. Padamnabhan, and P. Moncharoen. 2000. Global carbon stocks. In *Global climate change and pedogenic carbonates*, ed. R. Lal, J.M. Kimble, H. Eswaran, and B.A. Stewart, 15–25. Boca Raton: CRC/Lewis Publishers.

Evans, N., R.A. Gill, V.T. Eviner, and V. Bailey. 2017. Soil and belowground processes. In *Rangeland Systems*, ed. D.D. Briske, 131–168. Springer Series on Environmental Management.

Falkowski, P., R.J. Scholes, E.E.A. Boyle, J. Canadell, D. Canfield, J. Elser, N. Gruber, K. Hibbard, P. Högberg, S. Linder, and R.T. Mackenzie. 2000. The global carbon cycle: A test of our knowledge of earth as a system. *Science* 290: 291–296.

Fierer, N., M.S. Strickland, D. Liptzin, M.A. Bradford, and C.C. Cleveland. 2009. Global patterns in belowground communities. *Ecology Letters* 12: 1238–1249.

Flach, K.W., T.O. Barnwell Jr., and P. Crosson. 1997. Impacts of agriculture on atmospheric carbon dioxide. In *Soil organic matter in temperate agroecosystems. Long-term experiments in North America*, ed. E.A. Paul, K. Paustian, E.T. Elliot, and C.V. Cole, 3–13. Boca Raton, FL: CRC Press.

Foley, J.A. 2005. Global consequences of land use. *Science* 309: 570–574.

Frank, A.B., D.I. Tanaka, I. Hoffmann, and R.F. Follett. 1995. Soil carbon and nitrogen of the Northern Great Plains grasslands as influenced by long-term grazing. *Journal of Range Management* 48: 470–474.

Franzluebbers, A.J. 2005. Soil organic carbon sequestration and agricultural greenhouse gas emissions in the southeastern USA. *Soil and Tillage Research* 83: 120–147.

——. 2012. Grass roots of soil carbon sequestration. *Carbon Management* 3: 9–11.

Franzluebbers, A.J., and J.A. Stuedemann. 2009. Surface soil changes during twelve years of pasture management in the southern piedmont USA. *Soil Science Society of America Journal* 74: 2131–2141.

Franzluebbers, A.J., J.A. Stuedemann, and S.R. Wilkinson. 2001. Bermudagrass management in the Southern Piedmont USA. I. Soil and surface residue carbon and sulfur. *Soil Science Society of America* 65: 834–841.

Gaia, N. 1993. *An atlas of planet management*. Garden City: Anchor and Doubleday.

Gale, E.S., D.M. Sullivan, C.G. Cogger, A.I. Bary, D.D. Hemphill, and E.A. Myhre. 2006. Estimating plant-available nitrogen release from manures, composts, and specialty products. *Journal of Environmental Quality* 35: 2321–2332.

Generalic. 2019. Periodic table of the elements. Croatian-English Chemistry Dictionary & Glossary. https://glossary.periodni.com

Ghilarov, M.S. 1970. *Regularities in adaptions of arthropods to the terrestrial life*. Moscow: Nauka Publishing House.

Gliessman, S.R. 1998. *Agroecology: Ecological processes in sustainable agriculture*. Chelsea: Ann Arbor.

Gordon, D.M., S. Bauer, and J.R. Johnson. 2002. The genetic structure of Escherichia coli populations in primary and secondary habitats. *Microbiology* 148: 1513–1522.

Granger, D.E., J.W. Kirchner, and R. Finkel. 1996. Spatially averaged long-term erosion rates measured from in-situ-produced cosmogenicnuclides in alluvial sediment. *The Journal of Geology* 104: 249–257.

Guillame, T., D. Muhammad, and K. Yakov. 2015. Losses of soil carbon by converting tropical forest to plantations: Erosion and decomposition estimated by δ^{13} C. *Global Change Biology* 21: 3548–3560.

Guo, X., M.W. van Iersel, J. Chen, R.E. Brackett, and L.R. Beuchat. 2002. Evidence of association of salmonellae with tomato plants grown hydroponically in inoculated nutrient solution. *Applied Environmental Microbiology* 68: 3639–3643.

Gurtler, J.B., D.D. Douds, B.P. Dirks, J.J. Quinlan, A.M. Nicholson, J.G. Phillips, and B.A. Niemira. 2013. Salmonella and Escherichia coli O157: H7 survival in soil and translocation into leeks (*Allium porrum*) as influenced by an arbuscular mycorrhizal fungus (*Glomus intraradices*). *Applied Environmental Microbiology* 79: 1813–1820.

Haddix, M.L., A.F. Plante, R.T. Conant, J. Six, J.M. Steinweg, K. Magrini-Bair, R.A. Drijber, S.J. Morris, and E.A. Paul. 2011. The role of soil characteristics on temperature sensitivity of soil organic matter. *Soil Science Society of America Journal* 75: 56–68.

Haider, K.M., and G. Guggenberger. 2005. Organic matter: Genesis and formation. In *Encyclopedia of the soils in the environment*, ed. D. Hillel, vol. 3, 93–101. Oxford: Elsevier.

Hausenbuiller, R.L. 1978. *Soil science principles and practices*. Dubuque: W.C. Brown Co.

Hewins, D.B., M.P. Lyseng, D.F. Schoderbek, M. Alexander, W.D. Willms, C.N. Carlyle, S.X. Chang, and E.W. Bork. 2018. Grazing and climate effects on soil organic carbon concentration and particle-size association in northern grasslands. *Scientific Reports* 8 (1): 1336.

Hicks Pries, C.E., J.A. Bird, C. Castanha, P.J. Hatton, and M.S. Torn. 2017. Long term decomposition: The influence of litter type and soil horizon on retention of plant carbon and nitrogen in soils. *Biogeochemistry* 134: 5–16.

Hoagland, O.K., H.W. Miller, and A.L. Hafenrichter. 1952. Application of fertilizer to aid conservation. *Journal of Range Management* 5: 55–61.

Houghton, R.A. 1995. Changes in storage of terrestrial carbon since 1850. In *Soils and global change*, ed. R. Lal, J. Kimble, E. Levine, and B.A. Stewart. Boca Raton: CRC/Lewis Publishers.

——. 2007. Balancing the global carbon budget. *Annual Review of Earth and Planetary Sciences* 35: 313–347.

Hudson, H.J., and J. Webster. 1958. Succession of fungi on decaying stems of *Agropyron repens*. *Transactions of the British Mycological Society* 41: 165–177.

Huffine, W.W., and W.C. Elder. 1960. Effect of fertilizers on native grass pastures in Oklahoma. *Journal of Range Management* 13: 34–36.

Ingram, L.J., G.E. Schuman, P.D. Stahl, J.M. Welker, G.F. Vance, and G.K. Gunjegunte. 2004. The influence of grazing on microbial activity in a northern mixed-grass prairie. In *Agronomy abstracts*. Madison, Wisconsin: American Society of Agronomy.

IPCC. 2007. Climate Change 2007: The Physical Science Basis. Contribution of Working Group I to the Fourth Assessment Report of the Intergovernmental Panel on Climate Change Solomon. eds. S., D. Qin, M. Manning, Z. Chen, M. Marquis, K.B.M.Tignor and H.L. Miller. United Kingdom: Cambridge University Press.

———. 2013. *Climate change 2013: The physical science basis. Contribution of working group I to the fifth assessment report of the Intergovernmental Panel on Climate Change.* eds. T.F. Stocker, D. Qin, G.K. Plattner, M. Tignor, S.K. Allen, J. Boschung, A. Nauels, Y. Xia, V. Bex and P.M. Midgley. Cambridge/New York: Cambridge University Press. 1535 p.

Jabbagy, E.G., and R.B. Jackson. 2000. Below-ground processes and global change. *Ecolological Applications* 10: 423–436.

Jenkinson, D.S. 1990. The turnover of organic carbon and nitrogen in soil. *Philosophical Transactions of the Royal Society* 329: 361–368.

Jenny, H. 1941. *Factors of soil formation: A system of quantitative Pedology.* New York: McGraw-Hill.

Jobbagy, E.G., and R.B. Jackson. 2000. The vertical distribution of soil organic carbon and its relation to climate and vegetation. *Ecological Applications* 10: 423–436.

Joergensen, R.G., and F. Wichern. 2008. Quantitative assessment of the fungal contribution to microbial tissue in soil. *Soil Biological Biochemistry* 40: 2977–2991.

Johnson, C., G. Albrecht, Q. Ketterings, J. Beckman, and K. Stockin. 2005. *Nitrogen basics-the nitrogen cycle. Cornell University fact sheet 2.* Cornell: Cornell University Cooperative Extension.

Johnston, A., J.F. Dormaarand, and S. Smoliak. 1971. Long-term grazing effects on fescue grassland soils. *Journal of Range Management*: 185–188.

Juma, N.G. 1998. *The pedosphere and its dynamics: A systems approach to soil science.* Vol. 1. Edmonton: Quality Color Press Inc.

Kindler, R., A. Miltner, H.H. Richnow, and M. Kästner. 2006. Fate of gram-negative bacterial biomass in soil – Mineralization and contribution to SOM. *Soil Biological Biochemistry* 38: 2860–2870.

Kindler, R., A. Miltner, M. Thullner, H.H. Richnow, and M. Kästner. 2009. Fate of bacterial biomass derived fatty acids in soil and their contribution to soil organic matter. *Organic Geochemistry* 40: 29–37.

Kirchman, D.L., ed. 2010. *Microbial ecology of the oceans 36.* New York:Wiley and Sons.

Kirchmann, H., and E. Witter. 1989. Ammonia volatilization during aerobic and anaerobic manure decomposition. *Plant and Soil* 115: 35–41.

Kirschbaum, M.U.F., B. Harms, N.J. Mathers, and R.C. Dalal. 2008. Soil carbon and nitrogen changes after clearing mulga (*Acacia aneura*) vegetation in Queensland, Australia: Observations, simulations and scenario analysis. *Soil Biology and Biochemistry* 40: 392–405.

Klausner, S.D. 1997. *Nutrient management: Crop production and water quality (No. 101).* Northeast Regional Agricultural Engineering Service. Ithaca, New York: Cornell University.

Klausner, S.D., V.R. Kanneganti, and D.R. Bouldin. 1994. An approach for estimating a decay series for organic nitrogen in animal manures. *Agronomy Journal* 86: 897–903.

Klipple, G.E., and J.L. Retzer. 1959. Response of native vegetation of the central Great Plains to application of corral manure and commercial fertilizer. *Journal of Range Management* 12: 239–243.

Kucera, C.L., L.C. Dahlman, and M.R. Koelling. 1967. Total net productivity and turnover on an energy basis for tallgrass prairie. *Ecology* 48: 536–541.

Lal, R. 1998. Soil erosion impact on agronomic productivity and environment quality. *Critical Reviews in Plant Sciences* 17: 319–464.

———. 2003. Soil erosion and the global carbon budget. *Environment International* 29: 437–450.

———. 2004. Soil carbon sequestration impacts on global climate change and food security. *Science* 304: 1623–1627.

———. 2005. World crop residues production and implications of its use as a biofuel. *Environment International* 31: 575–584.

———. 2008. Carbon sequestration. *Philosophical Transactions of the Royal Society* 363: 815–831.

———. 2010. Managing soils and ecosystems for mitigating anthropogenic carbon emissions and advancing global food security. *Bioscience* 60: 708–721.

———. 2018. Accelerated soil erosion as a source of atmospheric CO_2. *Soil and Tillage Research.*

Lapina, K., R.E. Honrath, R.C. Owen, M. Val Martin, and G. Pfister. 2006. Evidence of significant large–scale impacts of boreal fires on ozone levels in the midlatitude northern hemisphere free troposphere. *Geophysical Research Letters* 33: L10815.

Larcher, W. 1983. *Physiological plant ecology*. Berlin: Springer.

Larney, F.J., K.E. Buckley, X. Hao, and W.P. McCaughey. 2006. Fresh, stockpiled, and composted beef cattle feedlot manure: Nutrient levels and mass balance estimates in Alberta and Manitoba. Technical reports: Waste management. *Journal of Environmental Quality* 35: 1844–1854.

Lewis, S.L. 2005. Tropical forests and atmospheric CO2: Current conditions and future scenarios. In *Avoiding dangerous climate change*, ed. H.J. Schellnhuber, 147–153. Cambridge, United Kingdom: Cambridge University Press.

Liang, C., J.P. Schimel, and J.D. Jastrow. 2017. The importance of anabolism in microbial control over soil carbon storage. *Nature Microbiology* 2: 1–6.

Magdoff, F., and H. Van Es. 2009. Building soils for better crops. Sustainable soil management. In *Handbook series book 10*, 3rd. ed. Sustainable Agriculture Research and Education (SARE) program, with funding from the National Institute of Food and Agriculture, U.S. Department of Agriculture.

Maher, B., J.M. Prospero, D. Mackie, D. Gaiero, P.P. Hesse, and Y. Balkanski. 2010. Global connections between aeolian dust, climate and ocean biogeochemistry at the present day and at the last glacial maximum. *Earth-Science Reviews* 99: 61–97.

Manley, J.T., G.E. Schuman, J.D. Reeder, and R.H. Hart. 1995. Rangeland soil carbon and nitrogen responses to grazing. *Journal of Soil and Water Conservation* 50: 294–298.

Mann, L.K. 1986. Changes in soil C storage after cultivation. *Soil Science* 142: 279–288.

Marschner, B., and K. Kalbitz. 2003. Controls of bioavailability and biodegradability of dissolved organic matter in soils. *Geoderma* 113: 211–235.

McKenzie, R.H., J.F. Dormarr, B. Adams, and W. Willms. 2003. *Manure application and nutrient balance on rangeland. Agdex 538–1*. Alberta: Agriculture, Food and Rural Development.

Mian, L.S., H. Maag, and J.V. Tacal. 2002. Isolation of Salmonella from muscoid flies at commercial animal establishments in San Bernardino County, California. *Journal Vector Ecology* 27: 82–85.

Miltner, A., R. Kindler, H. Knicker, H.H. Richnow, and M. Kästner. 2009. Fate of microbial biomass-derived amino acids in soil and their contribution to soil organic matter. *Organic Geochemistry* 40: 978–985.

Miltner, A., P. Bombach, B. Schmidt-Brücken, and M. Kästner. 2012. SOM genesis: Microbial biomass as a significant source. *Biogeochemistry* 111: 41–55.

Minderman, G. 1968. Addition, decomposition and accumulation of organic matter in forests. *Journal of Ecology* 56: 355–362.

Moebius-Clune, B.N., D.J. Moebius-Clune, B.K. Gugino, O.J. Idowu, R.R. Schindelbeck, A.J. Ristow, H.M. van Es, J.E. Thies, H.A. Shayler, M.B. McBride, K.S.M. Kurtz, D.W. Wolfe, and G.S. Abawi. 2016. *Comprehensive assessment of soil health – The Cornell framework, edition 3.2*. Geneva: Cornell University.

Mooney, H.A. 1972. The carbon balance of plants. *Annual Review of Ecology and Systematics* 3: 315–346.

Moorman, T.B. 1989. A review of pesticide effects on microorganisms and microbial processes related to soil fertility 1. *Journal of Production Agriculture* 2: 14–23.

———. 2018. Pesticide degradation by soil microorganisms: Environmental, ecological, and management effects. In *Soil biology*, 121–153. Boca Raton, Florida: CRC Press.

Müller-Nedebock, D., and V. Chaplot. 2015. Soil carbon losses by sheet erosion: A potentially critical contribution to the global carbon cycle. *Earth Surface Processes and Landforms* 40: 1803–1813.

NRAES. 1992. Resource, N. Agriculture, Engineering Service. On-farm composting. Ithaca, N.Y: NRAES Cooperative Extension.

NRAES. 1999. Resource, N. Agriculture, and Engineering Service. Poultry waste management handbook. Publ. No. NRAES-132. Ithaca, N.Y: Cooperative Extension, 152.

Natvig, E.E., S.C. Ingham, B.H. Ingham, L.R. Cooperband, and T.R. Roper. 2002. *Salmonella enterica* serovar *Typhimurium* and *Escherichia coli* contamination of root and leaf vegetables grown in soils with incorporated bovine manure. *Applied Environmental Microbiology* 68: 2737–2744.

NIACS. 2007. Carbon sequestration. http://nrs.fs.fed.us/niacs/

Norton, L.D. 1986. Erosion-sedimentation in a closed drainage basin in Northwest Indiana. *Soil Science Society of America Journal* 50: 209–213.

NRC (National Research Council). 2004. *Committee on air quality management in the United States, board on environmental studies and toxicology, board on atmospheric sciences and climate, division on earth and life studies, air quality management in the United States.* Washington, D.C.: National Academies Press.

Olson, K., A. Gennadiyev, A. Zhidkin, and M. Markelov. 2012. Impacts of land-use change, slope, and erosion on soil organic carbon retention and storage. *Soil Science* 177: 269–278.

Pacala, S., and R. Socolow. 2004. Stabilization wedges: Solving the climate problem for the next 50 years with current technologies. *Science* 305: 968–972.

Pan, Y., R.A. Birdsey, J. Fang, R. Houghton, P.E. Kauppi, W.A. Kurz, O.L. Phillips, A. Shvidenko, S.L. Lewis, J.G. Canadell, and P. Ciais. 2011. A large and persistent carbon sink in the world's forests. *Science* 333: 988–993.

Parham, J.A., S.P. Deng, H.N. Da, H.Y. Sun, and W.R. Raun. 2003. Long-term cattle manure application in soil. II. Effect on soil microbial populations and community structure. *Biology and Fertility of Soils* 38: 209–215.

Parker, W.F., and B.J. Mee. 1982. Survival of *Salmonella adelaide* and fecal coliforms in coarse sands of the swan coastal plain, Western Australia. *Applied Environmental Microbiology* 43: 981–986.

Paustian, K., H.P. Collins, and E.A. Paul. 1997. Management controls on soil carbon. In *Soil organic matter in temperate agroecosystems*, ed. E.A. Paul, E.T. Elliott, K. Paustian, and C.V. Cole, 15–49. New York: CRC Press.

Persson, T., E. Baath, M. Clarholm, H. Lundkvist, B.E. Söderström, and B. Sohlenius. 1980. Trophic structure, biomass dynamics and carbon metabolism of soil organisms in a scots pine Forest. *Ecological Bulletins* 32: 419–459.

Pietikainen, J., M. Pettersson, and E. Baath. 2005. Comparison of temperature effects on soil respiration and bacterial and fungal growth rates. *FEMS Microbiologial Ecology* 52: 49–58.

Pimentel, D. 2006. Soil erosion: A food and environmental threat. *Environment, Development and Sustainability* 8: 119–137.

Pimentel, D., C. Harvey, P. Resosudarmo, K. Sinclai, D. Kurz, M. McNair, et al. 1995a. Environmental and economic costs of soil erosion and conservation benefits. *Science* 267: 1117–1123.

Pimentel, D., C. Harvey, P. Resosudarmo, K. Sinclair, D. Kurz, and M. McNair. 1995b. Environmental and economic costs of soil erosion and conservation benefits. *Science New Series* 267 (5201): 1117–1123.

Povirk, K. 1999. Carbon and nitrogen dynamics of an alpine grassland: effects of grazing history and experimental warming on CO2 flux and soil properties. MS thesis, Laramie, Wyoming: Dept. Renewable Resources.

Pratt, P.F., F.E. Broadbent, and J.P. Martin. 1973. Using organic wastes as nitrogen fertilizers. *California Agriculture* 27: 10–13.

Pumphrey, F.V., and R.D. Hart. 1973. Fertilizing rangeland in Northeast Oregon. In *Special Report 378*. Corvallis, Oregon: Agriculture Experiment Station, Oregon State University.

Rasmussen, P.E., and H.P. Collins. 1991. Long-term impacts of tillage, fertilizer and crop residue on soil organic matter in temperate semi-arid regions. *Advances Agronomy* 45: 93–134.

Rasmussen, P.E., and W.J. Parton. 1994. Long-term effects of residue management in wheat-fallow. I. Inputs, yield and soil organic matter. *Soil Science Society America Journal* 58: 523–530.

Rauzi, F., R.L. Lang, and L.I. Painter. 1968. Effects of nitrogen fertilization on native rangeland. *Journal of Range Management*: 287–291.

Rhodes, M.W., and H. Kator. 1988. Survival of Escherichia coli and Salmonella spp. in estuarine environments. *Applied Environmental Microbiology* 54: 2902–2907.

Robertson, G.P., and E.A. Paul. 2000. Decomposition and soil organic matter dynamics. In *Methods in ecosystem science*, ed. O.E. Sala, R.B. Jackson, H.A. Mooney, and R.W. Howarth. New York: Springer.

Ruddiman, W.F. 2003. The anthropogenic greenhouse era began thousands of years ago. *Climatic Change* 61: 261–293.

Ruhe, R.V. 1969. Quaternary Landscapes of Iowa. Ames: Iowa State University Press, 255 pp.

Ruhe, R.V., and R.B. Daniels. 1965. Landscape erosion-geologic and historic. *Journal of Soil and Water Conservation* 20: 52–57.

Sakirkin, S.L.P., C.L.S. Morgan, and B.W. Auvermann. 2010. Effects of sample processing on ash content determination in solid cattle manure with visible/near-infrared spectroscopy. *Transactions of the ASABE* 53: 421–428.

SARE. 2012. Why soil organic matter is so important. Sustainable agriculture research and education. https://www.sare.org/Learning-Center/Books/Building-Soils-for-Better-Crops-3rd-Edition/Text-Version/Organic-Matter-What-It-Is-and-Why-It-s-So-Important/Why-Soil-Organic-Matter-Is-So-Important

Sawyer, J.E. and A.P. Mallarino. 2007. Carbon and nitrogen cycling with corn biomass harvest. *Integrated Crop Management News*.

Schimel, D.S. 1995. Terrestrial ecosystems and the carbon cycle. *Global Change Biology* 1: 77–91.

Schimel, J., and S.M. Schaeffer. 2012. Microbial control over carbon cycling in soil. *Frontiers in microbiology* 3: 348.

Schimel, D.S., D.C. Coleman, and K.A. Horton. 1985. Soil organic matter dynamics in paired rangeland and cropland toposequences in North Dakota. *Geoderma* 36: 201–214.

Schimel, D.S., J.I. House, K.A. Hibbard, P. Bousquet, P. Ciais, P. Peylin, B.H. Braswell, M.J. Apps, D. Baker, A. Bondeau, and J. Canadell. 2001. Recent patterns and mechanisms of carbon exchange by terrestrial ecosystems. *Nature* 414: 169–172.

Schlesinger, W.H. 1977. Carbon balance in terrestrial detritus. *Annual Review of Ecology and Systematics* 8: 51–81.

———. 1997. *Biogeochemistry: An analysis of global change*. 2nd ed, 558. San Diego: Academic Press.

Schmidt, H.P., A. Anca-Couce, N. Hagemann, C. Werner, D. Gerten, W. Lucht, and C. Kammann. 2019. Pyrogenic carbon capture and storage. *GCB Bioenergy* 11: 573–591.

Schuman, G.E., H.H. Janzen, and J.E. Herrick. 2002. Soil carbon dynamics and potential carbon sequestration by rangelands. *Environmental Pollution* 116: 391–396.

Schuman, G.E., J.D. Reeder, J.T., Manley, R.H. Hart, and W.A. Manley. 1999. Impact of grazing management on the carbon and nitrogen balance of a mixed-grass rangeland. *Ecological applications* 9: 65–71.

Schuman, G.E., L.J. Ingram, P.D. Stahl, and G.F. Vance. 2005. Dynamics of long-term carbon sequestration on rangelands in the western USA. In *XX International grassland congress*, vol. 26, 590. Wageningen: Academic Publishers.

Schweigert, M., S. Herrmann, A. Miltner, T. Fester, and M. Kästner. 2015. Fate of ectomycorrhizal fungal biomass in a soil bioreactor system and its contribution to soil organic matter formation. *Soil Biology and Biochemistry* 88: 120–127.

Scurlock, J.M.O., and D.O. Hall. 1998. The global carbon sink: A grassland perspective. *Global Change Biology* 4: 229–233.

Semenov, V.M., L.A. Ivannikova, T.V. Kuznetsova, N.A. Semenova, and A.S. Tulina. 2008. Mineralization of organic matter and the carbon sequestration capacity of Zonal soils. *Eurasian Soil Science* 41: 717–730.

Sierra, C., S. Malghani, and H.W. Loescher. 2017. Interactions among temperature, moisture, and oxygen concentrations in controlling decomposition rates in a boreal forest soil. *Biogeosciences* 14: 703–710.

Silveira, M., E. Hanlon, M. Azenha, and H. da Silva. 2012. *Carbon sequestration in grazing land ecosystems. UF/IFAS extension SL373*. Gainesville, Florida.

Simpson, I.J., F.S. Rowland, S. Meinardi, and D.R. Blake. 2006. Influence of biomass burning during recent fluctuations in the slow growth of global tropospheric methane. *Geophysical Research Letters* 33: L22808.

Singh, B.P., and Z. Rengel. 2007. The role of crop residues in improving soil fertility. In *Nutrient cycling in terrestrial ecosystems*, ed. P. Marshner and Z. Rengel, 183–214. Berlin: Springer.

Six, J., S.D. Frey, R.K. Thiet, and K.M. Batten. 2006. Bacterial and fungal contributions to carbon sequestration in agroecosystems. *Soil Science Society of America Journal* 70: 555–569.

Smoliak, S. 1986. Influence of climatic conditions on production of *Stipa-Bouteloua* prairie over a 50-year period. *Journal of Range Management*: 100–103.

Smoliak, S., J.F. Dormaar, and A. Johnson. 1972. Long-term grazing effects on Stripa-Bouteloua prairie soils. *Rangeland Ecology and Management* 25: 246–250.

Spaeth, K.E. 2018. *Circle gardening: Growing vegetables outside the box*. College Station: Texas A&M University Press.

Stenstrom, T.A. 1989. Bacterial hydrophobicity, an overall parameter for the measurement of adhesion potential to soil particles. *Applied Environmental Microbiology* 55: 142–147.

Stirling, G., H. Helen, T. Pattison, and M. Stirling. 2016. Introduction: Soil health, soil biology and sustainable agriculture and evidence based information soil health. In *Soil biology, soil borne diseases and sustainable agriculture*, ed. P. Storer, 1st ed., 1–5. Queensland: Csiro Publishing.

Strickland, M.S., C. Lauber, N. Fierer, and M.A. Bradford. 2009. Testing the functional significance of microbial community composition. *Ecology* 90: 441–451.

Summerfield, M.A., and N.J. Hulton. 1994. Natural controls of fluvial denudation rates in major world drainage basins. *Journal of Geophysical Research* 99: 13,871–13,883.

Taiz, L., and E. Zeiger. 1998. *Plant physiology*. 2nd ed. Sunderland: Sinauer Associates, Inc. Publ.

Thomason, B.M., D.J. Dodd, and W.B. Cherry. 1977. Increased recovery of salmonellae from environmental samples enriched with buffered peptone water. *Applied Environmental Microbiology* 34: 270–273.

Thorup, R.M. 1984. *A practical guide to soil fertility and fertilizer use*. San Francisco: Chevron Chemical Co.

Trumbore, S.E. 2000. Age of soil organic matter and soil respiration: Radiocarbon constraints on belowground C dynamics. *Ecological Applications* 10: 399–411.

Urbanski, S.P., J.M. Salmon, B.L. Nordgren, and W.M. Hao. 2009. A MODIS direct broadcast algorithm for mapping wildfire burned area in the western United States. *Remote Sensing of Environment* 113: 2511–2526.

USDA-NRCS. 2000. U.S. Department of Agriculture Natural Resources Division Resources Assessment Division. Washington, D.C.

USDA-USFS. 2019. https://www.fs.fed.us/ecosystemservices/carbon.shtml.

USFS. 2015. U.S. forest service climate change advisor's office briefing paper https://www.fs.fed.us/climatechange/documents/CarbonAssessmentsBriefingPaper.pdf

Verheijen, F.G.A., R.J.A. Jones, R.J. Rickson, and C.J. Smith. 2009. Tolerable versus actual soil erosion rates in Europe. *Earth-Science Reviews* 94: 23–38.

Waksman, S.A. 1936. *Humus origin, chemical composition, and importance in nature*. Baltimore: Williams and Wilkins Co.

Waldrop, M.P., and M.K. Firestone. 2004. Microbial community utilization of recalcitrant and simple carbon compounds: impact of oak-woodland plant communities. *Oecologia* 138: 275–284.

Walker, P.H. 1966. *Postglacial environments in relation to landscape and soils on the Cary Drift (Research bulletin 549)*, 549. Ames: Agriculture and Home Economics Experiment Station, Iowa State University.

Weil, R.R., and N.C. Brady. 2017. *The nature and properties of soils*. New York: Pearson.

Welch, N.T., J.M. Belmont, and J.C. Randolph. 2007. Summer ground layer biomass and nutrient contribution to above-ground litter in an Indiana temperate deciduous forest. *The American Midland Naturalist* 157: 11–27.

Wetterstedt, M. 2010. Decomposition of soil organic matter. Uppsala. Doctoral thesis.

Whittaker, R.H. 1975. *Communities and ecosystems*. 2nd ed. New York: McMillan Publ. Co.

Whittaker, R.H., and G.E. Likens. 1973. The primary production of the biosphere. *Human Ecology* 1: 299–369.

Wiedinmyer, C., and J.C. Neff. 2007. Estimates of CO_2 from fires in the United States: Implications for carbon management. *Carbon Balance and Management* 2: 12p.

Wiedinmyer, C., B. Quayle, C. Geron, A. Belote, D. McKenzie, X. Zhang, S. O'Neill, and K. Klos Wynne. 2006. Estimating emissions from fires in North America for air quality modeling. *Atmospheric Environment* 40: 3419–3432.

Wienhold, B.J., J.R. Hendrickson, and J.F. Karn. 2001. Pasture management influences on soil properties in the northern Great Plains. *Journal of Soil and Water Conservation* 56: 27–31.

Wikipedia 2018. https://commons.wikimedia.org/wiki/File:Agromet-Pilmet_Tytan_10_plus_P9110022.jpg.

Wilkinson, B.H., and B.J. McElroy. 2007. The impact of humans on continental erosion and sedimentation. *Geological Society of America Bulletin* 119: 140–156.

Winfield, M.D., and E.A. Groisman. 2003. Role of nonhost environments in the lifestyles of Salmonella and Escherichia coli. *Applied Environmental Microbiology* 69: 3687–3694.

Wright, R.C. 1989. The survival patterns of selected fecal bacteria in tropical fresh waters. *Epidemiology and Infection* 103: 603–611.

Xiao, C. 2017. Soil organic matter storage (sequestration) principles and management. Potential role for recycled organic materials in agricultural soils of Washington State Dept. Of Ecology Pub. No. 15-07-005.

Yan, H., S. Wang, C. Wang, G. Zhang, and N. Patel. 2005. Losses of soil organic carbon under wind erosion in China. *Global Change Biology* 11: 828–840.

Yoo, K., R. Amundson, A.M. Heimsath, and W.E. Dietrich. 2005. Erosion of upland hillslope soil organic carbon: Coupling field measurements with a sediment transport model. *Global Biogeochem Cycles* 19: 1–17.

Yue, Y., J. Ni, P. Piao, T. Wang, M. Huang, A.G.L. Borthwick, T. Li, Y. Wang, A. Chappell, and K. Van Oost. 2016. Lateral transport of soil carbon and land-atmosphere CO_2 flux induced by water erosion in China. *PNAS* 113: 6617–6622.

Zhong, W., T. Gu, W. Wang, B. Zhang, X. Lin, Q. Huang, and W. Shen. 2010. The effects of mineral fertilizer and organic manure on soil microbial community and diversity. *Plant and Soil* 326: 511–522.

Zibilske, L.M. 1998. Composting of organic wastes. In: Principles and Applications of Soil Microbiology, p. 482–497, Eds. Sylvia, D. M., Fuhrmann, J. F., Hartel, P. G., and Zuberer, D. A. Upper Saddle River, NJ: Prentice-Hall, Inc.

Zomer, R.J., D.A. Bossio, R. Sommer, and L.V. Verchot. 2017. Global sequestration potential of increased organic carbon in cropland soils. *Scientific Reports* 7: 1–8.

Chapter 7
Soil-Hydrology-Plant Assessment Technologies for Cropland, Rangeland, Pastureland, and Gardens

Abstract Soil health is critical to sustaining agricultural productivity and environmental well-being and is a necessary endeavor due to increased pressure to produce food, feed, fiber, and fuel throughout the world. Regional, national, and global efforts by the Food and Agriculture Organization of the United Nations, governments, agricultural agencies, universities, and private industries are pursuing new scientific approaches, technologies, and modes of communication to promote soil health into mainstream agriculture. Soil health is a prominent subject due to public recognition and landuser interest and commitment to work toward sustaining soil productivity on cropland, rangelands, forests, and gardens.

The inherent characteristics of the soil are the result of long-term succession involving the interaction of climate, soils, plants, and natural disturbance regimes. In Chap. 2, discussion about ecological sites and descriptions explains the importance of the historic characteristics of local soils and their baseline physical and chemical characteristics—a necessary step in advancing soil health principles. In this chapter, the concept of evaluating soil and surface stability, hydrologic function (Pellant et al. 2020), and biotic integrity is key to providing an accurate assessment of overall conditions in any land unit (e.g., cropland, rangeland, and pastureland). The reason for partitioning these three assessments separately is because there are often examples where one assessment is properly functioning, whereas another assessment is not. For example, soil and site stability may be functional (normal infiltration capacity, intact soil aggregates and stability, minimal erosion, no rills or gullies), yet biotic integrity may be dysfunctional (e.g., loss of native plants, preponderance of invasive plants, lack of plant vigor due to a macro- or micro-nutrient limitation, invasive plants, or troublesome plant pathogen, etc.). Therefore, we introduce the concept of evaluating the preponderance of evidence of three assessments used in Intepreting Indicators of Rangeland Health (Pellant et al. 2020), via separate, yet related indicators (some indicators cross over for one or more of the assessments) for cropland systems. This idea has been used since the mid-1990s on rangeland and recently for pastureland assessments (Pellant et al. 2005; Spaeth 2020). An example of a cropland soil assessment matrix is provided later in this chapter; however, this matrix can be customized to fit local cropping and cultural, and environmental

© Springer Nature Switzerland AG 2020
K. E. Spaeth Jr., *Soil Health on the Farm, Ranch, and in the Garden*,
https://doi.org/10.1007/978-3-030-40398-0_7

settings. A particular soil assessment matrix in a cultivated irrigated barley field in Wyoming will not exactly fit a no-till corn field in Iowa.

It is important to recognize soil diversity, how they developed, function, and eventually sustain life in particular ecosystems. Cropland soils that were once prairies or rangeland soils in cold and warm deserts are recognizably different, and expectations about their functionality, especially in manipulated agricultural settings and response to management, must be uniquely considered (Brown and Herrick 2016). The land classification "ecological sites" used by many US government agencies can provide a reference for understanding soil capabilities and a guide for planning and achieving a realistic approach to soil health. The US Department of Agriculture is actively involved in developing ecological site descriptions in coordination with agencies in the Department of Interior, universities, and other partners.[1] Assessment and monitoring of soil health are of major interest to farmers/land managers; therefore, it is important to have information about historic physical and chemical characteristics to develop a realistic plan for sustaining or improving the land. For example, for a particular soil, historic level of soil organic matter is 3%, and a plan to increase it to 5% is not a valid goal from several perspectives; however, if depleted below 3%, a plan to increase soil organic matter is a worthy goal. Environmental conditions over millennia controlled soil development and the local organic carbon cycle, which ultimately determined the constraints of organic matter content in the soil. Artificially increasing the soil, which is at it's target equilibrium organic matter content in the soil is usually a short-lived endeavor without constant additions of organic amendments, e.g., manure and compost. In summary, to assess soil health parameters, a reference is needed for some key physical and chemical characteristics.

There are many versions of cropland assessment tools that are reliable and quick and easy for evaluating basic soil health parameters on the farm. Visual Soil Assessment (VSA) is one approach and is a simple method to assess soil health and plant performance. Examples of VSA assessment tools are available: The Food and Agriculture Organization of the United Nations (Shepherd 2008); USDA-Natural Resources Conservation Service (NRCS) (2015), and Cornell University (Moebius-Clune 2017); University of Wisconsin; Iowa Soil Health Field Indicators Assessment Card 2017). USDA-NRCS has links to numerous soil health scorecards[2] and individual State Soil Quality Indicator Guides.[3] Many of these methods are reliable, economical, give rapid results, and provide a "quick assessment of soil health conditions that are meaningful to farmers and landowners with minimum training." VSA methods are largely qualitative and can be performed in the field without specialized equipment and expense.

---

[1]Ecological sites have not been completed throughout the United States; however, the work is progressing rapidly. Contact USDA-NRCS field offices for information on current status of completed ecological sites. WEB site for USDA-NRCS Ecological Site Information System at

[2] https://www.nrcs.usda.gov/wps/portal/nrcs/detailfull/soils/health/assessment/?cid=nrcs142p2_053871

[3] https://www.nrcs.usda.gov/wps/portal/nrcs/detailfull/soils/health/assessment/?cid=nrcs142p2_053871

On US grazing lands, separate methodologies are used for field assessments. For rangeland, Interpreting Indicators of Rangeland Health (IIRH) (Pellant et al. 2005; 2020 V5) was designed to provide information about how ecological processes such as the water cycle, energy flow, and nutrient cycle are functioning at the ecological site level. Rangeland health focuses on evaluation of preponderance of evidence of three assessments: soil/site stability, hydrologic function, and biotic integrity at an ecological site level. The NRCS incorporated the rangeland health methodology into their National Range and Pasture Handbook (USDA-NRCS 1997; section 600.0402 C) (NRPH) for use in the inventory and assessment phases during the conservation planning process with private landowners (USDA NRCS 1997). Rangeland health is also a key protocol in USDA-NRCS rangeland National Resource Inventory (NRI) field studies from 2003 to present (Spaeth et al. 2003, 2005).

For pastureland, a field protocol, Pasture Condition Score Sheet (PCSS), was developed to (1) provide "a framework for planning and assessing management at a site" (Toledo et al. 2016), (2) evaluate current productivity and stability, and soil and water resources on pastureland, and (3) assist in identifying future conservation treatment needs required to maintain or improve pasture conditions (Cosgrove et al. 2001). The PCSS (Cosgrove et al. 2001) was incorporated into the Pastureland NRI protocols beginning in 2009. Some states have modified the Cosgrove et al.'s 2001 PCSS to fit individual state needs because of inconsistencies of particular indicators with varying ecological dynamics and constraints across the United States. An additional tool, Determining Indicators of Pasture Health (DIPH), has been developed by USDA-NRCS (Toledo et al. 2016; Spaeth et al. 2020), which utilizes the concepts of Interpreting Indicators of Rangeland Health, where soil/site stability, hydrologic function, and biotic integrity are assessed separately according to preponderance of evidence.

Questions

- The development of ecological sites now incorporates altered states or land uses such as cropland, pastureland, and agroforestry practices. Discuss how altered ecological states fit into the state and transition concept and why it is important to link altered states with the original reference state to assess conditions and health. Why is it important to put altered land uses into the ecological site concept?
- Many visual assessments on cropland use a scoresheet based on a numerical score to evaluate soil health, which is inevitably related to cropland health. What are the advantages and disadvantages of tallying a score for cropland visual assessments? Would visual assessments on cropland be enhanced by avoiding an overall score and partitioning the assessment into three environmental categories such as soils, hydrology, and biotic components—discuss?
- Local farmers can develop a visual assessment matrix for cropland based on local and pertinent parameters and indicators. How would a local farmer select indicators that would best evaluate field-by-field soil health assessments?

- Range and pastureland health assessments avoid tallying an overall score. Discuss why three assessments, soil site stability, hydrologic function, and biotic integrity, are evaluated separately.
- What are some common and unique indicators used to assess range and pasture-land health?

Getting to Know Your Soil

What are the dynamics of your soil—what are its strengths and limitations regarding crop production? Having a baseline of information about local soils is a good first step in the assessment of soil health. It is always a good idea to send soils to your local state agricultural soil lab or local private labs. Soil tests are fine if you want to obtain information about basic chemical analyses of your soil. Most soil test labs provide basic information on soil pH, sometimes organic matter, the main macronutrients (phosphorus, potassium, calcium, magnesium, sodium, and sulfur), salinity (soluble salts), and cation exchange capacity (CEC). CEC is the ability of negatively charged soil particles (clay and organic matter) to attract and retain positively charged ions called cations (calcium, Ca^{++}; magnesium, Mg^{++}; potassium, $K+$; sodium, $Na+$; ammonium, NH_4+; aluminum, Al^{+++}; and hydrogen, $H+$), and many other micronutrients. Clay and organic matter in the soil have net negative charges and will attract these positively charged cations, just as the poles of a magnet attract their opposite charge. Sands are typically low in colloids and, therefore, low in cations, whereas silt loams and clay loams have a higher CEC, and humus has a very high CEC. In addition, some labs will report some of the main micronutrients (parts per million, or ppm), including iron (Fe), boron, (B), manganese (Mn), copper (Cu), zinc (Zn), and molybdenum (Mo). A separte request and charge by a soil lab is usually required for micronutrient analysis. However, most soil reports will not give you information about the soil name, soil physical characteristics, the relationship of the soil to other adjacent soils on the landscape, general potential productivity of the soil to adapted crops, and information about how the soil is suited for engineering purposes. The USDA-Natural Resources Conservation Service county soil survey and Web Soil Survey can fill in the gaps.

Soil Surveys

The USDA-National Resources Conservation Service (NRCS) conducts the National Cooperative Soil Survey in the United States. The USDA National Soil Survey Program was initiated near the end of the nineteenth century (1896) and has been constantly refined as soil science advances. For a history of the US Soil Survey, see *Profiles in the History of the U.S. Soil Survey* (Helms et al. 2002). Soil surveys describe the characteristics of soils in a given area and are based on a standard system of classification. Soil surveys are used in a number of ways, such as in developing land use plans, providing information about adapted crops and potential production, assisting in designing engineering applications, and assisting in evaluating and predicting the effects of various land uses on the environment.

In a soil survey, soils are displayed on an aerial photograph with lines or polygons designating soil map units. Soil surveys provide descriptions for each soil: respective location on a landscape, layers or profile characteristics with depth ranges in inches and/or cm, relationships of one soil to another, soil suitability for various uses, and information about management needs. County soil survey reports are available free on the Web (Fig. 7.1). These publications are very informative and provide a thorough overview of soils and ecology at the county level. However, there are also new tools that have digitized this information in an easy-to-use format accessed via the World Wide Web discussed in the next section.

Web Soil Survey

Web Soil Survey (WSS) provides soil data and information produced by USDA-NRCS National Cooperative Soil Survey (https://websoilsurvey.sc.egov.usda.gov/App/HomePage.htm). It is the largest natural resource information system in the world, which contains soil maps and data about your soils. Soil information is available online for more than 95 percent of the nation's private land; however, the site

Fig. 7.1 Example of county soil survey reports throughout the United States

generally does not include public land owned and managed by the US Forest Service, Bureau of Land Management, or other federal agencies (there are some exceptions where public lands have been mapped through cooperative agreements).

In addition, there is an application (App) for iPhone and Android users where global positioning system (GPS) technology locates onsite coordinates and links with USDA-NRCS soil survey data. The "Soil Web: An Online Soil Survey Browser" app provides summaries of the soil type based on the iPhone's current location in the field. This app was developed by the California Soil Resource Lab at UC Davis and UC-ANR in collaboration with the USDA Natural Resources Conservation Service.[4]

Basic Steps to Get Started with Web Soil Survey

Go to the Web Soil Survey site and click on the Green Button (Start WSS).

1. Type in address, locate area of interest on map, or enter latitude and longitude. A map will appear for your location.
2. Icons at the top of the map will allow you to zoom the photo image, create an area of interest where you will draw a box around your area, make distance measurements, etc.
3. To obtain an overlay of the mapped soils, click on the yellow tab, Soil Map, in the upper left corner. Lines will appear in the image with numbers, and a corresponding legend will appear that defines the soil map units on the image. There are two important concepts to be familiar with: the soil map unit and the corresponding soil component names. A map unit symbol on a soil map represents one or more kinds of soils. A soil map unit is a "collection of soil areas or non-soil areas (miscellaneous areas) delineated in a soil survey. Each map unit is given a name that uniquely identifies the unit in a particular soil survey area" (SSS 2013). Soil scientists identify soil map units according to soil taxonomic classification principles. The soil map unit name represents the dominant soil; however, more finely delineated soil components with individual names are classified with the map unit. Map unit components are not delineated on a soil map, but they are described according to their geomorphology and association with the map unit. Remember that soils, including soil map unit and component delineations, are natural phenomena, and they are subject to variability. This means that the information about a particular soil map unit and components may vary, but keep in mind that the presence of minor components does not diminish the value or accuracy of the data. The objective of mapping soils is to identify soils in the context of the landscape, which have similar use and management requirements. When it is necessary to obtain detailed soil information for specialized construction projects, onsite investigations can be used to provide the needed detail. The soil component contains the level of detail needed to begin evaluations of soil health.

---

[4] https://casoilresource.lawr.ucdavis.edu/soilweb-apps/. Instructions for the app at https://play.google.com/store/apps/details?id=com.casoilresourcelab.soilweb

4. Click on the desired soil in the map unit legend, and a report will be created. The
   example soil report shown in Table 7.1 is from my garden location in Ada County,
   Idaho. Web Soil Survey is a powerful tool; other tabs such as "Intro to Soils,
   Suitability's and Limitations for Use, Soil Properties and Qualities, Ecological
   Site Assessments, and Soil Reports" are included. Explore all the possibilities
   and acquaint yourself with the information. This is the tool that land management professionals use in conservation planning and working with farmers and
   ranchers. Web Soil Survey is also a valuable tool for home gardeners.

Table 7.1 Example of soil report from Web Soil Survey and Jenness soil description (units in
English): https://soilseries.sc.egov.usda.gov/OSD_Docs/J/JENNESS.html

Ada County, Idaho 72—Jenness fine sandy loam, 2–4 percent slopes.

The Jenness series consists of very deep, well-drained soils on bottomlands, low terraces, and
alluvial fans. They formed in alluvium and colluvium from acid igneous rocks. Permeability is
moderate.

Slopes are 0–30 percent. The average annual precipitation is about 10 inches, and the average
annual air temperature is about 49 degrees F. TAXONOMIC CLASS: Coarse-loamy, mixed,
superactive, nonacid, mesic, Xeric Torriorthents.

Map Unit Setting

   Landscape: Valleys

   Elevation: 2480–3100 feet

   Mean annual precipitation: 9–12 inches

   Mean annual air temperature: 49–52 degrees F

   Frost-free period: 135–155 days

Map Unit Composition

   Jenness and similar soils: 85 percent

Description of Jenness

Setting

   Landform: Stream terraces, fan remnants

   Downslope shape: Linear

   Across-slope shape: Linear

   Parent material: Alluvium derived from igneous rock

Properties and Qualities

   Slope: 2–4 percent

   Depth to restrictive feature: More than 80 inches

   Drainage class: Well-drained

   Capacity of the most limiting layer to transmit water (Ksat): Moderately high to high
(0.20–1.9 in hr.$^{-1}$)

   Depth to water table: More than 80 inches

   Frequency of flooding: None

   Frequency of ponding: None

   pH, neutral: 6.8

   Calcium carbonate, maximum content: 5 percent

   Soil organic matter: 3–4%

(continued)

Table 7.1 (continued)

Maximum salinity: Nonsaline (0.0–2.0 mmhos/cm)
Available water capacity: Moderate (about 7.6 inches)
Interpretive Groups
Farmland classification: Prime farmland if irrigated
Land capability classification (irrigated): 2e
Land capability (nonirrigated): 6c
Hydrologic soil group: B
Ecological site: Originally a rangeland soil: LOAMY 8–12 ARTRW8/PSSPS-ACTH7[a] (R011XY001ID).

Soil Profile Description (report given in English Units)

0 to 9 inches: Fine sandy loam, light brownish gray (10YR 6/2) loam, dark grayish brown (10YR 4/2), moist; weak, very fine granular structure, slightly hard, friable, slightly sticky, and slightly plastic; common fine roots; many very fine pores; neutral (pH 6.8); clear smooth boundary (2–9 inches thick).
9–22 inches: Fine sandy loam
22–32 inches: Loam
32–40 inches: Loam
40–53 inches: Sandy loam
53–72 inches: Silt loam

Type Location: Gem County, Idaho; about 5 miles north of Star Idaho, 1200 feet east of the NW cor. of section 15, T. 5 N., R. 1 W.
Competing Series: These are the Lolalita and Luckyrich series. Lolalita soils are dominantly coarse sandy loam or sandy loam in the 10- to 40-inch control section. Luckyrich soils have an average annual soil temperature of 54–59 degrees F.
Geographic Setting: Jenness soils are on bottomlands and very low terraces along small intermittent streams, very gently to steeply sloping alluvial fans. The soil formed in micaceous, recent local alluvium, and colluvium from acid and intermediate igneous rocks. The average annual precipitation is 9–12 inches, including 1–2 feet of snow.
Geographically Associated Soils: These are the Chilcott, Elijah, *Lanktree, Power, and Purdam soils*. All of these soils have argillic horizons. Chilcott, Elijah, Lanktree, and Purdam soils are on higher terraces. Power soils are on similar landscapes.
Use and Vegetation: Used for rangeland and irrigated cropland. Small grains, corn, sugar beets, alfalfa, and improved pasture are grown on irrigated land. Vegetation in the potential natural plant community is sagebrush (*Artemisia*), bluebunch wheatgrass (*Pseudoroegneria spicata*), Sandberg bluegrass (*Poa sandbergii*), Thurber needlegrass (*Achnatherum thurberianum)*, and giant wildrye (*Leymus condensatus*).
Distribution and Extent: Southwestern Idaho. The series is moderately extensive.

This information can provide a baseline to conduct soil health assessments. Where to start with all this information? Let us look at some of the most pertinent information:
1. Name of the soil and slope percent range.
2. Data from map unit setting, which contains information about elevation ranges, average precipitation, mean annual air temperature, and frost-free period.

(continued)

Table 7.1 (continued)

3. Map unit composition. The Jenness and similar soils make up 85% of this map unit. Other minor map unit components can exist (15% of the map unit). Your soil or part of it could be one of these minor components. If your soil does not match the description, type in the other minor soil component names in a web browser to see if there is a better match. In the Jenness report, the other minor soil components were not listed, but in many reports, the names and percentage of the map unit are given.
4. Evaluate the information of properties and qualities such as baseline pH, organic carbon, soil texture, soil physical properties, and chemistry.
5. What is the land capability class? A detailed discussion of this concept follows.
6. Lastly, check the typical profile information. The profile identifies layers or soil horizons, and the respective texture and depth in inches.

[a]ATARW8 Wyoming big sagebrush (*Artemisia wyomingensis*), PSSPS bluebunch wheatgrass (*Pseudoroegneria spicata*), and ACTH Thurber needlegrass (*Achnatherum thurberianum*).

In soil survey reports and laboratory data from Web Soil Survey, the horizons are defined by symbols, e.g., O, A, E, B, C, and R. Soil morphology is a science in itself, and it can be quite complex. The O horizon is the uppermost surface layer of the soil that comprise organic matter, which can be undecomposed or partially decomposed litter such as hay, leaves and needles, moss, and lichens. In some instances, the O horizon may be very thin or nonexistent for several reasons. This layer may decompose rapidly depending on local climatic conditions, be raked away during normal lawn and ground maintenance, or eroded. As you progress from the O to the C horizon, organic matter typically decreases. In most soils, the next is the A horizon, which is the uppermost mineral horizon. In laymen's terms, this layer typically is your topsoil layer and can contain decomposed organic material. During the mapping of soils, soil color is an easily discernible soil characteristic, and it is used to identify and classify soils. Soil scientists use a Munsell color chart, which contains color chips, much like you find at paint store to determine hue (yellow to redness), value (lightness or darkness), and chroma (color intensity or brightness). Following the A horizon, an E horizon could exist. The E horizon is a zone where silicate clay, iron, and aluminum oxides or a combination of these may have leached downward and are lighter in color than the A horizon above or the B horizon below. The B horizon typically forms below the A horizon. The B horizon contains less organic matter, and in laymen's terms, this layer is often referred to as subsoil. The C horizon is unconsolidated material (broken up rock). Plant roots usually do not easily penetrate through this layer, except through cracks and fissures. Very little organic material is found in the C horizon. The R-layer is hard bedrock with very little evidence of being weathered. Granite, basalt, quartzite, limestone, and sandstone are examples.

Text Box 7.1 Soil Color and Soil Organic Matter (Fig. 7.2)

One important visual property that is useful in determining soil carbon is soil color. Munsell color charts can be used to visually estimate soil organic matter content; however, limitations include individual perceptions of color, soil type, and water content. USDA-NRCS soil descriptions include Munsell color information (see Jenness soil profile example for A horizon). Soil color determinations using Munsell color charts are available for most soil components. For example, mollisols are dark in color, but many other soil orders are not; therefore, do not make the mistake of incorrectly assuming that soil organic matter has been depleted because a dark soil color is not observed. Use of Munsell color charts requires some training and practice but can be useful in determining color changes in soils—which can be an indicator of soil organic matter loss.

Assessment of carbon storage in field settings, indicators such as soil texture, clay mineralogy, soil color, soil flora observed (earthworms, insects, arthropods, etc.), visual breakdown of plant materials, and smell are often used. One good visual property associated with soil carbon in VSA utilizes soil color. Munsell soil color charts are developed in the early 1900s and are used by soil scientists in describing physical characteristics of soils during mapping and reporting. Soil organic matter and conversion to soil organic carbon content [soil organic matter = (SOC) (1.72)] can be approximated using soil color. Generally, the darker brown the soil, the higher the OM content, but other soil characteristics such as soil texture, moisture, carbonate, and mineral contents on soil color should be considered (Escadafal et al. 1989). For example, Fe oxides and hydroxides have characteristic colors: hematite alpha (Fe_2O_3), brownish red; magnetite (Fe_3O_4), blackish gray; wustite (FeO), grayish blue; goethite alpha (FeOOH), bright yellowish brown; lepidocrocite gamma (FeOOH), orange; and ferrihydrite (Fe_2O_3) reddish brown (Schwertmann 1993).

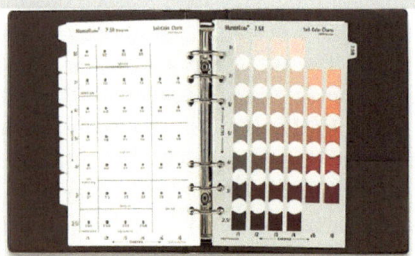

Fig. 7.2 Munsell color book and comparison of soil ped with color chart. Courtesy of USDA-NRCS

Land Capability Classification System
Land capability classes (LCC) are published for the inherent soils in county soil surveys and are used in the Web Soil Survey. There are other systems to evaluate land capability and highly erodible lands; however, LCC is straightforward and useful for group soils on the basis of their capability to produce cultivated crops and pasture plants without deteriorating over a long period of time. Land capability classification is subdivided into capability class and capability subclass" (USDA-NRCS 2013).

Some of the assumptions for grouping soils into capability classes are as follows:

(a) **How the soil is classified based on its characteristics**
(b) **Degree of limitations for agricultural purposes or hazard when used**
(c) Ratios of output to input
(d) Assumption of a high level of management
(e) Limitations such as the presence of surface standing water or excess water within the soil; droughty conditions, lack of water for adequate crop production; presence of stones; and presence of soluble salts or exchangeable sodium, or both

A "capability class is the broadest category in the land capability classification system. Class codes I (1), II (2), III (3), IV (4), V (5), VI (6), VII (7), and VIII (8) are used to represent both irrigated and non-irrigated land capability classes" (USDA-NRCS 2013).

Classes and Definitions (USDA-NRCS 2013)

Class I (1) soils have slight limitations that restrict their use.
Class II (2) soils have moderate limitations that reduce the choice of plants or require moderate conservation practices.
Class III (3) soils have severe limitations that reduce the choice of plants or require special conservation practices or both.
Class IV (4) soils have very severe limitations that restrict the choice of plants or require very careful management or both.
Class V (5) soils have little or no hazard of erosion but have other limitations, impractical to remove, that limit their use mainly to pasture, range, forestland, or wildlife food and cover.
Class VI (6) soils have severe limitations that make them generally unsuited to cultivation and that limit their use mainly to pasture, range, forestland, or wildlife food and cover.
Class VII (7) soils have very severe limitations that make them unsuited to cultivation and that restrict their use mainly to grazing, forestland, or wildlife.
Class VIII (8) soils and miscellaneous areas have limitations that preclude their use for commercial plant production and limit their use to recreation, wildlife, or water supply or for aesthetic purposes.

The capability subclass is the second category in the land capability classification system. Class codes e, w, s, and c are used for land capability subclasses" (USDA-NRCS 2013). Subclasses are not assigned to soils in capability class I (1), and subclass "e" is not used in class V (5).

Subclass e is made up of soils for which the susceptibility to erosion is the dominant problem or hazard affecting their use. Erosion susceptibility and past erosion damage are the major soil factors that affect soils in this subclass.

Subclass w is made up of soils for which excess water is the dominant hazard or limitation affecting their use. Poor soil drainage, wetness, high water table, and overflow are the factors that affect soils in this subclass.

Subclass s is made up of soils that have soil limitations within the rooting zone, such as shallowness of the rooting zone, stones, low moisture-holding capacity, low fertility that is difficult to correct, and salinity or sodium content.

Subclass c is made up of soils for which the climate (the temperature or lack of moisture) is the major hazard or limitation affecting their use.

Visual Soil Assessments: Cropland

Visual soil assessments are useful in that they provide a quick and simple way to assess soil condition and cropland plant performance. There are many examples of VSA assessment tools: the Food and Agriculture Organization of the United Nations (Shepherd 2008); University of Wisconsin; Iowa Soil Health Field Indicators Assessment Card (2017); USDA-Natural Resources Conservation Service (NRCS) (2015); and Cornell University (Moebius-Clune 2017).[5] More comprehensive analyses are available through state university soil labs and private labs to assess the status of carbon and microorganisms (Moebius-Clune 2017). In "Comprehensive Assessment of Soil Health – The Cornell Framework" (Moebius-Clune 2017), a comprehensive assessment protocol is presented to address and emphasize the integration of soil biological, physical, and chemical measurements. Complete instructions are given for sampling with discussions on field and laboratory analyses.

Before a VSA is conducted, some preliminary information is needed: The key to a successful VSA is to have good reference information. Prerequisite information needed for VSA are as follows:

- Soil profile information
- Baseline pH of soil component
- Baseline soil physical properties (texture, structure, aggregate stability)
- Baseline soil chemistry and nutrients
- Historic information: was the soil originally grassland, rangeland, or forestland? Has an ecological site and description been developed for the soil map unit of interest?[6]

---

[5] USDA-NRCS has links to numerous soil health scorecards and individual state soil quality indicator guides.

[6] USDA-NRCS is working to complete the development of ecological sites and descriptions for the United States. If an ESD is not available in Web Soil Survey, contact your local NRCS office for information.

- Potential production capabilities for the soil
- Monthly precipitation and temperature (minimum to maximum)

Many US states, agencies, and governments have developed VSA score sheets. Table 7.2 provides a comparison of some VSA scorecards. Many of these VSA scorecards tally a numeric rating. The total score corresponds to a health category. Care must be used to rely on these overall scores, e.g., a score may reflect a healthy category; however, there may be indicators of concern that could be serious, irrespective of an overall score. On range and pastureland health assessment protocols, an overall rating score is not calculated, and the assessments are separated into hydrologic function, soil and surface stability, and biotic integrity. A similar approach could also be used on cropland VSAs (see Text Box 7.2). Instead of using existing VSA tools, cropland producers and managers can design custom health matrices based on key parameters and elements (see Table 7.2) that are of interest on a local field scale (see Text Box 7.3). Then perhaps, it would be more precise to label this approach a "soil-hydrology-plant health matrix."

Text Box 7.2 Comparing VSA Scores with Conventional, Laboratory-Based Measures of Soil

Properties
Visual soil assessment (VSA) methods have been developed to provide farmers, land managers, and regulatory agencies with a simple tool to assess and monitor soil health quickly, cheaply, and effectively. Shepherd (2003) evaluated VSA scores with measured soil properties to demonstrate that VSA can be used to complement quantitative laboratory measurements of soil to characterize and monitor soil properties. The ranking of correlated variables of VSA with laboratory-based characteristics were dry aggregate-size distribution, saturated hydraulic conductivity, air permeability, macroporosity, bulk density, aggregate stability, dry aggregate-size distribution, total carbon, and anaerobic mineralizable N. Shepherd (2003) found strong relationships between VSA scores and key quantitative measurements of soil characteristics. He summarized that "visually assessed soil characteristics can, in a number of instances, provide a more reliable indication of soil conditions that predominate throughout the year than measured values."

Table 7.2 Comparison of descriptive properties used in various visual assessment scorecards

Descriptive properties	Iowa State	Northeastern Illinois	Maryland Soil Quality Assessment (1997)	Wisconsin Soil Health Scorecard	USDA-NRCS in Field Soil Health Assessment Worksheet
Topography		✓			
Soil type		✓			
Soil structure	✓	✓		✓	✓
Soil aggregate stability	✓				✓
Compaction (bulk density) (surface and/or subsurface	✓	✓	✓	✓	
Penetration resistance					✓
Erosion and erodibility		✓	✓	✓	
Surface crust				✓	✓
Ease of tillage				✓	
Tilth, friability, mellowness (feel)			✓	✓	
Organic matter soil color			✓	✓	
Soil color		✓		✓	✓
pH			✓	✓	
Soil fertility				✓	
Soil hardness				✓	
Soil texture				✓	
Topsoil depth				✓	
Soil test (N, P, K)				✓	
Soil micronutrients				✓	
Nutrient deficiencies				✓	
Soil air and water					
Infiltration capacity (retention)	✓		✓	✓	
Drainage		✓	✓	✓	
Drainage moisture (ponding)	✓				✓
Water holding capacity	✓		✓		
Nutrient holding capacity			✓		
Soil aeration				✓	
Soil life					

(continued)

Table 7.2 (continued)

	Iowa State	Northeastern Illinois	Maryland Soil Quality Assessment (1997)	Wisconsin Soil Health Scorecard	USDA-NRCS in Field Soil Health Assessment Worksheet
Soil biotic components (earthworms, arthropods, insects), biological diversity	✓	✓	✓	✓	✓
Biopores					✓
Smell	✓			✓	
Plant life					
Plant residue (surface cover)			✓	✓	✓
Plant decomposition	✓			✓	✓
Plant emergence (seed germination)	✓			✓	
Plant health and vigor	✓		✓	✓	
Root growth	✓		✓		
Root health				✓	✓
Stem health				✓	
Stem health				✓	
Overall crop appearance				✓	
Plant drought resistance				✓	
Mature crop condition				✓	
Analytical properties				✓	
Crop yield				✓	
Grain test weight				✓	
Cost of production and profit				✓	
Animals					
Human health				✓	
Animal health				✓	
Wildlife				✓	
Chemicals in groundwater				✓	
Surface water conditions				✓	

Cornell University has evaluated cost-effective means for determining soil health. They have listed many potential indicators for rapid quantitative assessments (Table 7.3). The publication, "Comprehensive Assessment of Soil Health" (Moebius-Clune et al. 2016), is an in-depth training manual including a compilation of pertinent material on soil health.

Table 7.3 Potential indicators for the evaluation of soil health (Moebius-Clune et al. 2016)

Physical	Biological	Chemical
Texture	Root pathogen pressure assessment	Phosphorus
Bulk density	Beneficial nematode population	Nitrate nitrogen
Macro-porosity	Parasitic nematode population	Potassium
Meso-porosity	Potentially mineralizable nitrogen	pH
Micro-porosity	Cellulose decomposition rate	Magnesium
Available water capacity	Particulate organic matter	Calcium
Residual porosity	Active carbon	Iron
Penetration resistance at 10 kPa	Weed seed bank	Aluminum
Saturated hydraulic conductivity	Microbial respiration rate	Manganese
Dry aggregate size (<0.25 mm)	Soil proteins	Zinc
Dry aggregate size (0.25–2 mm)	Organic matter content	Copper
Dry aggregate size (2–8 mm)		Exchangeable acidity
Wet aggregate stability (0.25–2 mm)		Salinity
Wet aggregate stability (2–8 mm)		Sodicity
Surface hardness with penetrometer		Heavy metals
Subsurface hardness with penetrometer		
Field infiltrability		

Text Box 7.3 Example of a Cropland and Garden Health Assessment Matrix Using Three Attributes: Soil and Site Stability, Hydrologic Function, and Biotic Integrity (Table 7.4).

Table 7.4 Customized example of health matrix using some key elements from Table 7.2

	Cropland soil health attribute		
Indicators	Soil and site stability	Hydrologic function	Biotic integrity
1. Rills	X	X	
2. Water flow patterns	X	X	
3. Soil cover (bare ground %)	X	X	
4. Gullies	X	X	
5. Wind erosion	X		
6. Litter movement	X		
7. Soil structure	X	X	
8. Soil surface resistance to erosion (soil aggregate stability)	X	X	X
9. Soil surface loss or degradation	X	X	X
10. Soil compaction layer	X	X	X
11. Surface crusts	X	X	X

(continued)

Table 7.4 (continued)

Indicators	Cropland soil health attribute		
	Soil and site stability	Hydrologic function	Biotic integrity
12. Ponding and drainage of water	X	X	
13. Dead or dying plants or plant parts			X
14. Litter/residue cover and depth (cover ___ %)		X	X
15. Plant emergence (seed germination)			X
16. Overall plant health and vigor			X
(a) Roots			X
(b) Stem			X
(c) Fruit			X
17. Plant color (related to chlorosis or deficient macro- and/or micro-nutrients)			X
18. Soil biotic components (earthworms, arthropods, insects), biological diversity			X
19. Plant decomposition after harvest			X
20. Annual production (units used _____) Estimated: ___ ÷ expected ___ = ___ %			X
21. Invasive plants-weeds			X

The "X" indicates that the indicator is applied to the respective attribute (soil and site stability, hydrologic function, and biotic integrity). The descriptions below may change and be relevant or non-relevant based on local environmental conditions
Note: This matrix is a guide for growers to consider what indicators may be important in their operations

1. *Rills.* Associated with water erosion, small channels, usually a few centimeters deep, formed by runoff. Inter-rill erosion includes soil loss by raindrop splash and erosion from shallow overland flow. *Relevance to soil health*: SSS, transport and movement of soil, soil redistribution, and loss onsite; and HF, rapid loss of water, reduced infiltration, and water storage onsite.

2. *Water flow patterns.* Path(s) of moving water across the soil surface during periods of rain or snowmelt. Sometimes referred to as sheetflow or overland flow. In cropland, water flow patterns are highly dependent upon natural drainage and altered drainage paths caused by practices associated with land leveling, drainage, and irrigation systems, construction of terraces and waterways, and cropland planting patterns. *Relevance to soil health*: SSS, associated with inter-rill erosion and sediment transport; HF, accelerated water loss and erosion, and increased length and number of water flow paths associated with reduced infiltration and water storage onsite.

3. *Soil cover and corresponding bare ground percent.* A critical factor in promoting soil health and reducing runoff and erosion during the growing season and fallow periods. Estimate percent of soil surface covered with dead plant material, litter, organic mulch, or live plants (principal crop or cover crop). *Relevance*

to soil health: SSS, increased detrimental effects of raindrop splash, highly correlated with reduced infiltration capacity, soil movement and loss, reduction in aggregate stability and increased incidence of soil crusting; HF, reduction in infiltration, increased runoff and erosion, correlated with changes in plant community dynamics and production.

4. *Gullies.* Ephemeral gullies are deeper than rills and are small erosion channels caused by concentrated overland flow from runoff between two adjacent slopes (a natural drainage) after a single rainstorm event. Ephemeral gullies can appear on cropland each year but can be filled in with tillage operations. Gullies can form from rills and/or ephemeral gullies; they consist of deeper channels that by definition cannot be crossed during tillage operations. Gullies normally follow natural drainage channels. *Relevance to soil health*: SSS, soil loss erosion and landscape degradation; HF, accelerated runoff and transport of water off-site, and water table affects.

5. *Wind erosion.* Soil loss, movement, and deposition. Highly correlated with major site deterioration and desertification effects. Windblown particles causing abrasion to plants, exposed roots. *Relevance to soil health*: SSS, soil erosion and deposition, change in soil surface dynamics, and buried plants.

6. *Litter movement.* Includes plant litter, mulch, crop residues, compost material, etc. Movement of litter due to wind and water is an indication of current wind and water erosion. Visual indications can include debris dams and lodged litter against obstructions. Distance of litter movement is associated with active erosional processes and nutrient redistribution (Abrahams et al. 1995). *Relevance to soil health*: SSS, correlated with accelerated runoff and erosion and sediment transport. Secondary effects-associated nutrient loss.

7. *Soil structure.* Soil structure is a physical property of soil and is characterized by how the individual particles (sand, silt, and clay) and organic matter are arranged into larger aggregates called peds. If you were to excavate a clod of dry soil with a shovel and gently break it apart, notice how the soil tends to break apart into various sizes and/or shapes—these are the peds. The peds break apart along natural weak points which have low tensile strength. Soils covered by natural vegetation and undisturbed by previous tillage will exemplify the inherent structure of the soil. Pore space in the soil is another soil physical property that determines how fast and how much air and water are transmitted throughout the soil and influences rooting and the activities of certain soil organisms (e.g., earthworms). Soil porosity is highly influenced by soil structure and texture. Water percolation is affected by the respective soil structural types and together these factors determine ease or difficulty in cultivation, susceptibility to erosion (within limits), and use in engineering.

   Soil structure is identified by shape and appearance. The easiest structure to identify is single grained where the soil particles consist of non-aggregated sand or loose windblown loess. Granular structure appears as small, rounded aggregates that are easily separated from the mass. Granular soil structure is characteristic of a soil surfaces (A horizons) high in organic matter with earthworm activity. Platy structure has particles which are arranged in flattened planes and are usually arranged horizontally. During the soil-forming processes, soil parent materials deposited by water and ice can develop a platy structure. Clay soils that are compacted by machinery can also develop a platy structure. If soil peds are blocky, irregular, or cube like, then the structure is blocky. There are two types of blocky structures: angular, where the block edges are sharp and distinct, and subangular, where the planes are rounded in appearance. Blocky structure is usually seen in B horizons but may also occur in A horizons. Columnar and prismatic structures appear as vertically oriented columns, which vary in height (subsurface horizons in arid and semiarid regions; poorly drained areas in humid regions). Columnar is differentiated from prismatic as the top of the peds are more rounded. Prismatic/columnar structure is often associated with swelling clays and some subsoils where sodium is present. The last structural type, massive, represents a soil condition where there is no evidence of the above structural types. The soil particles appear as a mass with no apparent pattern.

What happens to soil structure when you cultivate or otherwise disturb the soil? Several things begin to occur: (1) soil structure is reduced and/or damaged and (2) soil aggregates are broken up and aerated, exceeding the norm, which exposes, oxidizes, and accelerates the breakdown of soil organic matter. Farming operations that use heavy machinery (deep plows, heavy discs, chisels) can penetrate into the subsoil, disturb, and alter soil structure. Cultivation practices that mix surface and subsoil can lead to a loss of organic matter and may bring heavier clay particles and sodium (if in the subsoil horizon) to the surface and cause a loss in soil quality. Soil surfaces of frequent and heavily cultivated soils tend to crust over after wet-dry cycles.

Relevance to soil health: SSS and HF, soil structures are important factors in determining soil health as they affect infiltration, water holding capacity, and drainage. They also have a physical impact on plant roots and growth. The principal factors associated with deteriorating structure are cultivation, compaction, and absence of vegetation.

8. *Soil surface resistance to erosion (soil aggregate stability).* Associated with vegetative cover, rooted plants, bare ground, and soil aggregate stability (internal correlations with organic matter and microbial activity) (Morgan 1986). Note: soil aggregate stability is highly correlated with soil texture; some light textured soils naturally do not form stable aggregates. *Relevance to soil health*: SSS, higher stability of soil aggregates correlated with reduced impact of raindrops, increased infiltration, and reduced runoff (Goff et al. 1993). If soils are exposed to repeated significant raindrop impact with scarce plant cover, soil aggregate stability will deteriorate. HF, associated with higher infiltration capacity and greater water storage; BI, stable water balance and greater soil water availability and storage for plant growth and microbial activity.

9. *Soil surface loss or degradation.* A result of wind and/or water erosion. Indications include loss of thickness of surface soil horizon, loss of organic matter, changes in soil color, soil textural, and structural changes. Depending on inherent soil texture by horizon, subsurface layers commonly exhibit less

infiltration capacity and loss of the soil pores and porosity (Spaeth et al. 1996a, b). *Relevance to soil health*: SSS, significant impact on all three attributes. Indication of past and/or current erosion, loss of organic matter, and decline in overall soil function; HF, significant effects, common reductions of infiltration and water holding capacity, increases in runoff and erosion, which have concomitant effects on biotic integrity and plant growth and production; BI, impact on the function of living organisms due to loss of nutrients and organic matter.

10. *Soil compaction layer.* A result of soil disturbance due to past or current repeated compaction from farm machinery and other vehicles, livestock trampling, foot traffic, and raindrop impact with loss or absence of plant or litter cover. *Relevance to soil health*: Affects all three soil health attributes due to changes in hydrologic cycle, including reduced infiltration and water storage, increased runoff, soil erosion, and sedimentation. Compaction can significantly restrict root development and penetration.

11. *Surface crusts* (see discussion in Chap. 3). Form on the soil surface by raindrop impact on bare soil, evaporative processes forming chemical crusts, and compaction from various sources such as grazing and other related vehicular traffic. Vesicular crusts are common where aggregate stability breaks down and gas bubbles form a crusty layer 1–10 mm thick. *Relevance to soil health*: Impacts all three soil health attributes (see Chaps. 3, 4, 5).

12. *Ponding and drainage of water.* Ponding is a natural occurrence on heavy textured soils. However, with soil surface disturbance such as tillage and the development of physical soil crusts, ponding of water may remain longer than on undisturbed soils or soils with high surface reside and organic matter. Physical crusts are more common on silt, clay, and loam soils, and they can also occur on lighter textured soils and are relatively thin and weak. *Relevance to soil health*: SSS, HF, physical crusts are generally associated with low organic matter content and reduced biological activity. As physical crusts develop, infiltration rates are reduced, ponding can occur, and overland water flow increases as runoff. Erosion rates are exacerbated as well. Longer ponding intervals are an indication of reduced infiltration capacity, which can be caused by a variety of indicators in this listing. On rangeland, ponding can be magnified on heavier textured soils with heavy trampling by livestock.

13. *Dead or dying plants or plant parts.* Plant stems, branches, leaves, and roots of annual plants cease their life cycle at the end of the growing season; however, signs of mortality during the growing periods can be due to many factors such as water stress, disease, nutrient deficiencies, and other aspects relating to a lack of soil health. *Relevance to soil health*: BI, soil health factors are inextricably linked to biological activity and many species of fungi and bacteria play important roles in maintaining soil structure and cycling nutrients, and provide competitive effects with pathogenic organisms (see Stirling et al. 2016 for in-depth discussions of soil health and biology relationships with soil-borne diseases).

14. *Litter/residue cover and depth.* Litter, ground residue, or mulches are dead plant material, including leaves, stems, and branches that are detached from the

plant. Stems and seed heads that are dead or dormant but standing upright and attached to the plant are considered standing dead material that will eventually become litter/residue. As litter decomposes, it becomes a source of soil organic material and raw materials for nutrient cycling, helps moderate the soil microclimate and water evaporation, and provides a barrier against raindrop impact, excessive runoff, and wind and water erosion. *Relevance to soil health*: HF, litter intercepting raindrop energy and soil splash, assists in obstructing overland flow, promotes infiltration, and reduces evapotranspiration, wind, and water erosion. BI, affects biotic integrity via nutrient cycling and organic matter cycle.

15. *Plant emergence (seed germination).* Seed germination and emergence are important to overall crop health and is a vulnerable stage of the cropping cycle. Many factors can be responsible for poor germination and survivability (poor quality seed, field and seedbed conditions, sowing problems, climate, and disease organisms). It is important to identify the causal agents associated with this indicator. *Relevance to soil health*: SSS and BI.

16. *Overall plant health (roots, stem, fruits).* There is a high correlation between healthy soils and healthy plants. The two are very closely related and plant health is a good indicator of soil health, although, there are always exceptions, especially in vegetable crops. Example 1: tomatoes can develop a hardened white-to-gray papery irregular-shaped patch where fruit is exposed to sun. Sunscald spots may eventually turn into more hardened gray patches. Example 2: the occurrence of small-to-large necrotic spots on bottom of tomato fruit, a concave rotting spot that can penetrate into fruit, and occurrence of a black end on fruit. This is a physiologic non-parasitic disease called "blossom end rot." Irregular rainfall (sometimes a lack of calcium in soil or lack of calcium transfer to fruit, which would then be soil related) due to extreme fluctuations in moisture, infrequent and then sudden heavy rainfall, or irrigation. Example 3: a common fungal disease, fusarium wilt (*Fusarium oxysporum*) can occur any time in the season and growth stage and is associated with warm humid air and soil temperatures (78–90 °F). *Relevance to soil health*: BI, judgment needs to be applied to this indicator as there many causal agents to plant health; however, including it in an overall soil-plant health matrix is useful as it identifies existing problems.

17. *Plant color.* Related to indicator 16; however, this indicator provides an extra indicator for biotic integrity. Are their adequate plant nutrients in the soil, is nutrient transport functioning, which is then also related to mycorrhizae, other microorganisms, photosynthesis and disease organisms—a very complex process. *Relevance to soil health*: BI.

18. *Soil biotic components (earthworms, arthropods, insects), biological diversity.* Soil organisms contribute greatly to the physical and chemical properties of the soil, which in turn affects plant growth (Stirling et al. 2016). The biotic organisms in the soil are associated with decomposition and recycling nutrients, improving soil structure and aggregation. Beneficial biotic components also help reduce the effect of soil-borne diseases. Specificity is needed to identify

what soil organisms naturally reside in a particular soil. For example, a VSA sheet may list earthworms, but these organisms may not be a component in many semiarid and arid environments. Matching the key indicator soil organisms with a particular soil component is important in making a proper determination of this indicator. Soil-borne pathogens can be identified and quantified in several different ways (plate counts, spore counts, and bioassay methods), and the methods used are dependent on the particular organisms. *Relevance to soil health*: BI, this indicator has a wide range of application, beneficial soil organisms to problematic pathogens. Plant health is a good first indicator used to identify the status of soil biotic components.

19. *Plant decomposition after harvest.* Decomposition involves both abiotic and biotic processes. Microbial action begins on standing live vegetation as different fungal species colonize different parts of plants. A myriad of organisms (insects, earthworms, mites, millipedes, centipedes—detritivores and saprophages, etc.) fragment the debris, open plant cuticles exposing parenchyma cells to microbial invasion, and consume raw material, which through excreta, is again acted upon by microorganisms. The basic components of organic matter are dry matter-water content, elemental composition, and biologically derived components. The components in organic matter breakdown at different rates; fungal mycelium and non-spore-forming bacteria quickly utilize carbohydrates (sugars) and simple proteins in the organic material. Bacteria and fungi provide more than 95% of the biotic contribution to organic matter decomposition; however, they are short lived, and as they expire, they are consumed by other microbial organisms. As this process continues, nutrients are immobilized in microbial tissue, and upon death, nutrients are released or mineralized—nutrients (nitrogen, phosphorous, sulfur, etc.) are again available for use by microbes and primary producers. Enzymatic biochemical processes of soil microorganisms reduce decomposed organic matter into mineral compounds that may be utilized by plant roots. Mineralization is the reduction of inorganic mineral compounds, and as plant residues are being mineralized by microorganisms, they become part of the soil. A considerable amount of the original carbon (65–85% of what was part of plant litter and animal detritus) is released back into the atmosphere as CO_2 via respiration. Only about 15–35% of this remaining carbon may remain in the soil as live biomass (~2–5%), non-living labile carbon compounds (~3–10%), or stabilized humus (~10–30%). Stabilization and the accumulation of organic matter in the soil are regulated by the soil environment and microbial activity (Weil and Brady 2017).

Plant litter will decompose at different rates according to crop residue management. No-till, reduced tillage, and various cultivation methods will result in different decomposition rates. Standing dead material will require longer periods of time to decompose since the process is expedited when the material comes in contact with the ground layer and soil. The concept here focuses on the organic matter cycle, are the soil microbes decomposing the plant litter. In some cultivation systems, where microbial action is compromised, check to see if undecomposed litter remains intact in buried soil longer than expected (given that soil temperatures are warm enough to foster decomposition). *Relevance to soil health*: BI.

20. *Annual production.* Productivity of plants is a direct representation of the energy captured by plants through photosynthesis. As an indicator, annual production is closely tied with many variables in this matrix and is considered a latent variable that is observed, but it represents many unmeasured variables (climatic variability, physical and chemical properties of soil, and healthy biological activity). *Relevance to soil health*: BI, through the process of photosynthesis, solar energy is converted into chemical energy used by plants. The functionality of photosynthesis can be affected by many factors in the environment and is a good indicator of the functionality of biotic integrity. *Relevance to soil health*: BI.

21. *Invasive plants-weeds.* This indicator is an important aspect of biotic integrity in perennial range and pasture systems. In cropland, weeds can provide cover and roots help stabilize soil, provide plant cover, redistribute plant nutrients, adsorb excess nutrients, and provide some diversity to the ecosystem. However, when weeds compete with seedling emergence and productivity of the crop, this indicator can be used to help focus on issues relating to pest management. Inclusion of this indicator is a decision to the grower—the aspect of expected crop production and weed encroachment is a biotic concern. *Relevance to soil health*: BI (Tables 7.5 and 7.6).

Determining Rangeland Conditions and Health

Throughout the history of the range management profession, various methodologies have been used to assess rangeland condition. Rangeland condition classes were used since the late 1930s up until the early 1990s and categorized conditions based on similarity from the "historic climax plant community (HCPC)": four groupings represented excellent (SI = 75–100%), good (SI = 50–74%), fair (SI = 26–49%), and poor (Si = 0–25%).[7] The percentage values represent the mathematical similarity of the existing native vegetation with a reference community. Historically, the reference community was a range site description; however, ecological sites have replaced the older range sites. Currently, a reference community is an ecological site (USDA-NRCS) or a similar concept, such as habitat types (USDA-USFS). The terminology of rangeland condition appears in the USDA Soil Conservation Range Handbook 1938, but there were no specific guidelines for use. Rangeland Condition classes again are included in the 1942 SCS Range Conservation Handbook. In the

---

[7] The use of the similarity index at this time veered from the mathematical comparison of existing plant community composition with HCPC. Assumptions were made about diminished native plant composition with replacement or the occurrence of less desirable native plants and/or native invasive plants, or exotic introduced species. The occurrence of this situation supposed that range condition was trending or was in fair-to-poor condition. This was a misnomer as many examples of changing plant composition are not necessarily associated with deteriorating soil surface stability and/or hydrologic function (Pellant et al. 2005, 2020; Spaeth 1990; Spaeth et al. 1996a, b; Boxell 2007, Boxell and Drohan 2009; Bagchi et al. 2013; Weltz et al. 1998, 2012, 2014).

Table 7.5 Example of soil health matrix with indicator criteria for none to slight, slight to moderate, moderate, moderate-to-extreme, and extreme-to-total departure from expected field conditions

Indicators	Ex-T	M-Ex	M	S-M	N-S
1. Rills					None
2. Water flow patterns	Numerous, very long, active erosion, flow paths connected	Very common, usually connected, erosion and deposition very common	Common flow paths, often connected, erosion and deposition common	Short flow paths, some connected, erosion and deposition minimal	None
3. Bare ground %	>95%		<70%		Minimal
4. Gullies	Several active gullies in drainage paths with unstable headcuts in field		Downcutting and erosion in major drainage areas in field, headcuts may be evident, waterways have downcuts and erosion, vegetation not supporting overland flow of water		None Note: waterways may have been constructed to provide stable drainage
5. Wind erosion	Widespread >50% of field	Common wind erosion throughout season	Occasional signs of wind erosion	Minimal wind erosion	None
6. Litter movement					
7. Soil structure	Soil structure pulverized and does not resemble structure associated with mapped soil component				Soil structure concomitant with mapped soil component
8. Soil surface resistance to erosion (soil aggregate stability)	Soils that inherently associated with stable soil aggregates not apparent				Based on expected aggregation for soil textural classes
9. Soil surface loss or degradation	Apparent soil erosion with depositional areas occurs with every storm event	Erosion observable throughout most of field with common depositional areas throughout	Some observable erosion and soil deposition areas during season	Slight or very occasional soil loss during season	Soil erosion not evident over the long term

10. Soil compaction layer	Extensive compaction restricting infiltration and root penetration and development	Widespread compaction throughout field with partial root restriction, root development affected	Moderate compaction with reduced infiltration, some plant root penetration partially restricted	Infrequent or very minimal, infiltration adequate with no root restriction	None
11. Surface physical crusts	Compacted hard crust forming after precipitation, seedling emergence sig Affected		Thin crust forming after rainfall event, some effect on seedling emergence	Minimal crust, not inhibiting seedling germination	None
12. Ponding and drainage of water	Ponding common and lasts longer than expected for field soil texture				Infiltration capacity as expected with soil texture. On heavily textured soils, ponding is transient
13. Dead of dying plants or plant parts	Extensive signs of plant mortality and disease symptoms	Widespread plant mortality and disease symptoms		Very slight or rare signs of plant mortality or disease symptoms	None
14. Litter/residue cover and depth	<5%	5–20%	20–30%	30–60%	>70%
15. Plant emergence (seed germination)	Very poor germination and seedling emergence	Poor	Fair, uneven stand	Good	Very good germination and seedling emergence
16. Overall plant health and vigor (a) Roots (b) Stem (c) Fruit	Plant vigor low, pathogens effect on crop is high		Plant vigor moderate, pathogens effect on crop is moderate		Plant vigor high, none or minor signs of disease
17. Plant color (related to chlorosis or deficient macro- and/or micro-nutrients)	Stunted growth, chlorosis or yellowing of plant tissue		Growth potential is variable throughout field, variation in color		Vibrant robust growth of plants, green color of plants signifies adequate nutrients

(continued)

Table 7.5 (continued)

Indicators	Ex-T	M-Ex	M	S-M	N-S
18. Soil biotic components (earthworms, arthropods, insects), biological diversity					
19. Plant decomposition after harvest	Organic carbon cycle is severely compromised due to lack of microbial action. Decomposition of litter on soil surface during warm temperatures is severely delayed		Organic carbon cycle is moderately compromised due to lack of microbial action. Decomposition of litter on soil surface during warm temperatures is moderately delayed		Organic matter cycle fully functional. When plant material comes in contact with soil, decomposition readily occurs
20. Annual production	< 50%	50–75%	75–80%	80–95%	95–100% of production capacity
21. Invasive plants-weeds (focus is on impact to crop productivity, not organic matter production or plant cover for hydrology and erosion). Healthy weed populations can be a sign of good soil health as they assist in transferring nutrients throughout the soil profile	Invasive plants/weeds severely affecting crop productivity				Minimal effect of invasive plants/weeds on crop productivity

The criteria can be adjusted to meet local conditions and customized according to site dynamics. Cells have been deliberately left blank to provide the reader with an opportunity to consider local factors relating to the respective indicators. Some of the indicator interpretations may not fit particular cropping systems; therefore, this matrix should be customized to fit local conditions for assessing soil health

Table 7.6 A soil-hydro-plant field evaluation based on the custom matrix from Table 7.4

Indicators	Attribute	Rating	Notes
1. Rills	SSS, HF	SM	Some rilling after last storm
2. Water flow patterns	SSS, HF	SM	Water flow patterns degraded
3. Soil cover (bare ground %)	SSS, HF	M	About 50%, residue reduced by tillage
4. Gullies	SSS, HF	NS	None
5. Wind erosion	SSS	NS	None
6. Litter movement	SSS	M	Litter movement and debris lodged in plants
7. Soil structure	SSS, HF	ME	Soil structure altered by tillage
8. Soil surface resistance to erosion (soil aggregate stability)	SSS, HF, BI	M	Soil aggregates loose and disintegrate easily upon wetting
9. Soil surface loss or degradation	SSS, HF, BI	SM	Some soil deposition in low areas and outside field edges
10. Soil compaction layer	SSS, HF, BI	SM	Some compaction and surface soil is hard when dry
11. Surface crusts	SSS, HF, BI	NS	No crusting after rain events
12. Ponding and drainage of water	SSS, HF	NS	Good drainage, no ponding
13. Dead of dying plants or plant parts	BI	NS	Healthy
14. Litter/residue cover and depth	HF, BI	SM	Litter was reduced by tillage, use less tillage
15. Plant emergence (seed germination)	BI	NS	Good seedling emergence and survival
16. Overall plant health and vigor	BI	NS	Good
17. Roots			Good
18. Stem			Good
19. Fruit			To be determined at harvest
20. Plant color (related to chlorosis or deficiency of macro-and/or micro-nutrients)	BI	NS	Good color, adequate N, P, K with soil test
21. Soil biotic components (earthworms, arthropods, insects), biological diversity	BI	SM	Some earthworm activity and beetle grubs, diversity is lower than reference conditions
22. Plant decomposition after harvest	BI	NS	Normal
23. Annual production	BI	NS	To be determined
24. Invasive plants-weeds	BI	SM	Bindweed scattered in field

E-T	M-E	M	S-M	N-S	E-T	M-E	M	S-M	N-S	E-T	M-E	M	S-M	N-S
			1					1			8			
			2					2				9		
		3					3					10		
			4						4					13
			5		7								14	

(continued)

Table 7.6 (continued)

E-T	M-E	M	S-M	N-S	E-T	M-E	M	S-M	N-S	E-T	M-E	M	S-M	N-S
		6					8							15
	7						9							16
		8					10							17
			9					11					18	
		10						12						19
			11				14						21	
			12											
Soil & Site Stability Attribute Rating: Scattered between SM to M					Hydrologic Function Attribute Rating: Scattered between SM to M					Biotic Integrity Attribute Rating: Predominantly NS to SM				

Notes: SSS has some critical issues that need to be addressed: surface cover, litter movement during storms, soil aggregates disturbed by tillage, compaction and tightness of soil structure after rain event

HF has some issues related to higher bare ground than planned, soil aggregates discussed in SSS, and soil structure was damaged during tilling when too wet causing hard clods.

BI is satisfactory, except for poor aggregate stability and possible lower soil porosity, which may be affecting soil microbial activity.

Notes: Complete with Cropland Evaluation matrix. Record observer name, field location and ID, date, status of crop (phenology), e.g., corn at three-leaf stage, 10 cm (4 in. tall), soil map units, current soil moisture, topsoil texture, current crop, last three-year rotation, tillage and field preparation, soil test (yes, no), and fertilizer records (yes, no)

1976 SCS National Range Handbook, similarity index appears again with a discussion of rangeland condition classes and examples of calculating rangeland condition. In the USDA-NRCS National Range and Pasture Handbook (1997), "Evaluating and Rating Ecological Sites" included discussions on rangeland trend, similarity index, and rangeland health. These three methods (although uniquely different) were included as ways to access and evaluate rangeland. A revision of the handbook (USDA-NRCS 2020) will maintain these evaluation tools with specific instructions on when and how they should be used. Regarding the similarity index, it is no longer used to evaluate overall rangeland condition or health. The value of the similarity index is to provide a mathematical computation of how similar the existing plant community is to a reference or a designated stand of choice. The reference plant community can be the historic plant community with expected native plants or any other sub-state reference community as long as the plant community is defined and objectives for using a sub-state reference are given.

History of Similarity Indices

The mathematical treatment of using similarity indices in plant ecology has been used since the early 1900s (Jaccard 1908; Gleason 1920; Sorensen 1948; Ellenberg 1956; Motyka 1950; Bray and Curtis 1957; Grieg-Smith 1964; Pielou 1969; Sneath and Sokal 1973; Mueller and Ellenberg 1974; Clifford and Stephenson 1975; Whittaker, 1975; Chambers 1983; Legendre and Legendre 1998; Egghe 2010;

Chiclana et al. 2013). Many different conceptual similarity indices have been proposed to quantitatively measure the degree to which species composition between quadrats, releves, stands of vegetation, communities, and sites is alike or conversely different. The simplest indices are based on species presence/absence, and number of species (species richness), while more complex indices compare individual species and respective composition or abundance measures (can be production, canopy or foliar cover, species density, frequency).

Similarity Coefficients Based on Presence (Tables 7.7 and 7.8)

$$\text{Jaccard's index SIJ} = (100)\frac{a}{a+b+c}(100)\frac{7}{7+3+2} = 58.3\%.$$

where

c = species unique to Stand 1 = 2
b = species unique to Stand 2 = 3
a = species common to Stand 1 and 2 = 7

$$\text{Sorensen's Index Sis} = \frac{2(a)}{b+c} \times 100 \, \text{Sis} = \frac{2(7)}{9+10} \times 100 = 73.6\%$$

where

a = species common to Stand 1 and 2 = 7
b = total species in Stand 1 = 7
c = total species in Stand 2 = 9

Table 7.7 Presence =1, absence =0 for two vegetation stands

Species	Stand 1	Stand 2	Common to 1 and 2
1	1	1	1
2	1	1	1
3	**1**	0	
4	1	1	1
5	0	**1**	
6	1	1	1
7	**1**	0	
8	1	1	1
9	0	**1**	
10	1	1	1
11	0	**1**	
12	1	1	1

The stands could be from the same or different ecological sites depending on the objective study.
Note: unique species in bold

Table 7.8 Example of similarity indices with species quantities

Species	Stand 1 (lbs)	Stand 2 (lbs)	Unique spp. Stand 1	Unique spp. Stand 2	Spp. presence common to 1 and 2	Spp. common to 1 and 2	Mw
A	60	25			1	85	25
B	1	0.5			1	1.5	0.5
C	0.5	2			1	2.5	0.5
D		**5**	5				
E	2	3			1	5	2
F		**2**	2				
G	**1**		1				
H		**30**	30				
I		**1**	1				
J	2	8			1	10	2
K	5	6			1	11	5
L	**28.5**		28.5				
	100	82.5			6	115	35

Note: Species unique to stand in bold

Similarity Relationships with Species Quantities

Motyka index $\dfrac{2\,\text{Mw}}{\text{MA}+\text{MB}} \times 100 \quad \dfrac{2(35)}{100+82.5} \times 100 = 38.36\,\%$ based on species composition

where

Mw = sum of the smaller values of common spp. in Stands 1 and 2
MA = sum of values in Stand 1 = 100
MB = sum of values in Stand 2 = 82.5

Jaccard's index $\text{SI}_J = (100)\dfrac{a}{a+b+c}$ $\text{SI}_J = (100)\dfrac{6}{6+2+4} = 50\%$ based on species presence

Similarity Index Based on Species Quantities (USDA-NRCS (1997) (Table 7.9)

Stand 1: production represents the reference plant community composition from which comparisons are made. Total average annual production in stand 1 = 1000 lbs. ac^{-1}. In this example, species F, H, and I are not native endemic species to the reference plant community. They may be native invasive species or exotic introduced species.

SI = (400/1000) × 100 = 40%

Stand 2: the existing stand of plants in the field is 40% similar to the reference plant community (the historic plant community composition given in the respective ecological site description). The Motyka index (calculated above) equaled 38.36%.

Table 7.9 Species composition by weight (lbs ac⁻¹) for reference plant community (Stand 1) and comparison community (Stand 2). Stand 2 production is based on reconstructed weights by species (see below)

Species	Stand 1 ref. ESD (lbs ac⁻¹)	Stand 2 actual dry wt. (lbs ac⁻¹)	Allowable lbs. ac⁻¹
A	600	250	250
B	10	5	5
C	5	20	5
D	0	50	0
E	20	75	20
F	0	20	0
G	10	0	0
H	0	300	0
I	0	10	0
J	20	80	20
K	50	60	50
L	285	50	50
	1000	920	400

Alternative: Reconstructing the Present Plant Community

In USDA-NRCS, a historic reference plant community is used with reconstruction factors for each species. In this example, similarity is calculated based on reconstructed factors that represent several environmental variables. Invasive species can be native plants from other ecological sites or exotic introduced plants. Even though these species exist in Stand 2, they are discounted and given an allowable production of zero. This is a common procedure used by USDA-NRCS (1997) (English units used by USDA) (Table 7.10).

Stand 2 at the time of evaluation is reconstructed to account for annual airdry production using three other factors (% ungrazed, % growth, and % climate effect by respective species). Each plant species is reconstructed using the formula below:

$$\text{Re Factor} = \frac{\%\text{Dry Weight}}{(\%\text{Ungrazed})(\%\text{Growth})(\%\text{Climate})}$$

where:

- Dry weight (%) = actual percent of air-dry weight for the plant species.
- Ungrazed (%) = percent of current plant growth of each species that has not been removed by harvest or grazing.
- Growth (%) = percent of growth of each species that has occurred for the current growing season.
- Climate (%) = climate adjustment, percent of growth of each species in comparison with average year, e.g., 0.75 that represents the current year's production is 75% of normal or is 25% below normal. A value of 1.25 is 125% of normal climate conditions or 25% above normal.

Table 7.10 Similarity index calculation with reconstruction factors

Species	Stand 1 Ref ESD (lbs ac⁻¹)	Stand 2 Green Wt. (lbs ac⁻¹)	Stand 2 Dry Wt. (%)	Stand 2 Ungrazed (%)	Stand 2 Growth (%)	Stand 2 Climate (%)	Re factor	Re wt. (lbs ac⁻¹)	Pounds Allowable
A	600	333	0.9	0.9	0.9	0.95	1.17	389	389
B	10	5.5	0.95	1	0.9	0.95	1.11	6	6
C	5	6.6	0.85	1	1	0.95	0.89	6	5
D	**0**	**20**	**0.9**	**1**	**1**	**0.95**	**0.95**	**19**	**0**
E	20	40	0.9	1	1	0.95	0.95	38	20
F	**0**	**6.6**	**0.9**	**1**	**1**	**0.95**	**0.95**	**6**	**0**
G	10	10	0.9	1	1	0.95	0.95	9	9
H	**0**	**4**	**0.95**	**1**	**1**	**0.95**	**1.00**	**4**	**0**
I	**0**	**4**	**0.85**	**1**	**1**	**0.95**	**0.89**	**4**	**0**
J	20	26.6	0.6	1	1	0.95	0.63	17	17
K	50	50	0.9	1	1	0.95	0.95	47	47
L	285	62.5	0.9	1	1	0.95	0.95	59	59
	1000	568.8							552

Stand 1 is the reference plant community. Species in bold are present in Stand 2, but not endemic to ecological site; therefore, pounds allowable = 0

Species in bold are not endemic to ecological site; therefore, allowable pounds = 0.

SI = (552/1000) × 100 = 55.2%

Stand 2 is 55.2% similar to Stand 1 of the reference plant community, or 44.8% dissimilar based on annual production by weight.

The expected average weight for each species is listed in the reference plant community. The reference site (stand 1) is the historic plant community composition listed in the ecological site description. For example, species A can equal up to 600 lbs. ac^{-1}, which is the determined total annual production weight allowed for that species. In stand 2, the green harvested weight of species A is 333 lbs. ac^{-1}. The reconstructed weight for species A is 389 lbs. ac^{-1} yr.$^{-1}$. Since the total allowable annual production for species A is 600 lbs. ac^{-1} dry weight, and the reconstructed annual production is 389 lbs. ac^{-1} yr.$^{-1}$, the full amount (389 lbs. ac^{-1} yr.$^{-1}$) is allowed in the similarity index calculation. Species D is an invasive plant and the reconstructed annual production is 19 lbs. ac^{-1}. Since the allowable production for this invasive species is zero, the allowable production used in the similarity index calculation is zero. The USDA-NRCS (1997) method of calculating similarity with reconstruction factors is not an exact science and subjectivity does enter in with respect to three of the reconstruction values. Dry weight can be determined by harvest and drying and weighing samples. Climate data can help in determining climate percent. Whatever decisions are made, consistency is key (this is not an exact science).

For percent similarity based on presence/absence of the given species, the Jaccard's index for this example equals:

$$SIJ = (100)\frac{a}{a+b+c} SIJ = (100)\frac{7}{7+4+0} = 63.6\% \text{ similarity based on species presence / absence}$$

where

- c = species unique to Stand 1 = 0
- b = species unique to Stand 2 = 4
- a = species common to Stands 1 and 2 = 7

Species	Stand 1	Stand 2	Common to 1 and 2
A	1	1	1
B	1	1	1
C	1	1	1
D		1	
E	1	1	1
G		1	
H	1	1	1
I		1	
J		1	
K	1	1	1
L	1	1	1

Species Importance Curves

Species importance curves can be used to visualize the ranking and importance of species for comparative purposes. Log scales are commonly used for the y axis to better depict data distribution; however, other data such as cover, frequency, and density could also be used. In Fig. 7.3, the reference community, that is species A and L, is the most dominant, with two species showing annual productivity greater than 100 lbs. ac^{-1} yr.$^{-1}$. Four species have production values between 10 and 100 lbs. ac^{-1} yr.$^{-1}$ and two species (B and C) between 1 and 10 lbs. ac^{-1} yr.$^{-1}$. Stand 2

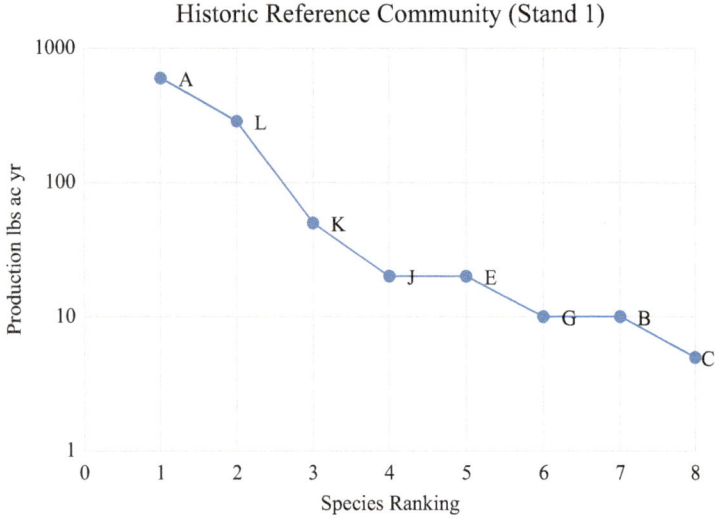

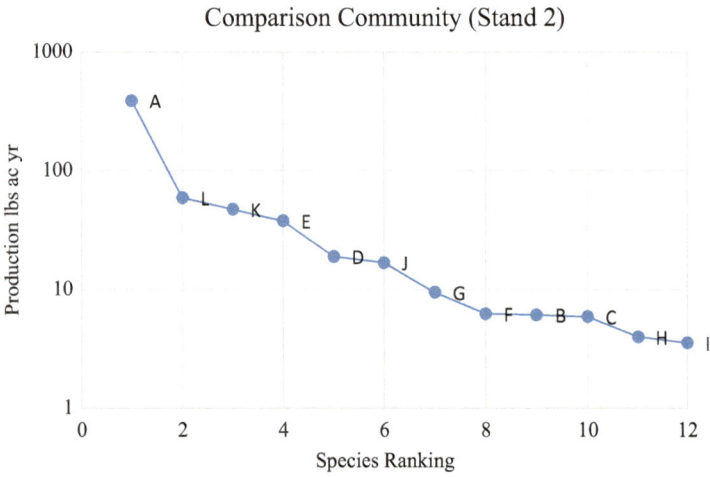

Fig. 7.3 Species importance curves for reference community (Stand 1) and comparison community (Stand 2). The y axis is log 10 scale and x axis is ranking of species from largest to smallest

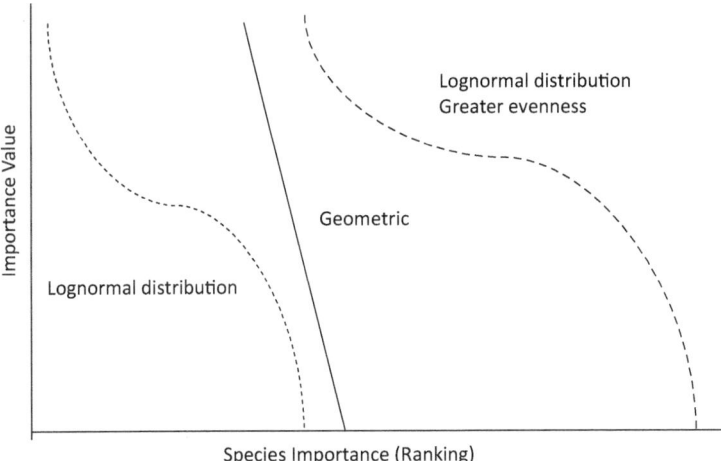

Fig. 7.4 Example of species importance curves, lognormal distribution with intermediate species diversity with one or few dominant species, and a geometric curve indicative of dominant species

shows that species A and L are still dominant, but species L has decreased in productivity. Species D, F, H, and I are new arrivals and happen to be an invasive species.

The shape of the curves in dominance diversity curves also provides information about degree of dominance and diversity (Whittaker 1965; Grieg-Smith 1983) (Figs. 7.3 and 7.4). A sharp linear line is depicted as a geometric curve, usually representing one or several species that are dominant with the remainder of the species with low importance values (production, cover, density, frequency values). Plant communities or stands with geometric curves typically are represented by lower species diversity and less evenness and perhaps associated rigorous environments, whereas the lognormal distribution implies that there are more species of intermediate importance values (evenness), with trace species trailing off. The species dominance curves are an indication of the degree of dominance: Are one or few species contributing a large importance value? Or, is there more evenness among the species importance values?

It should be noted that species with low importance values (biomass, cover, frequency, and density) does not indicate that the species is ecologically insignificant. Often species that occupy a trace in the plant community can be important indicator plants. For example, in North Idaho forests, e.g., western red cedar (*Thuja plicata*)/ wild ginger (*Asarum caudatum*) habitat type, wild ginger is the diagnostic indicator species (Cooper et al. 1987). Soils common to this habitat type are moister in the upper soil profile with a cryic soil temperature regime. A contiguous habitat type, the western red cedar/queencup beadlily (*Clintonia uniflora*) occurs in drier habitats with a frigid temperature regime. Both these indictor forb species typically occupy only trace to 1% foliar cover of the forest understory.

Interpreting Indicators of Rangeland Health

In 1994, a seminal book by the National Research Council NRC (1994) discussed the concept of rangeland health as an alternative method to assess rangelands. Choosing the term "health" was controversial when associated with natural systems (Rapport et al. 1998); however, it was maintained as it already was used on rangelands and agricultural soils (Brown and Herrick 2016). The NRC (1994) defines rangeland health as *"The degree to which the integrity of the soil and ecological processes of rangeland ecosystems are maintained"*. As the rangeland health concept evolved, its scope included such ecological processes as the water cycle, energy flow, and nutrient cycle that are functioning at the ecological site level. "Rangeland health is the degree to which the integrity of the soil, vegetation, water, and air as well as the ecological processes of the rangeland ecosystem are balanced and sustained. Integrity is defined as the maintenance of the functional attributes characteristic of a local, including normal variability" (Pyke et al. 2002). As described in Pellant et al. (2020): "Scientists and managers face continuing challenges to translate rangeland health into terms that the public can comprehend and that resource specialists can use to assist in identifying areas where ecological processes are or are not functioning properly. [The Interpreting Indicators of Rangeland Health] IIRH protocol does this using observable indicators. This protocol relies on a combination of qualitative and quantitative measures to assess the functional status of rangelands. Qualitative assessments provide relatively rapid techniques to rate site protection indicators, including both plant and soil components (Morgan 1986). The use of qualitative information to determine vegetation and soil conditions has a long history in land management inventory and monitoring. In some cases, qualitative assessments were used independently."

The rangeland health assessment evaluates 17 indicators to determine the preponderance of evidence for three ecosystem attributes (soil and site stability, hydrologic function, and biotic integrity) (Table 7.11). Each of the rangeland health assessments (Pellant et al. 2005, 2020) is defined as follows:

- *Soil/site stability*: "the capacity of an area to limit redistribution and loss of soil resources (including nutrients and organic matter) by wind and/or water, and to recover this capacity when a reduction does occur."
- *Hydrologic function*: "the capacity of an area to capture, store, and safely release water from rainfall, run-on, and snowmelt (where relevant), to resist a reduction in this capacity, and to recover this capacity when a reduction does occur."
- *Biotic integrity*: "the capacity of the biotic community to support ecological processes within the natural range of variability expected for the site, to resist a loss in the capacity to support these processes, and to recover this capacity when losses do occur. The biotic community includes plants (vascular and nonvascular), animals, insects, and microorganisms occurring both above and below ground."

The three ecosystem attributes (soil and site stability, hydrologic function, and biotic integrity) are determined from specific indicators (Table 7.1) that are specific to the ecological site. The 17 indicators are rated on expected conditions associated with the reference state for the ecological site (most often the historic plant community but can be other references with documentation). The 17 indicators include rills, water flow patterns, pedestals and terracettes, bare ground, gullies, wind scour and depositional areas, litter movement, soil resistance to erosion, soil surface loss or degradation, plant composition relative to infiltration, soil compaction, plant functional/structural groups, plant mortality, litter amount, annual production, invasive plants, and reproductive capability. Each of the 17 indicators is rated on a scale of departure from reference conditions (none-to-slight, slight-to-moderate, moderate, moderate-to-extreme, and extreme-to-total). Since there are many unique site-specific effects and nonlinear environmental relationships on rangeland, a single overall rating of rangeland health is not conducive to rating the status of a site. For example, some attributes may be functioning adequately, while other attributes are not (e.g., soil/site stability may be functioning according to what is expected for the site; however, biotic integrity may be compromised).[8]

Before a rangeland assessment can be conducted, a reference sheet is needed for the ecological site. The reference sheet is based on quantitative and qualitative data and expert knowledge of soil, hydrology, and plant relationships. Baseline reference information of the 17 indicators is used to gauge and evaluate current conditions. The expected conditions for each indicator are based on the natural range of variability for the historic plant community or other state if used for reference. The rangeland health tool has been used on private farms and ranches and to assess national rangeland conditions (federal and private) to help identify areas where conservation resources are needed or should be prioritized and provide "early warnings" of potential problems. Some of the major uses of IIRH[9] are (1) to provide an assessment of 17 indicators to determine the current status of soil and surface stability, hydrologic function, and biotic integrity at the ecological site level in conjunction with the corresponding reference sheet; (2) to communicate major ecological concepts to others, improve communication, and focus on ecosystem properties, functionality, and processes; and (3) help identify and prioritize resource needs and identify sites and areas that are at risk of degradation. Conversely, IIRH should not be used to (1) cite the cause of rangeland resource problems; (2) make independent management recommendations—supporting quantitative data should be used in conjunction with IIRH; and (3) independently monitor and determine trends without supporting quantitative data. In order to use IIRH effectively, knowledge of vegetation, soils, and ecological processes is required at the ecological site level and when two or more qualified individuals work together, the consistency of evaluating the indicators and attributes is improved.

---

[8] The USDI-Bureau of Land Management, US Geologic Survey, and USDA conduct one-week training sessions on rangeland health. The manual by Pellant et al. (2005) and the revised edition (2020) provide detailed theory, discussion, and full documentation on use and application of Interpreting Indicators of Rangeland Health.

[9] Consult Pellant et al. (2005, 2020) for specific uses and training needs of IIRH.

Text Box 7.4 USDA-NRCS Report on Rangeland Health Using Interpreting Indicators of Rangeland Health Model (USDA-NRCS 2018) (Fig. 7.5)

The current National Resource Inventory (NRI) began in 2003 and is based on a field survey that includes a fixed GPS points where data are collected identifying the soil type, ecological site, and site characteristics. Two transects (45.7 m; 150 ft) are positioned 45 deg. from magnetic north and data are collected on plant foliar and ground cover by species, annual production by species, plant height, vegetative canopy gaps, and soil aggregate stability. Within the microplot (45.7 m; 150 ft) diameter, applied conservation practices, resource concerns, rangeland health assessment, disturbances, and full plant census are recorded.

According to the NRI sample (2004–2010), nearly 82% of the nation's 405 million acres of non-federal rangeland is in relatively healthy condition and has no significant soil, hydrologic, or biotic integrity problems. Nationally, 18.3% (±0.8%) of the of non-federal rangeland was rated at moderate or worse departure from reference conditions for one of the three attributes (Fig. 7.6a) and 7.1% ((±0.5%) showed at least moderate or worse departure for all three attributes (Fig. 7.6b). Of the three rangeland health attributes, biotic integrity showed the greatest departure from reference conditions, with at least moderate and greater departure recorded on 15.2% (±0.7%) of non-federal rangeland. Hydrologic function ranked second at 11.9% (±0.8%), and third, soil and site stability with 9.3% (±0.8%) rated at moderate or worse departure from reference conditions.

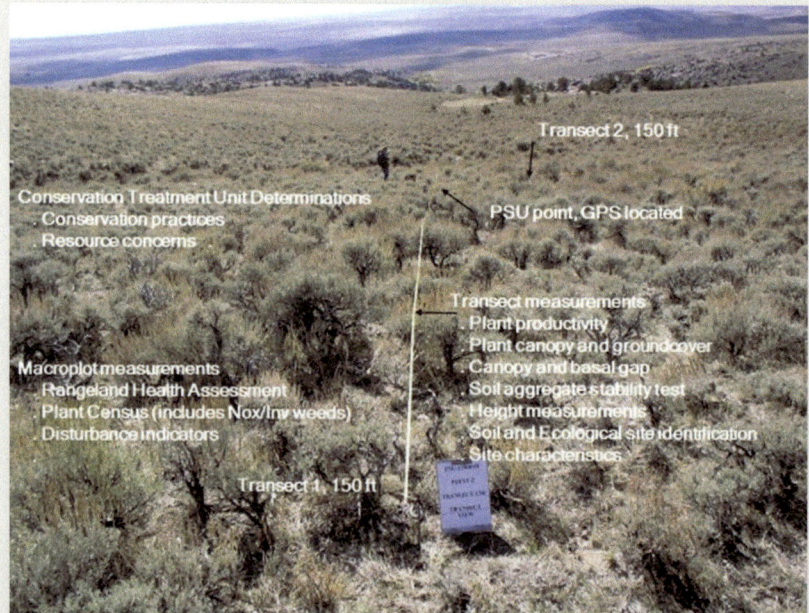

Fig. 7.5 Field layout of NE-SW and NW-SE transects and data field protocols of USDA-NRCS rangeland field NRI study

(continued)

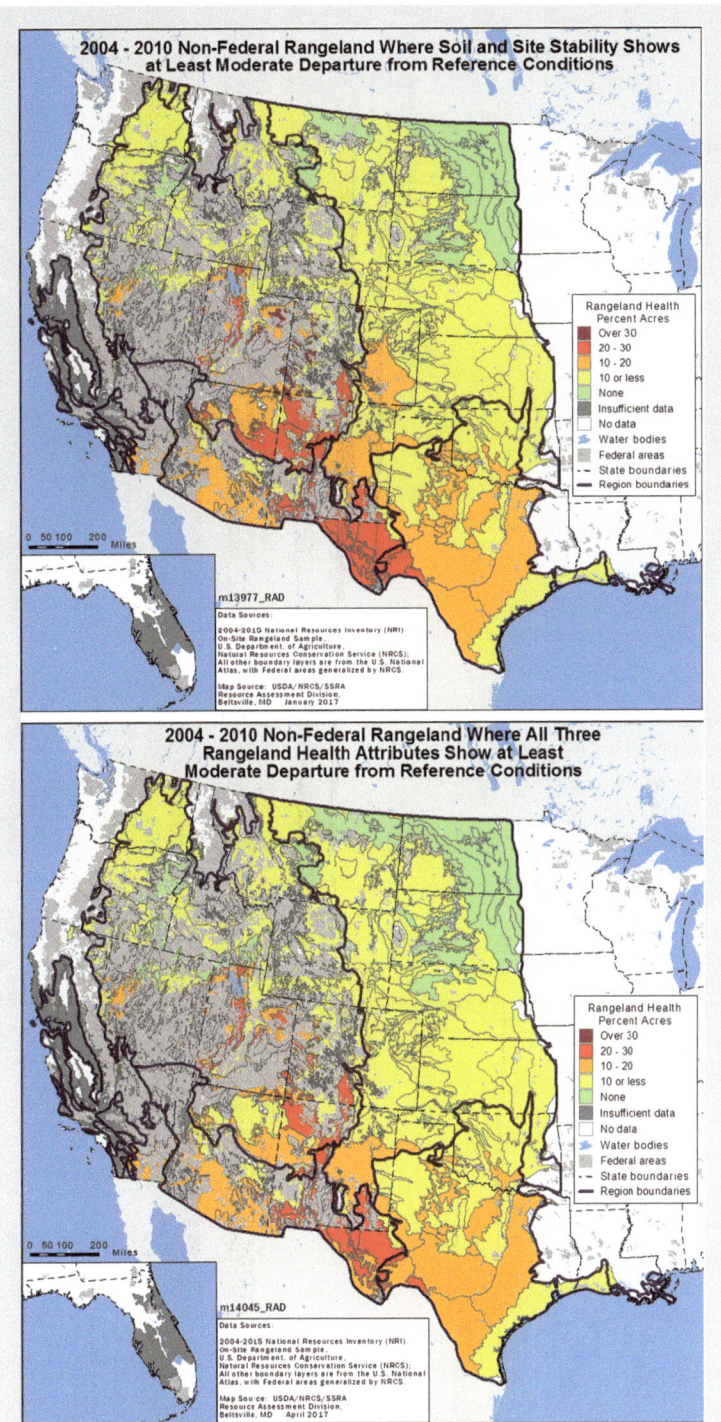

Fig. 7.6 (**a**) Non-federal rangeland where soil and site stability shows at least moderate departure from reference conditions. (**b**) Non-federal rangeland where all three rangeland health attributes show at least moderate departure from reference conditions. (USDA-NRCS 2018)

Text Box 7.5 Example of Rangeland Health Matrix and Interpretation Sheet (Tables 7.11, 7.12, 7.13, and 7.14).

Table 7.11 Standard indicators included for interpreting indicators of rangeland health with attributes (soil and site stability, hydrologic function, and/or biotic integrity) to which each indicator applies (Pellant et al. 2005, 2020)

	Rangeland health attribute		
	Soil and site stability	Hydrologic function	Biotic integrity
1. Rills.	X	X	
2. Water flow patterns.	X	X	
3. Pedestals and/or terracettes.	X	X	
4. Bare ground %.	X	X	
5. Gullies.	X	X	
6. Wind-scoured, blowouts, and/or depositional areas.	X		
7. Litter movement	X		
8. Soil surface resistance to erosion Interspace: ___ plant canopy: ___	X	X	X
9. Soil surface loss or degradation	X	X	X
10. Plant community composition and distribution relative to infiltration and runoff		X	
11. Compaction layer	X	X	X
12. Functional/structural groups Relative dominance ___ F/S groups not expected for the site ___ # F/S groups ___ Spp. # in dom. and subdom. F/S groups ___			X
13. Dead of dying plants or plant parts			X
14. Litter cover and depth Cover ___ %		X	X
15. Annual production (___ pounds or ___ kilograms) Estimated: ___ ÷ expected ___ = ___ %			X
16. Invasive plants			X
17. Vigor with an emphasis on reproductive capability of perennial plants			X

The "X" indicates that the indicator is applied to the attribute

Table 7.12 Example of a rangeland health rating sheet with field observed interpretations

Indicators	Attribute	Rating	Notes
1. Rills	SSS, HF	NS	Slope is 0 to 1%, no rills
2. Water flow patterns	SSS, HF	NS	Slope is 0 to 1%, no exacerbated water flow patterns
3. Pedestals and/or terracettes	SSS, HF	SM	Some pedestalled blue grama plant clumps from extreme heavy rainfall events
4. Bare ground %	SSS, HF	NS	Bare ground less than 2%, expected for site
5. Gullies	SSS, HF	NS	No gullies
6. Wind erosion-scoured, blowouts, and/or depositional areas	SSS	NS	No apparent signs of wind erosion or deposition
7. Litter movement	SSS	NS	Litter intact as ground cover, litter ≈ 25%
8. Soil surface resistance to erosion Interspace: ___ plant canopy: ___	SSS, HF, BI	NS	High plant cover, mostly 98–100% foliar cover, ground cover litter ≈ 25%
9. Soil surface loss or degradation	SSS, HF, BI	NS	No apparent loss of surface horizon (Berda loam 0–7 in), organic matter intact, soil color matches soil component description (7.5YR 6/3)
10. Plant community composition and distribution relative to infiltration and runoff	HF	M	Plant community composition has transitioned to higher percentage of buffalo grass (a sod, stoloniferous species associated with lower infiltration capacity (see Chap. 5).
11. Compaction layer	SSS, HF, BI	SM	Some areas seem slightly compacted by livestock
12. Functional/structural groups (a) Dominant plant comp. SM (b) FS not expected for site SM (c) # F/S groups NS (d) Spp. # in F/S groups SM	BI	M	Site transitioning from blue grama (bunch grass) to buffalo grass (sod forming spp.), some influx of broom snakeweed (perennial half shrub). Some decline in forb diversity. Increase in three awn grass.
13. Dead of dying plants or plant parts	BI	SM	Some blue grama plants have dead or declining plant centers
14. Litter cover and depth. Cover %	HF, BI	NS	Expected litter matches site expectations (≈ 25%)
15. Annual production (pounds or kilograms) Estimated ÷ expected = %	BI	SM	Plant production slightly reduced due to increase in buffalo grass composition and loss of blue grama plants (3000–3500 lbs. ac)

(continued)

Table 7.12 (continued)

Indicators	Attribute	Rating	Notes
16. Invasive plants	BI	M	Broom snakeweed, a native plant, but increasing on site
17. Vigor with an emphasis on reproductive capability of perennial plants	BI	SM	Perennial plants are producing seed; however, blue grama is stressed with some dead or dying plant centers

E-T	M-E	M	S-M	N-S	E-T	M-E	M	S-M	N-S	E-T	M-E	M	S-M	N-S
			3	1			10	3	1				11	8
				2					2			12	13	9
		11		4				11	4			16	15	14
				5					5				17	
				6					8					
				7					9					
				8					14					
				9										

Soil & Site Stability Attribute Rating: Predominantly NS, with some SM; monitor and reduce soil compaction	Hydrologic Function Attribute Rating: Scattered between NS to SM; moderate for plant community composition and distribution relative to infiltration and runoff	Biotic Integrity Attribute Rating: Predominantly NS to SM, with two M. Functional group changes from bunch to sod grass, increases in P half shrub and broom snakeweed (native invasive plant)

Ecological site: Deep Hardland Loamy 6.3–8.3 cm (16–21 in). PZ (R077CY022TX), major land resource area (MLRA): 077C-Southern High Plains, Southern Part. Site near Muleshoe, Texas. (See Chap. 5, Fig. 5.5, State 1.2)

NS none-to-slight, *SM* slight-to-moderate, *M* moderate, *ME* moderate-to-extreme, *ex* extreme-to-Total

Notes: SSS has some issues related to pedestalling of blue grama and compaction from moderate grazing intensity during wet periods

HF has some issues related to some pedestalling of blue grama and effect of higher buffalo grass composition resulting in reduced infiltration capacity

BI is most affected by changes in plant community composition and trends; use prescribed grazing to shift composition to blue grama dominance; broom snakeweed, a native invasive plant, is difficult to control without herbicides

Table 7.13 Evaluation matrix used to rate the 22 indicators and 5 departure categories of the 3 attributes of pastureland health (Spaeth 2020)

Indicators	Extreme-to-total	Moderate-to-extreme	Moderate	Slight-to-moderate	None-to-slight
1. Erosion (sheet and rill)	Numerous and frequent throughout. Nearly all rills are wide, deep and long. Occur in exposed and vegetated areas	Moderate in number at frequent intervals. Many rills are wide, deep, and long. Occur in exposed areas and in some adjacent vegetated areas	Moderate in number at infrequent intervals. Moderate rill width, depth, and length. Occur mostly in exposed areas, and steeper slopes	Scarce and scattered. Minimal rill width, depth, and length. Occur in exposed areas, and steeper slopes	Current or past formation of rills – none.
2. Erosion (gullies)	Sporadic or no vegetation on gully banks and/or bottom. Numerous nick points. Significant active bank and bottom erosion, including downcutting. Substantial depth and/or width. Active headcuts may be present	Intermittent vegetation on gully banks and/or bottom. Nick points common. Moderate active bank and bottom erosion, including downcutting. Significant width and/or depth. Active headcuts may be present	Occasional vegetation on gully banks and/or bottom. Occasional nickpoints and/or slight downcutting. Moderate depth and/or width. Active headcuts absent	Vegetation on most gully banks and/or bottom. Few nickpoints and/or minimal downcutting. Minimal gully depth and/or width. Headcuts absent	None
3. Erosion, wind-scoured and/or depositional areas	Extensive. Wind blowouts/scours usually connected. Large soil depositions around obstructions	Common. Wind scours frequently connected. Moderate soil depositions around obstructions	Occasionally present. Wind scours infrequently connected. Minor soil deposition around obstructions	Infrequent and few. Wind scours rarely connected. Trace amounts of soil deposition around obstructions.	None or as expected in reference ESD

(continued)

Table 7.13 (continued)

Indicators	Extreme-to-total	Moderate-to-extreme	Moderate	Slight-to-moderate	None-to-slight
4. Erosion (streambank or shoreline)	Banks bare, major vertical downcutting, major sloughing, little or no bank vegetation. Hydrology of riparian system severely altered	More than half the expected bank vegetation absent, vegetation trampled; sloughing and vertical banks active erosion. Hydrology of riparian system highly altered.	About half the bank vegetation trampled; active sloughing and downcutting. Hydrology of riparian system moderately altered	Some indication of trampled bank vegetation, active sloughing downcutting, or vertical slopes are minimal. Hydrology of riparian system slightly altered	Bank vegetation intact, minimal trampling and/or sloughing
5. Water flow patterns	Extensive. Long and wide. Erosional and/or depositional areas widespread. Usually connected	More numerous and widespread. Longer and wider than expected. Erosional and/or depositional areas common. Occasionally connected	Lengths and/or widths slightly to moderately higher than expected. Minor to moderate erosional and/or depositional areas. Infrequently connected	Length and width nearly match expected. Some minor erosional and/or depositional areas. Rarely connected.	Natural, well vegetated, or as described in ESD
6. Bare ground (%)	Substantially higher than expected. Bare ground patches are large and frequently connected	Much higher than expected. Major bare ground patches throughout stand are large and occasionally connected	Moderately higher than expected. Bare ground patches are moderate in size and sporadically connected	Slightly higher than expected. Bare ground patches are small and rarely connected	Amount and size of bare areas match that expected for the site. Else, no bare ground in stand
7. Pedestals and/or terracettes	Pedestals extensive; roots frequently exposed. Terracettes, if present, are widespread	Pedestals widespread; roots commonly exposed. Terracettes, if present, are common	Pedestals common; roots occasionally exposed. Terracettes, if present, are uncommon	Pedestals uncommon; roots rarely exposed. Terracettes scarce	None Terracettes, none

Indicators	Extreme-to-total	Moderate-to-extreme	Moderate	Slight-to-moderate	None-to-slight
8. Litter movement (wind or water)	Extreme movement of all size classes, (including large). Significant accumulations around obstructions or in depressions	Moderate to extreme movement of small to moderate size classes. Moderate accumulations around obstructions or in depressions.	Moderate movement of mostly small size classes. Small accumulations around obstructions or in depressions	Slight movement of small size classes. Minimal or no accumulations around obstructions or in depressions.	None or as described in ESD
9. Effects of plant community composition and distribution on infiltration and runoff *Assume that decreased infiltration causes a corresponding increase in runoff. Indicator 9 is correlated with indicator 10	Changes in plant community (functional/ structural groups) composition and/or distribution are associated with severe reduction in infiltration and a significant increase in runoff	Changes in plant community (functional/ structural groups) composition and/or distribution are associated with significantly or greatly decreased infiltration and a large increase in runoff	Changes in plant community (functional/ structural groups) composition and/or distribution are associated with moderate reduction in infiltration and a moderate increase in runoff	Community (functional/ structural groups) composition and/or plant distribution are associated with moderate reduction in infiltration and slight to moderate increase in runoff.	Infiltration and runoff are as expected for pasture state in S&T model. Plant composition and corresponding soil physical properties are not impeding infiltration
10. Soil surface loss or degradation	Soil surface horizon is very thin to absent throughout. Soil surface structure is similar to or more degraded than subsurface. No distinguishable difference between surface and subsurface organic matter content	Severe soil loss and/or degradation throughout. Minor differences in soil organic matter content and structure between surface and subsurface layers	Moderate soil loss and/ or degradation in plant interspaces with some degradation beneath plant canopies. Soil organic matter content is markedly reduced	Slight soil loss and/or soil structure shows slight signs of degradation, especially in plant interspaces. Minor change in soil organic matter content	No apparent soil loss or degradation (reference ESD narrative)

Indicators	Extreme-to-total	Moderate-to-extreme	Moderate	Slight-to-moderate	None-to-slight
11. Compaction layer	Extensive and/or strongly developed (thickness and density); may severely restrict root penetration and infiltrability	Widespread and/or moderately to strongly developed (thickness and density); may greatly restrict root penetration and infiltrability.	Moderately widespread and/or moderately developed (thickness and density); may moderately restrict root penetration and infiltrability	Not widespread and/or weakly developed (thickness and density); may weakly restrict root penetration and infiltrability.	No apparent compaction
12. Live plant foliar cover (hydrologic and erosion benefits)[a]	Less than 40% live foliar cover. Remaining is either dead standing material or bare ground	40–60% live foliar cover. Remaining is either dead standing material or bare ground	60–75% live foliar cover. Remaining is either dead standing material or bare ground.	75–95% live foliar cover. Remaining is either dead standing material or bare ground	More than 95% live foliar cover. Remaining is either dead standing material or bare ground
13. Forage plant diversity Note: (legumes adaptability based on what is expected for site in ESD)	Diversity severely lacking in comparison with site potential and/or with management objectives	Low diversity in comparison with site potential and/or plant diversity not in accordance with management objectives	Moderate diversity in comparison with site potential and/or plant diversity is not optimum with management objectives	Diversity slightly decreased in comparison with site potential and/or plant diversity is somewhat lacking with management objectives	High diversity of desirable forage plants in stand and/or plant diversity in full accordance with management objectives
14. Percent desirable forage plants (for identified livestock class)	Desirable forage species <20% dry weight	Desirable forage species 20–40% dry weight	Desirable forage species 40–60% dry weight	Desirable forage species 60–80% dry weight	Desirable forage species exceed 80% dry weight
15. Invasive plants	Invasive species dominate the site	Invasive species common throughout the site	Invasive species scattered throughout the site	Invasive species present in infrequent disturbed areas within the site	Invasive species rare, except in very infrequently disturbed areas

Indicators	Extreme-to-total	Moderate-to-extreme	Moderate	Slight-to-moderate	None-to-slight
16. Annual production	Less than 20% of potential production. Considering recent weather conditions	21–40% of potential production. Considering recent weather conditions	41–60% of potential production. Considering recent weather conditions	61–80% of potential production. Considering recent weather conditions	Annual production >80% of potential. Considering recent weather conditions
17. Plant vigor with an emphasis on reproductive capability of perennials	Plant reproduction and/or recovery after use is extremely reduced. Pale, yellow or brown, or severely stunted plants	Plant reproduction and/or recovery after use is greatly reduced. Yellowish green forage, or moderately or slightly stunted plants	Plant reproduction and/or recovery after use is moderately reduced. Adequate recovery. Yellowish and dark green areas due to manure and urine patches	Plant reproduction and/or recovery is slightly-to-moderately reduced after use Good recovery. Light green and dark green plants present	Plant reproduction and/or recovery is what is expected for the site. Rapid recovery. All healthy green plants.
18. Dead or dying plants or plant	Extensive mortality and/or dying plants/plant parts concentrated in one or more functional groups	Widespread mortality and/or dying plants/plant parts concentrated in one or more functional groups	Moderate mortality and/or dying plants/plant parts concentrated in one or more functional groups	Occasional mortality and/or dying plants/plant parts concentrated in one or more functional groups	No apparent mortality and/or dying plants/plant or plant parts
19. Litter cover and depth	Accumulation of litter cover and depth, and decomposition extremely out of balance with current weather conditions	Accumulation of litter cover and depth, and decomposition moderately-to-extremely out of balance with current weather conditions	Accumulation of litter cover and depth, and decomposition moderately out of balance with current weather conditions	Accumulation of litter cover and depth, and decomposition slightly out of balance with current weather conditions	Accumulation of litter cover and depth, and decomposition as expected for the site, and with current weather conditions

Table 7.13 (continued)

Indicators	Extreme-to-total	Moderate-to-extreme	Moderate	Slight-to-moderate	None-to-slight
20. Percentage of nontoxic legumes[b] Note: If bloating legumes dominate the stand by weight, rating = extreme to Total. Substantial risk to livestock with and without bloat prevention protocols. Fields with high legume composition should be considered for hayland.	If ES altered pasture state supports legumes, stands have less than 2% by weight and/or legume composition extremely out of balance with management objectives	If ES altered pasture state supports legumes, stands have 2–5% by weight and/or legume composition moderately-to-extremely out of balance with management objectives	If ES altered pasture state supports legumes, stands have 5–15% by weight And/or Legume composition moderately out of balance with management objectives	If ES altered pasture state supports legumes, stands have 15–30% by weight and/or legume composition slightly out of balance with management objectives	If ES altered pasture state supports legumes, stands have 30–35% by weight and/or legume use in accordance with management objectives
21. Uniformity of use	Little-grazed or ungrazed patches where forage species are rejected cover over 50% of the area. Rejected patches are generally connected or uniform use due to overutilization	Little-grazed or ungrazed patches where forage species are rejected cover 26 to 50% of the area Patches are occasionally connected	Little-grazed or ungrazed patches where forage species are rejected cover 10 to 25% of the area. Patches sporadically connected	Light-grazed or ungrazed and unconnected patches where forage species rejected are small and isolated (<10% cover). Urine and dung patches avoided	Uniform grazing throughout pasture. Areas where forage species are rejected only present at urine and dung patches

Indicators	Extreme-to-total	Moderate-to-extreme	Moderate	Slight-to-moderate	None-to-slight
22. Grazing and utilization Note: Utilization percentages can be temporarily adjusted in grazing rotation systems given that rest and/ or deferment are planned	Pasture severely overgrazed (>70% utilization), plant height continually below recommended grazing height for spp. Livestock concentration areas >10% of the pasture and transport contaminated runoff can directly drain into water channels unbuffered	Pasture utilization 65–70%, plant height is continually below recommended grazing height for spp. Livestock concentration areas and trails cover 5–10% of the area and drain into water channels unbuffered	Pasture utilization 60–65%; current utilization is temporary and not representative of continual management. Isolated and unconnected livestock concentration areas and trails (<5% of area) can potentially drain into water channels unbuffered	Pasture utilization 50–60%; plant height generally meets recommended grazing height for spp. Some livestock trails and one or two small unconnected concentration areas	Pasture utilization = <50%; plant height meets recommended grazing height for spp. No presence of livestock concentration areas or heavy use areas

[a] To define all possible undesirables (invasives, shrubs, and other weedy herbaceous forbs would be difficult). 60 percent cover has been shown to be the breakpoint of foliar cover where soil surface is relatively protected (Gifford 1985; Thurow 1986)

[b] Note: some literature mentions maximum legume comp. at 40–50 percent to minimize bloat potential

Table 7.14 Example field worksheet for determining indicators of pasture health (Spaeth 2020)

Indicators (Spaeth 2020)	Attribute	Rating	Notes
Erosion (Sheet and Rill)	SSS, HF		
Erosion (Gullies) if present	SSS, HF		
Erosion (Wind) if present	SSS, HF		
Erosion (Streambank/shoreline) if present	SSS, HF		
Water-flow Patterns	SSS, HF		
Bare ground %	SSS, HF		
Pedestals and Terracettes	SSS, HF		
Litter movement	SSS, HF		
Soil surface resistance to erosion	SSS, HF, BI		
Soil surface loss or degradation	SSS, HF, BI		
Compaction Layer	SSS, HF, BI		
Plant foliar cover (hydrologic and erosion benefits	SSS, HF		
Forage Plant Diversity	BI, LMQF		
Percent Desirable Forage Plants (for identified livestock class)	LMQF		
Invasive plants	BI, LMQF		
Annual production	BI, LMQF		
Plant Vigor with an emphasis on reproductive capability of perennial plants	BI		
Dead or Dying Plants or Plant Parts	BI		
Litter cover and depth	HF, BI		
Percent non-toxic legumes (based on adaptability of Ecol. Site and/or what is expected stand for the site)	BI, LMQF		
Uniformity of use	HF, BI, LMQF		
Livestock Concentration Areas	SSS, HF, LMQF		

E-T	M-E	M	S-M	N-S

Soil & Site Stability Attribute Rating

E-T	M-E	M	S-M	N-S

Hydrologic Function Attribute Rating

E-T	M-E	M	S-M	N-S

Biotic Integrity Attribute Rating

E-T	M-E	M	S-M	N-S

Livestock Management Quality Factor

Notes

Salt placement by watering area. Livestock use heavy along fence lines. Forage plant diversity could be improved by controlling undesirable weedy plants. Invasive shrubs scattered throughout pasture. SOM somewhat depleted from past cropping history and water erosion events. About 50% of soil aggregates dispersed in water.

Determining Indicators of Pasture Health (Spaeth 2020)

Determining Indicators of Pasture Health (DIPH) is a detailed assessment tool and includes a matrix of 22 indicators that can be used to determine the preponderance of evidence for three separate pastureland ecosystem attributes: biotic integrity, soil/site stability, and hydrologic function. DIPH is a similar methodology to IIRH V5 (Pellant et al. 2020), although there are specific indicators that are relevant to pastureland systems in DIPH. The premise associated with IIRH and DIPH is that many unique site-specific effects and nonlinear environmental relationships exist in grazing land ecosystems, and these methodologies provide a means of detecting changes in ecological attributes relative to a site's ecological potential.

DIPH may be used as a standardized approach similar to IIRH to conduct a comprehensive pasture assessment of hydrologic function, soil and surface stability, and biotic integrity. Three health assessments are evaluated in both IIRH and DIPH and are designed to provide information about how well ecological processes – such as the water cycle, energy flow, and nutrient cycling – are functioning at a site. The three ecosystem attributes (soil and site stability, hydrologic function, and biotic integrity) are determined from specific indicators (some indicators are used for one or more of the three assessments) (Table 7.14). The methodology, DIPH, is centric to the dynamics of the ecological site (ES). Various soil and plant variables may be different across the continuum of pasturelands in the USA where some pasture environments are capable of sustaining high species diversity and many different adapted forage species (including legumes) and soil biota such as earthworms, while some pasture systems are limited in these respects by various environmental constraints. For example, a wide variety of cool season grasses and legumes may be grown and maintained successfully in humid cold temperate climates in New England states (USA), whereas a semiarid subtropical climate in Louisiana may only support a maximum diversity of two warm season pasture grasses [bermudagrass (*Cynodon dactylon*) and Bahia grass (*Paspalum notatum*)], with no inherent introduced long-term sustainability of non-toxic legumes (which act as annuals).

Ecological site descriptions (if available) can provide valuable information about environmental parameters and reference conditions for specific indicators related to adaptability of certain forage species, legumes, invasive plants, as well as hydrology and erosion properties such as drainage, flooding, water flow paths, and propensity for rills, gullies, and erosion. Although ecological site descriptions can be valuable documents that provide reference information related to climate-soils-plants-hydrology-management, both IIRH (Sect. 7.1.4, Pellant et al. 2020) and DIPH can be used when ecological site information is not available (Tables 7.13 and 7.14).

Review for Preponderance of Evidence (Table 7.14, example of field notes, Spaeth 2020)

A. Soil and Site Stability

Slight-to-moderate with two moderate concerns. The critical indicators related to erosion are rated slight-to-moderate; however, bare ground was rated moderate—moderately higher than expected with patches sporadically connected.

Many of the problems related to soil stability can be largely be corrected with prescribed grazing and improvement of biotic integrity factors with pest management practices of weedy and woody species invasion.

B. Hydrologic Function

  1. Some of the key indicators such as bare ground, annual production, litter cover and depth, invasive plants, and grazing and utilization were rated moderate. The first three indicators above are important in proving protective cover to the soil surface, which is directly correlated with rainfall interception and reducing raindrop soil splash and sheet and rill erosion. Invasive plants such as shrubs can result in loss of understory cover, but this is not a problem on this site as invasive plants were largely Canada thistles. The main concern with bare ground, annual production, and litter cover and depth at moderate rating is that evaporation rates are higher than expected, and the result is depleted water-holding capacity which will affect plant growth and production. Overall water balance is now compromised but can be remedied with the planned rest schedule.
  2. Uniformity of use was rated moderate to extreme (little-grazed or ungrazed patches where forage species are rejected cover 26–50 percent of the area). Patches are occasionally connected, and grazing and utilization were rated moderate (pasture utilization 60–65 percent; current utilization is temporary and not representative of continual management). The pasture will be rested from July 15 to end of August, so there is no real concern about overgrazing at the present time.

C. Biotic Integrity

  3. Biotic integrity indicators ranged from slight to moderate to moderate-to-extreme. The overall attribute rating is moderate. Where indicators are rated moderate or worse, there is cause for concern. Since plant community shifts affect the quality of forage availability, species changes also affect soil surface stability and hydrology; e.g., shifts from bunchgrass to sod grass result in lower infiltration capacity and the prevalence of higher runoff. Improvement is needed regarding the indicators rated as moderate for BI.
  4. Annual productivity was moderate as was litter cover and depth. Uniformity of use was rated moderate-to-extreme (little-grazed or ungrazed patches where forage species are rejected cover 26–50 percent of the area). Patches are occasionally connected because of stands of undesirable weedy species (Canada thistle and yellow mustard).
  5. Forage plant diversity, invasive species, plant vigor, dead or dying plants or plant parts, litter cover and depth, and uniformity of use were rated moderate or worse. Legumes are not adapted based on Ecological site. Two dominant grass spp. and Canada thistles in overgrazed areas with yellow mustard. Overall plant diversity is low, compared to site potential and species that are adapted to this site. Invasive weed management is needed.

Text Box 7.6 Example Visual Assessment Scoresheet for Gardens (Spaeth 2018)

The garden "Soil Health/Quality Indicator Scorecard" can be used for new and existing gardens for evaluating soil quality and health. The indicator scorecard is a subjective evaluation but can provide a relative baseline based on number of indicators in each column. The indicator columns are undesirable and/or unhealthy and desirable and/or healthy, with the middle column being intermediate or transitional between the two. Note: not all the indicators represent an unhealthy situation, for example, texture, color, and infiltration capacity may be dependent on the inherent characteristics of your soil—the "norm" for your soil type. In many home garden situations, the soil has been severely disturbed, especially in subdivisions, where the topsoil has been removed or haphazardly mixed with the subsoil. In such situations, consider bringing in topsoil to establish the garden. Many of the indicators below can be improved and managed, i.e., pH, organic matter, biological activity, infiltration, water holding capacity, and overall soil health (Table 7.15).

Table 7.15 Example scorecard for garden soil indicators

Garden soil indicator scorecard	Date:_____ Site name: _____ Assessment completed by: _____ _____		
		(circle appropriate category)	
Indicators	**Undesirable and/or unhealthy**		**Desirable and/or healthy**
1. Is there a sufficient topsoil layer?	Subsoil is exposed or very near the surface	Topsoil is thin, about 4–6 inches. Some topsoil may be needed	Greater than 6 inches deep
2. Depth of surface and immediate subsoil layer	Less than 1 ft.	1–2 ft.	Greater than 2 ft.
3. During hot and dry climatic conditions, are there soil cracks?	Obvious and many	Occasional	None
4. Color of topsoil layer (moist)	Yellow, gray, light gray, mottled color	Light brown	Black, dark brown, dark red, color uniform
5. Ease of digging hole 1 ft. deep	Not possible or very difficult	Moderately difficult	Relatively easy
6. Soil compaction	Dense, tight, and compacted. Hard layer inhibits plant roots, roots growing laterally	Moderately compacted, few roots growing laterally	Not compacted, many vertical roots with adequate side branching

Table 7.15 (continued)

Garden soil indicator scorecard	Date:_____ Site name: _____ Assessment completed by: _____		
7. Soil structure	Hard, dense, or dusty powdery when dry. Surface crust forms after hard rain. Difficult to prepare smooth seedbed	Soil is lumpy, crumbles with pressure. Moderate crusting after hard rain	Very crumbly, no crusting after hard rain. Soil is mellow, and seedbed is easy to prepare
8. Odor of soil	Oily, chemical, gasoline, rotten eggs, other foul odors	No chemical smell, but slightly disagreeable	Fresh earthy smell

Warning: Certain contaminants, lead, and other pollutants may not have an odor. If you are suspicious of chemical pollutants, or if the soil has a noticeable disagreeable odor, do not do the smell test. Send a soil sample to an appropriate lab.

9. Existing plant cover under normal undisturbed conditions (if not mowed or clipped)	Low	Medium	High
10. Vigor of existing vegetation (plant coloration and growth)	Many dead or stressed plants, yellow color, and bare patches	Some dead or stressed plants, growth is sporadic, a few bare patches	High plant vigor, good green coloration, growth is rapid, no bare patches
11. Plants reproducing	None	Some	High
12. Plant roots in upper 4–12 inches	None	Few	Many
13. Current plant health	Yellow, discolored, blighted, streaky, spotty, or many dead, dry leaves and stems. Plants spindly, thin, and weak	Yellow-to-green coloration. Plant color falls off or discolors as season progresses, and with hot weather	Green-to-dark green coloration. Plants upright with healthy stems
14. Bare ground	Patchy and connected areas of bare ground	Some bare ground and patchiness	No bare ground, plant cover is entire throughout area
15. Biological activity, earthworms, insects	None	Few	Many

Table 7.15 (continued)

Garden soil indicator scorecard	Date:_____ Site name: _____ Assessment completed by: _____		
16. Decomposition, biological activity, microbes (litter and organic material decompose when soil conditions are moist)	Old plant litter, very slow or almost no decomposition of organic materials	Moderate decomposition of plant litter and organic materials (months)	Rapid decomposition of plant litter and organic materials (several weeks to month)
17. Water-holding capacity	Soil dries out quickly, droughty. Plants wilt quickly after water events in hot weather	Some plant wilting or curling in hot weather, but recover quickly	Soil retains moisture. Plants maintain turgor and structure in hot weather
18. Infiltration (estimate based on observations during rainfall)	Rapid to immediate or very slow (over 1 hour)	Most water infiltrates in about 20 minutes to 1 hour	Not rapid or very slow, most water infiltrates between 10 and 20 minutes
Data from Web Soil Survey or soil test (if no data, then omit)			
19. Organic matter (obtain from Web Soil Survey and/or soil test)	Less than 2%	2–4%	4–6%
20. pH (obtain general info from Web Soil Survey and/or soil test)	Less than 6.4 or greater than 7.2		
21. N, P, K	Low level	Medium level	Adequate or high level
22. Micronutrients	Low level	Medium level	Adequate or high level
Preponderance of evidence (total indicators in each column).			

References

Abrahams, A.D., A.J. Parsons, and J. Wainwright. 1995. Effects of vegetation change on interrill runoff and erosion, walnut gulch, southern Arizona. *Geomorphology* 13 (1–4): 37–48.

Bagchi, S., D.D. Briske, B.T. Bestelmeyer, and X. Ben Wu. 2013. Assessing resilience and state transition models with historical records of cheatgrass *Bromus tectorum* invasion in north American sagebrush steppe. *Journal of Applied Ecology* 50: 1131–1141.

Boxell, J.J, 2007. Changes in soil physical and hydrological properties due to *Bromus tectorum* (cheatgrass) invasion. UNLV Retrospective Theses and & Dissertations. 2216. https://digitalscholarship.unlv.edu/rtds/2216.

Boxell, J., and P.J. Drohan. 2009. Surface soil physical and hydrological characteristics in *Bromus tectorum* L. (cheatgrass) versus *Artemisia tridentata* Nutt. (big sagebrush) habitat. *Geoderma* 149: 305–311.

Bray, J.R., and J.T. Curtis. 1957. An ordination of the upland forest communities of Southern Wisconsin. *Ecological Monographs* 27: 325–349.

Brown, J.R., and J.E. Herrick. 2016. Making soil health a part of rangeland management. *Journal of Soil and Water Conservation* 71: 55A–60A.

Chambers, J.C. 1983. Measuring species diversity on revegetated surface mines: an evaluation of techniques. United States Department of Agriculture, Forest Service, Mines and Mineral Resources. Paper 1.

Chiclana, F., J.M. Tapia García, M.L. Moral, and E.A. Herrera-Viedma. 2013. A statistical comparative study of different similarity measures of consensus in group decision making. *Informatiopn Science* 221: 110–123.

Clifford, H.T., and W. Stephenson. 1975. *An introduction to numerical classification*, 49–197. New York: Academic Press.

Cooper, S., K. Neiman, and R. Steele. 1987. Forest habitat types of northern Idaho. In *UDSA Forest Service, intermountain forest and range experiment station general technical report INT-236*. Ogden: USA.

Cosgrove, D., D.J. Undersander, and J. Cropper. 2001. *Guide to pasture condition scoring*. Fort Worth: US Dept. of Agriculture, Natural Resources Conservation Service, Grazing Lands Technology Institute.

Egghe, L. 2010. Good properties of similarity measures and their complementarity. *Journal of the American Society for Information Science and Technology* 61: 2151–2160.

Ellenberg H. 1956. Grundlagen der Vegetationsgliederung. Part 1: Aufgabe and Methoden der Vegetationskunde. Einfuhrung in die Phytosoziologie 4.

Escadafal, R., M.C. Girard, and D. Courault. 1989. Munsell soil color and soil reflectance in the visible spectral bands of Landsat MSS and TM data. *Remote Sensing of Environment* 27: 37–46.

Gifford, G.F. 1985. Cover allocation in rangeland watershed management (a review). In *Watershed management in the eighties, proceedings of a symposium*, ed. E.B. Jones and T.J. Ward, 23–31. New York: ASCE.

Gleason, H.A. 1920. Some applications of the quadrat method. *Torrey Botanical Club Bulletin* 47: 21–33.

Goff, B.F., G.C. Bent, and G.E. Hart. 1993. Erosion response of a disturbed sagebrush steppe hillslope. *Journal of Environmental Quality* 22: 698–709.

Greig-Smith, P. 1964. *Quantitative plant ecology*. 2nd ed. London: Butterworth.

Grieg-Smith, P. 1983. *Quantitative plant ecology*. 3rd ed. Berkeley: University of California Press.

Helms, D., A.B.W. Effland, and P.J. Durana. 2002. *Profiles in the History of the U.S. Soil Survey*. Ames: Iowa State Press.

Iowa Soil Health. 2017. Iowa soil health assessment card. https://store.extension.iastate.edu/product/Iowa-Soil-Health-Assessment-Card.

Jaccard, P. 1908. Nouvelles recherches sur la distribution. *Oorale Bulletin Societe Vaudoise des Sciences Naturelles* 44: 223–270.

Legendre, P., and L. Legendre. 1998. *Numerical ecology*. 2nd ed. Amsterdam: Elsevier.

Moebius-Clune, B.N., D.J. Moebius-Clune, B.K. Gugino, O.J. Idowu, R.R. Schindelbeck, A.J. Ristow, H.M. van Es, J.E. Thies, H.A. Shayler, M.B. McBride, K.S.M. Kurtz, D.W. Wolfe, and G.S. Abawi. 2016. *Comprehensive assessment of soil health – The Cornell framework*. 32nd ed. Geneva, New York: Cornell University.

Morgan, R.P.C. 1986. *Soil erosion and conservation*. New York: Longman Scientific and Technical, Wiley.

Motyka, J. 1950. Wstçpne badania nad lakami paludniowo-wschodnilj Lubel-szczyzny. Preliminary studies on meadows in the south-east of the province Lublin. *Annales Universitatis Mariae Curie-Sklodowska* 5: 367–447.

Mueller Dombois, D., and H. Ellenberg. 1974. *Aims and methods of vegetation ecology*, 547. New York: Wiley.

National Research Council. 1994. *Rangeland health. New methods to classify, inventory, and monitor rangelands*. Washington, D.C.: National Academy Press.

Pellant, M., P. Shaver, D.A. Pyke, and J.E. Herrick. 2005. *Interpreting indicators of rangeland health, version 4*. Technical Reference 1734-6. US Department of the Interior, Bureau of Land Management, National Science and Technology Center, Denver. BLM/WO/ST-00/001+ 1734/REV05.

Pellant, M., P.L. Shaver, D.A. Pyke, J.E. Herrick, F.E. Busby, G. Riegel, N. Lepak, E. Kachergis, B.A. Newingham, and D. Toledo. 2020. *Interpreting indicators of rangeland health, version 5. Tech Ref 1734–6*. Denver: U.S. Department of the Interior, Bureau of Land Management, National Operations Center.

Pielou, E.C. 1969. *An introduction to mathematical ecology*. New York: Wiley-Interscience.

Pyke, D.A., J.E. Herrick, P. Shaver, and M. Pellant. 2002. Rangeland health attributes and indicators for qualitative assessment. *Rangeland Ecology and Management* 55: 584–597.

Rapport, D.J., R. Costanza, and A.J. McMichael. 1998. Assessing ecosystem health. *Trends in Ecology and Evolution* 13: 397–402.

Schwertmann, U. 1993. Relations between iron oxides, soil color, and soil formation. *Soil Color*: 51–69.

Shepherd, T.G. 2003. Assessing soil quality using visual soil assessment. In *Tools for nutrient and pollutant management: Applications to agriculture and environmental quality*, Occasional report, ed. L.D. Currie and J.A. Hanly, vol. 17, 153–166. PalmerstonNorth: Fertilizer and Lime Research Centre, Massey University.

Shepherd, G. 2008. Visual soil assessment (VSA). https://www.carbonfarming.org.nz/wp-content/uploads/VSA SummaryStatement.pdf.

Sneath, P.H.A., and R.R. Sokal. 1973. *Numerical taxonomy*, 141–145. San Francisco: W. H. Freeman.

Soil Survey Staff, Natural Resources Conservation Service, United States Department of Agriculture. 2013. Web soil survey. Available online at http://websoilsurvey.nrcs.usda.gov/.

Sørensen, T.J. 1948. *A method of establishing groups of equal amplitude in plant sociology based on similarity of species content and its application to analyses of the vegetation on Danish commons*. I kommission hos E: Munksgaard.

Spaeth, K.E. 2018. *Circle gardening: Growing vegetables outside the box*. College Station: Texas A&M University Press.

———. 2020. *Determining indicators of pasture health. USDA-NRCS national range and pasture handbook*. Texas: Fort Worth.

Spaeth, K.E., F.B. Pierson, M.A. Weltz, and G. Hendricks, eds. 1996a. *Grazingland hydrology issues: Perspectives for the 21st century*. Denver: Society for Range Management.

Spaeth, K.E., F.B. Pierson, M.A. Weltz, and J.B. Awang. 1996b. Gradient analysis of infiltration and environmental variables as related to rangeland vegetation. *Transactions of the ASAE* 39: 67–77.

Spaeth, K.E., F.B. Pierson, J.E. Herrick, P.L. Shaver, D.A. Pyke, M. Pellant, and B. Dayton. 2003. New proposed national resources inventory protocols on nonfederal rangelands. *Journal of Soil and Water Conservation* 58: 18A–21A.

Spaeth, K.E., G.L. Peacock, J.E. Herrick, P. Shaver, and R. Dayton. 2005. Rangeland field data techniques and data applications. *Journal of Soil and Water Conservation* 60: 114A–119A.

Stirling, G., H. Hayden, T. Pattison, and M. Stirling. 2016. *Soil health, soil biology, soilborne diseases and sustainability agriculture*. Clayton South VIC: Cisero Publishing.

Thurow, T.L., W.H. Blackburn, and C.A. Taylor. 1986. Hydrologic characteristics of vegetation types as affected by livestock grazing systems, Edwards Plateau, Texas. *Journal of Range Management* 39: 505–509.

Toledo, D., M. Sanderson, J. Herrick, and S. Goslee. 2016. An integrated approach to grazingland ecological assessments and management interpretations. *Journal of Soil and Water Conservation* 69: 110A–114A.

USDA-NRCS. 1997. *National range and pasture handbook*. Texas: Fort Worth.

———. 2013. National Soil Survey Handbook Part 622. Land capability classification (622.02). Washington, D.C. http://soils.usda.gov/technical/handbook/contents/part622.html

———. 2015. Soil health literature. https://www.nrcs.usda.gov/wps/portal/nrcs/detailfull/soils/health/mgnt/?cid=stelprdb1257753

———. 2018. NRI rangeland resource assessment-2018. https://www.nrcs.usda.gov/wps/portal/nrcs/detail/national/technical/nra/nri/results/?cid=nrcseprd1343031

———. 2020. *National range and pasture handbook*. Texas: Fort Worth.

Weil, R.R., and N.C. Brady. 2017. The nature and properties of soils. NY: Pearson.

Weltz, M. and K. Spaeth. 2012. Estimating effects of targeted conservation on nonfederal rangelands. Rangelands 34: 35–40.

Weltz, M.A., M.R. Kidwell, and H. Dale Fox. 1998. Influence of abiotic and biotic factors in measuring and modeling soil erosion on rangelands: State of knowledge. Soil erosion on rangeland. *Journal of Range Management* 51: 482–495.

Weltz, M.A., K. Spaeth, M.H. Taylor, K. Rollins, F. Pierson, L. Jolley, M. Nearing, D. Goodrich, M. Hernandes, S.K. Nouwakpo, and C. Rossi. 2014. Cheatgrass invasion and woody species encroachment: Benefits of conservation. *Journal of Soil and Water Conservation* 69: 39A–44a.

Whittaker, R.H. 1965. Dominance and diversity in land plant communities. *Science* 147: 250–260.

Whittaker, R.H. 1975. Communities and ecosystems 2nd ed. New York: Macmillan Publ. Co.

Index

© Springer Nature Switzerland AG 2020   363
K. E. Spaeth Jr., *Soil Health on the Farm, Ranch, and in the Garden*,
https://doi.org/10.1007/978-3-030-40398-0